Metals and Minerals: Science and Technology

Metals and Minerals: Science and Technology

Editor: Keith Liverman

NY RESEARCH
P R E S S

New York

Published by NY Research Press
118-35 Queens Blvd., Suite 400,
Forest Hills, NY 11375, USA
www.nyresearchpress.com

Metals and Minerals: Science and Technology
Edited by Keith Liverman

International Standard Book Number: 978-1-63238-562-8 (Hardback)

The publisher's policy is to use permanent paper from mills that operate a sustainable forestry policy. Furthermore, the publisher ensures that the text paper and cover boards used have met acceptable environmental accreditation standards.

Trademark Notice: Registered trademark of products or corporate names are used only for explanation and identification without intent to infringe.

Cataloging-in-Publication Data

Metals and minerals: science and technology / edited by Keith Liverman.
    p. cm.
Includes bibliographical references and index.
ISBN 978-1-63238-562-8
1. Metals. 2. Minerals. 3. Metallurgy. 4. Mineralogy. I. Liverman, Keith.
TN607 .M48 2017
620.16--dc23

Printed in the United States of America.

Contents

Preface

Metals and minerals are both elements which have multiple industrial applications. Most metals are extracted from a mineral ore by the process of mining. This book outlines the processes of extraction and applications of metals and minerals in detail. It elucidates new techniques and their applications in a multidisciplinary approach. The book presents researches and studies performed by experts across the globe. The various studies that are constantly contributing towards advancing technologies and evolution of this field are examined in detail. Scientists and students actively engaged in this field will find this book full of crucial and unexplored concepts.

The information contained in this book is the result of intensive hard work done by researchers in this field. All due efforts have been made to make this book serve as a complete guiding source for students and researchers. The topics in this book have been comprehensively explained to help readers understand the growing trends in the field.

I would like to thank the entire group of writers who made sincere efforts in this book and my family who supported me in my efforts of working on this book. I take this opportunity to thank all those who have been a guiding force throughout my life.

Editor

Effect of Cold Rolling and Heat Treatment on the Mechanical Properties of GH4169 Alloy Sheet at Room Temperature

Shi-Hong Zhang [1,†], Neng-Yong Ye [1,2,†], Ming Cheng [1,2,*], Hong-Wu Song [1], Hong-Wei Zhou [2] and Ping-Bo Wang [2]

Academic Editor: Johan Moverare

[1] Institute of Metal Research, Chinese Academy of Sciences, Shenyang 110016, China; shzhang@imr.ac.cn (S.-H.Z.); nyye12b@imr.ac.cn (N.-Y.Y.); hwsong@imr.ac.cn (H.-W.S.)
[2] State Key Laboratory of Advanced Processing and Recycling of Non-ferrous Metals, Lanzhou University of Technology, Lanzhou 730050, China; zhouhongwei126ct@126.com (H.-W.Z.); wpb_lut@126.com (P.-B.W.)
*   Correspondence: mcheng@imr.ac.cn
†   These authors contributed equally to this work.

Abstract: The mechanical properties of GH4169 alloy sheet after cold rolling (at 0%, 10%, 30%, 50% and 70%) and solid solution were investigated. The textures and Taylor factors were characterized using electron backscattering diffraction (EBSD). The fractions of δ phase were measured by X-ray diffraction. The contributions of δ phase, grain size, texture, and work hardening on the mechanical properties were also discussed. The results showed increases in the yield strength (YS) (0.2%) as well as the ultimate tensile strength (UTS) of GH4169 superalloy sheet after cold rolling, when rolling reduction was increased. In contrast, following solid solution treatment, YS and UTS were increased then subsequently decreased. The changes of yield strength of GH4169 superalloy were attributed to the texture and work hardening, followed by the grain refinement and precipitation of δ phase. When the rolling reduction was below 30%, the influence of δ phase was greater than grain refinement and when the rolling reduction was larger than 50%, the controversial results occur. The precipitation of δ phase promoted the improvement of yield strength, the relationship between the fraction of δ phase and improved yield strength satisfactory fit to the following equation: $\sigma_\delta = 15.9W_\delta + 59.7$.

Keywords: GH4169 alloy; cold rolling; mechanical property; phase fraction; heat treatment

1. Introduction

GH4169 alloy is a nickel base superalloy (modified IN718) strengthened by the precipitation of ordered body centered cubic (BCC) γ''-Ni_3Nb and ordered face centered cubic (FCC) γ'-Ni_3(Al, Ti) [1,2]. It is extensively used in aerospace, nuclear, and petrochemical industries because of its excellent mechanical properties, good oxidation resistance, and corrosion at elevated temperature. The strengthened phase of γ'' is metastable [3–6], which can be transformed to the orthogonal thermodynamic equilibrium δ-Ni_3Nb phase below 900 °C. However, when the temperature is >900 °C, the δ phase can also be precipitated directly from the matrix [7]. The δ phase existed in the GH4169 alloy plays important effects on the creep properties and microstructure's evolution [8,9]. It reduces the notch sensibility and controls the grain growth through precipitation on the grain boundaries [10–12]. The effect of δ phase on the hot deformation behavior and plasticity of GH4169 superalloy has been widely analyzed. The existence of δ phase results in the decreasing of flow stress and makes the flow stress reached a peak value at small strain. Besides, the δ phase increases the strain rate sensitivity exponent, it decreases the strain hardening exponent [13]. Zhang [14] discussed the initial δ phase on the hot tensile deformation behavior of a Ni-based superalloy. The results showed that δ phase causes obvious work hardening at the beginning of hot deformation and accelerates flow softening via promoting the dynamic recrystallization with further strain. Therefore, the Delta Process became the most widely used components in aerospace applications. It successfully utilizes the pre-precipitated δ phase prior hot deformation to obtain fine grains through the pinning effect on the grain boundaries. Recently, cold rolling technology was firstly adopted to form blades, based on the requirements of precise forming and fatigue strength improvement.. Liu *et al.* [15] analyzed the phases' fraction of cold rolled Inconel 718. They found that the process of cold rolling promoted the precipitation of δ phase. Although, it increased the δ phase's fraction, the fraction of γ'' phase was decreased oppositely. So, adjusting the δ phase's fraction is important to get the best performance of GH4169 alloy at room-temperature [11].

There were some results acquired in previous research, but they were mainly focus on the phases' precipitation after cold deformation, little attention was paid on the mechanical properties of GH4169 alloy after cold deformation or heat treatment. For the blade forming, these two processes were both important to the final properties of blades. Especially the period of annealing at 980 °C, it not only determined the phase content, but also has a significant effect on the grain refinement [8,10,11]. Besides, the texture and its effect on the mechanical property of alloy were also needed to be paid attentions [15,16]. The final mechanical properties of GH4169 alloy would be affected by the factors mentioned above. Moreover, to understand how much they would provide a theoretical guidance for further process optimization of cold rolling technology. So, the grain size, texture, phase fraction after cold rolling and heat treatment and their effects on the mechanical properties of GH4169 alloy at room temperature were needed to be analyzed.

2. Experimental Procedures

The cold rolled GH4169 superalloy sheet with the size of $70 \times 21 \times 2.5$ mm was used in this paper. It was annealed at 990 °C/1 h. The chemical compositions include: Ni 52.90, Cr 17.96,Mo 3.04, Nb 5.00, Al 0.51, Ti 1.02, C 0.042, P 0.005, S 0.003,Si 0.17 (wt. %), and Fe Bal. OM micrograph is shown in Figure 1.The annealing sheets were firstly cold rolled to different reductions in thickness, *i.e.*, 0%, 10%, 30%, 50% and 70% (named cold rolled specimen, CRS), then solid solution treated at 980 °C/1 h (named solid solutioned specimen, SSS). The tensile tests were carried on the MTS E45 (MTS, Eden Prairie, MN, USA) at the velocity of 1 mm/min to acquire the 0.2% yield strength (YS) and ultimate tensile strength (UTS). The microstructures were observed on Zeiss Observation Z1m optical microscope (Carl Zeiss AG, Baden, German). The Electron Backscattered Diffraction (EBSD) data were acquired on the Hitachi S3400N scanning electron microscope (Hitachi Ltd, Tokyo, Japan) with Oxford Instrument HKL system (Oxford Instruments, Oxfordshire, UK). The scan step was 1 μm and the voltage was 20 kV. Phases of the alloy with different cold rolling reductions were measured by RigakuD/max-γA diffractometer (Hitachi Ltd, Tokyo, Japan) and the weight percentages of δ phase were calculated according to Liu *et al.*, methodology [17] which the error can be controlled within 10% [18–20]. Additionally, other annealing experiments were carried out to obtain the values of K and σ_0 in the Hall-Petch relationship. They were heated rapidly at 1000 °C and 1040 °C for 1 h separately to avoid the influence of δ phase.

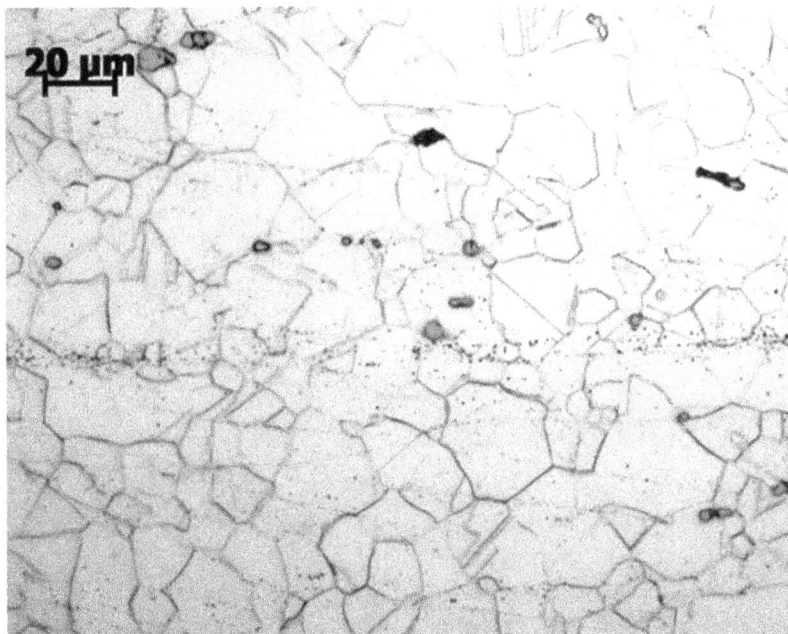

Figure 1. Microstructure of GH4169 alloy after annealing (990 °C/1 h).

3. Results and Discussion

3.1. Microstructure Evolution

　　The microstructure evolution of cold rolling and solid solution are shown in Figure 2. As can be seen, the increases of rolling reduction elongate the grains along the rolling direction. Meanwhile, plenty of deformation bands were formed, as shown in Figure 2a,b. After solid solution, δ phase gradually precipitated and its morphology transformed from needle-shaped to particle-shaped. The morphology of δ phase (when the rolling reduction = 30%) was needle-shaped and preferentially precipitated on the grain and twin boundaries. δ phase was also precipitated on the deformation bands and it was relative to small content, as shown with the black arrows in Figure 2c. On the other hand, when the rolling reduction reached 50%, the morphology of δ phase was mainly particle-shaped and largely precipitated both on the boundaries and deformation bands of grains, as shown with the black arrows in Figure 2d. Clearly, it reveals that large rolling reduction not only promotes the precipitation of δ phase, but also leads to the transformation of its morphology. However, distinguishing from the one formed in the hot deformation or heat treatment temperature below 900 °C, Notably, the particle-shaped δ phase in this paper neither originated from the phase transformation of $\gamma'' \rightarrow \delta$ nor spheroidized by deformation breakage or dissolution breakage. Obviously, it is related to the deformation bands and vacancies formed in the cold rolling process and the static recrystallization occurred in the heat treatment. When the rolling reduction reached 50%, static recrystallization was occurred at 980 °C which can be seen in later analysis. It weakens the orientation relationship between δ phase and matrix. Besides, high-angle grain boundaries were increased due to static recrystallization. It reveals that the deformation energy is high on these grain boundaries, which can result in the δ phase preferentially precipitated with a high energy (but not stable state), *i.e.*, particle-shaped.

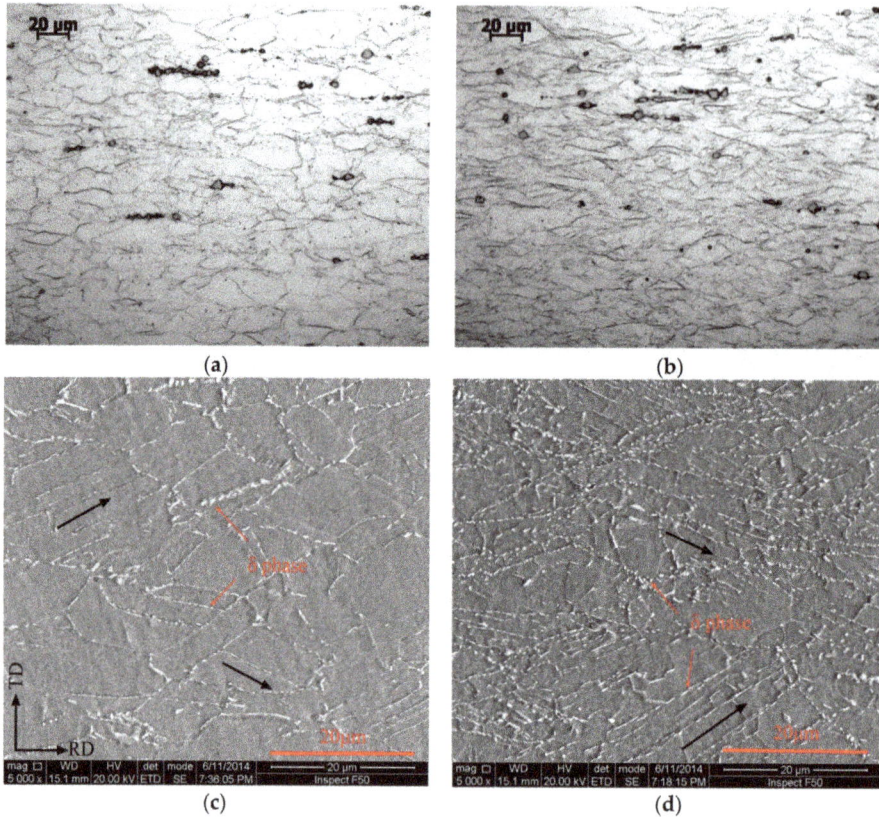

Figure 2. Microstructure evolution under different rolling reductions (980 °C 1 h) (a) $\varepsilon = 30\%$ (b) $\varepsilon = 50\%$ after cold rolling (c) $\varepsilon = 30\%$ (d) $\varepsilon = 50\%$ after solid solution.

3.2. Phase Fraction

Figure 3 demonstrates the diffraction patterns and phase fractions (wt. %) of SSS with different rolling reductions. It shows that the diffraction peaks of δ phase were gradually developed with the increases of rolling reduction, as shown with the arrow in Figure 3a. With the rolling reduction increasing from 0% to 30%, fraction of δ phase increases from 1.82% to 3.52%. When the rolling reduction reached to 70%, the fraction increased to 6.93%. The increase of δ phase's fraction was due to the large deformation during cold rolling process. In the process of cold rolling, the super vacancies were firstly formed with dislocations. However, in the heat treatment, Nb atoms were easily attracted by the super vacancy and finally form the vacancy-Nb atom pairs. Subsequently, the vacancy-Nb atom pairs diffuse to the defect rich on the deformation bands. It provides the necessary condition for the mass nucleation of δ phase, because of Nb element's non-equilibrium segregation. So the fraction of δ phase was increased with the rolling reduction after heat treatment.

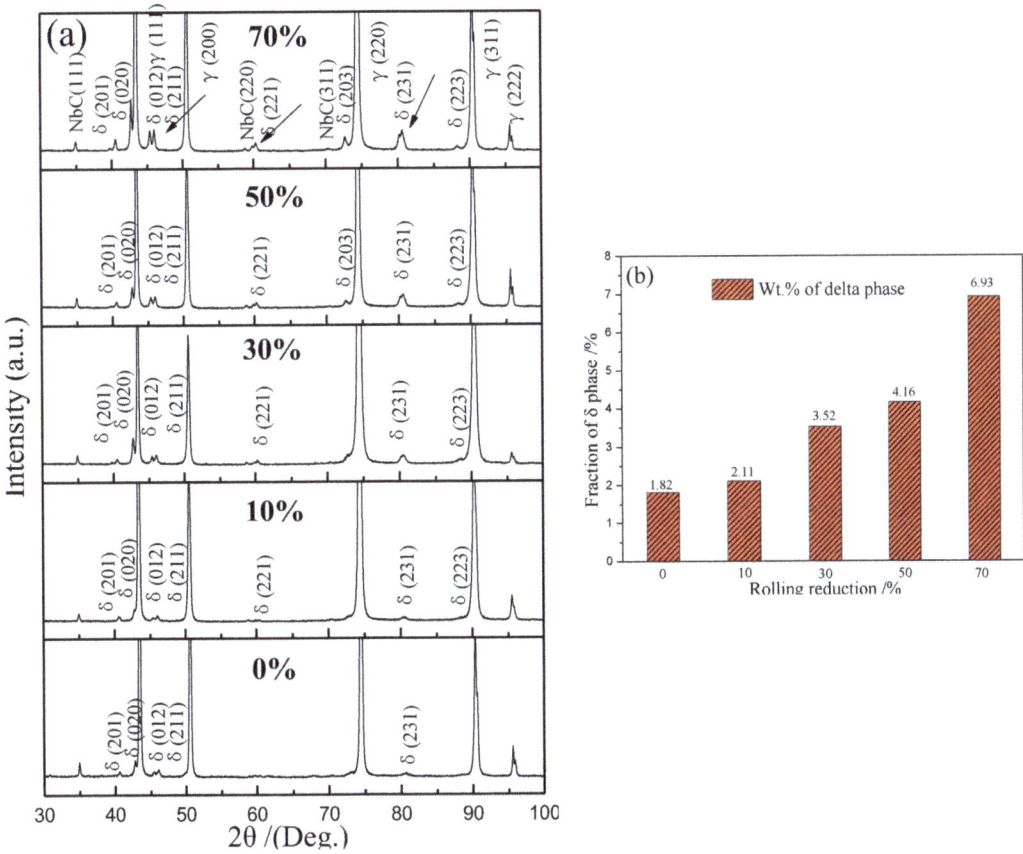

Figure 3. Diffraction patterns and phase fractions after solid solution under different cold rolling reduction (**a**) Diffraction patterns (**b**) Phase fractions.

3.3. Mechanical Properties

The mechanical properties will are changed due to the microstructure evolution and phase precipitation. The YS and UTS of CRS and SSS are shown in Figure 4. From Figure 4a,b, it can be seen that the YS and UTS of CRS increase with the increase of rolling reduction. On the contrary, the YS and UTS of SSS firstly increased then slightly decreased, when the rolling reductions increased. The YS and UTS of un-deformed CRS (0%) were 379.4 and 858.5 MPa, respectively. After cold rolling, the YS and UTS increased by 195.2 and 183.2 MPa, when the rolling reduction increases from 0% to 30%, respectively. When the rolling reduction reaches 70%, they further increases by 156.3 and 219.6 MPa separately comparing to the rolling reduction of 30%. After solid solution, the YS and UTS of un-deformed SSS0% were 457.3 and 923.2 MPa, respectively. They increased by 77.9 and 64.7 MPa compared to un-deformed CRS, as shown with the arrow in Figure 4. With increasing the rolling reduction to 30%, the YS and UTS of SSS increased to 624.1 and 1136.1 MPa; and 166.8 MPa and 212.9 MPa, respectively. However, when the rolling reduction reaches to 70%, the YS and UTS decreased to 593.6 and 1111.3 MPa, and 30.5 MPa and 24.8 MPa, respectively. These variations are relative to phase precipitation, static recrystallizaiton, texture and work hardening.

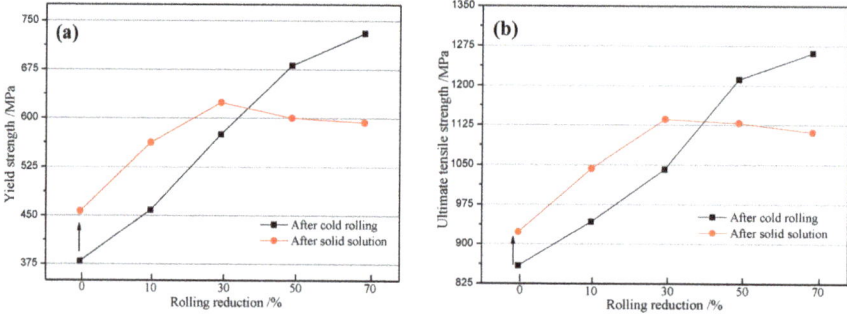

Figure 4. Mechanical properties after cold rolling and solid solution (**a**) Yield strength (**b**) Ultimate tensile strength.

3.4. Discussion

3.4.1. Strength Variation after Cold Rolling

In the process of cold rolling, the movement and tangle of dislocations formed a lot of deformation bands and sub-grain structure, as shown with black arrows in Figure 5a,b. The texture of {110} <112> C type was formed simultaneously, as shown in Figure 5d,e. When reduction ratio reached to 50%, the sub-grain structure and the texture were slightly reduced and weakened respectively as shown in Figure 5c,f.

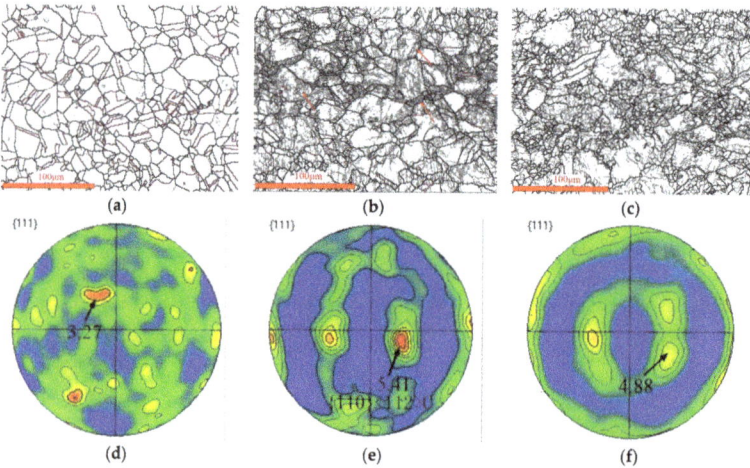

Figure 5. Microstructure and pole maps under different cold rolling reduction (**a**)/(**d**) $\varepsilon = 0\%$ (**b**)/(**e**) $\varepsilon = 30\%$ (**c**)/(**f**) $\varepsilon = 50\%$.

From Figure 5a–c, it can be seen that the grain size were changed with the increases of rolling reduction. From 0% to 30%, the grain size decreases from 11.9 to 11.7 μm. Further decline (9.8 μm) was recorded, when the rolling reduction increases to 50%.

To get the values of K and σ_0 in Hall-Petch relationship, size of 33.8 and 82.5 μm grains were acquired by annealing to the original sheet at 1000 °C and 1040 °C for 1 h, separately. Their corresponding YS were 347.6 and 313.84 MPa, separately. By linear fitting, the values of K and σ_0 were 547.1 MPa· mm$^{1/2}$ and 253.6 MPa, respectively. In turn, the Hall-Petch relationship based on the GH4169 superalloy can be expressed:

$$\sigma_d = 253.6 + 547.1 d^{-1/2} \tag{1}$$

where σ_d is the YS' variable value due to grain size, d is the average grain size. According to Equation (1), it can be implied that the grain size after cold rolling just has a small effect on the YS after cold rolling. Therefore, it is neglected in this part.

Except the grain sizes, the textures are also changed, as shown in Figure 5d–f. Taylor factor is a function of grain orientation and applied stress field. Therefore, the effect of texture evolution on the YS of CRS can be expressed [16]:

$$\Delta\sigma_T = M_i\tau_i - M_j\tau_j \tag{2}$$

where $\Delta\sigma_T$ is the YS' variable value due to texture evolution, M_i, M_j are the average values of Taylor factor under different rolling reductions, τ_i, τ_j are critical shear stress which can be calculated:

$$\tau = \tau_0 + \alpha Gb\rho^{1/2} \tag{3}$$

where τ_0 is the stress necessary to move a dislocation in the absence of other dislocations. It is also neglected due to its small value [21]. $\alpha = 0.5$ is a constant value between 0.3 and 0.6 [21]. $G = 76.3$ GPa is the shear modulus. $b = 36.3^{-8}$ cm is the Burgers vector. ρ is the dislocation density, which describes the level of work hardening and can be calculated as following [22]:

$$\rho = \frac{\beta^2}{4.35b^2} \tag{4}$$

where β is the full wave at half maximum (FWHM) for diffraction peak at halof GH4169 superalloy. From Equations (2)–(4), it can be seen that Equation (2) describes the coordinate effect of texture and working hardening on the YS.

Figure 6 is the evolution of Taylor factors under different rolling reduction of CRS. It shows that it firstly increases from 3.017 to 3.083, then decreased to 3.018. The Taylor factor under different rolling reductions do not change so much due to its FCC structure.

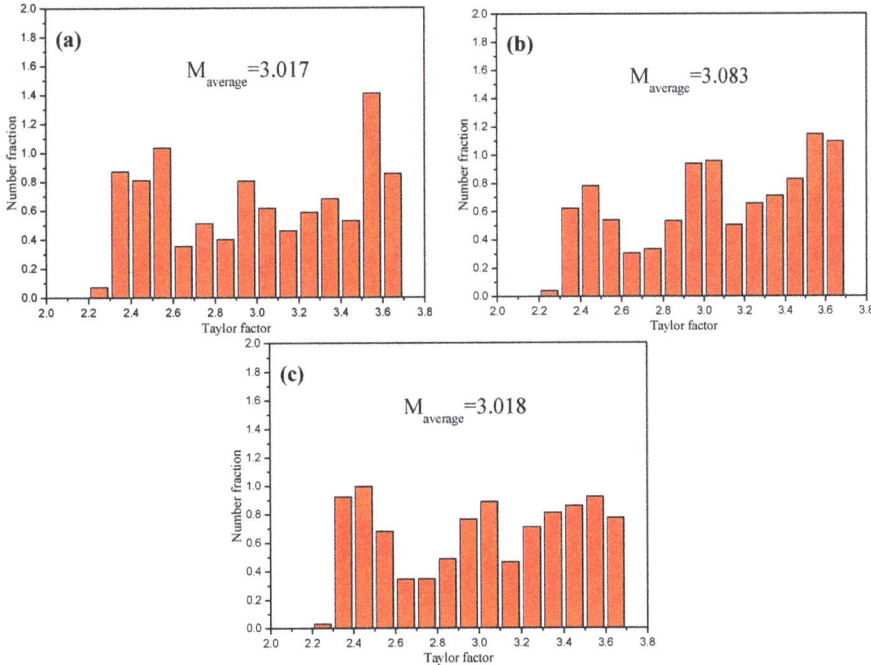

Figure 6. Taylor factor under different rolling reduction.

According to Equations (2)–(4), the effects of texture and work hardening on the YS of CRS are shown in Table 1. When cold rolling reduction increases from 0% to 30%, the contribution of texture and work hardening on the YS was approximately 233.4 MPa. When the rolling reduction reaches to 50%, the contribution was further increased by 109.9 MPa. It has the same trend with experiment results, as $\Delta\sigma_c$ shown in Table 1.

Table 1. Effects of texture and work hardening on the yield strength (YS) of cold rolled specimen (CRS).

$\beta \times 10^{-3}$/rad			τ_i/MPa			$\Delta\sigma_T$/Mpa		$\Delta\sigma_c$/Mpa	
0%	30%	50%	0%	30%	50%	0% → 30%	30% → 50%	0% → 30%	30% → 50%
7.1	11.1	13.3	130.1	203.0	243.8	233.4	109.9	195.2	155.5

Although the Taylor factors did not change so much during cold rolling process, the texture does have a great influence on the YS of CRS with the help of work hardening, including deformation bands and sub-grain structures.

3.4.2. Strength Variation after Solid Solution

After solid solution, δ was precipitated, except grain and texture's evolutions, which can be seen from Figures 2 and 3.

When the rolling reduction was below 30%, the dislocation density was obviously reduced, due to static recovery during solid solution and the appeared annealing twins (red lines in Figure 7a,b. In turn, the influence of work hardening was weakened, however, the texture of {110} <112> C type was reserved (Figure 7d,e). The maximum values of pole density are nearly the same as the situation of CRS. However, when the rolling reduction reached 50%, static recrystallization occurs, as shown in Figure 7c. Not only, it leads to the decreases of dislocation density and grain size, but also decreases the maximum values of pole density from 4.88 to 2.70 (Figure 7f).

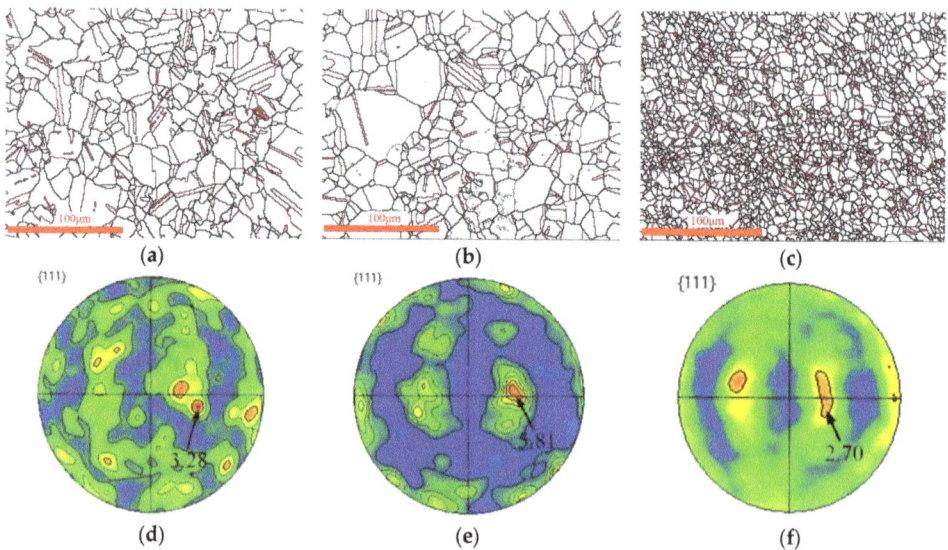

Figure 7. Microstructure and pole maps under different cold rolling reduction (**a**)/(**d**) ε = 0% (**b**)/(**e**) ε = 30% (**c**)/(**f**) ε = 50%.

After solid solution, the grain size of SSS with rolling reduction 0%, 30%, and 50% were 11.6, 11.5 and 5.1 μm respectively. For the situation of rolling reduction under 0 and 30%, the effect of grain size on the YS of SSS was ignored because of small differences. However, when the rolling reduction reached 50%, the grains were refined a lot due to static recrystallization. The contribution of grain size on the YS of SSS reach 81.4 MPa compared to rolling reduction under 30%. There might be a critical rolling reduction for static recrystallizaiton of cold rolled GH4169 superalloy sheet; the assumption that needs further research works.

During the solid solution, the texutre was also changed due to static recrystallizaiton. The evolution of Taylor factor under different rolling reduction of SSS was shown in Figure 8. The Taylor factor was firstly increased from 2.877 to 3.012 then it decreased to 2.924.

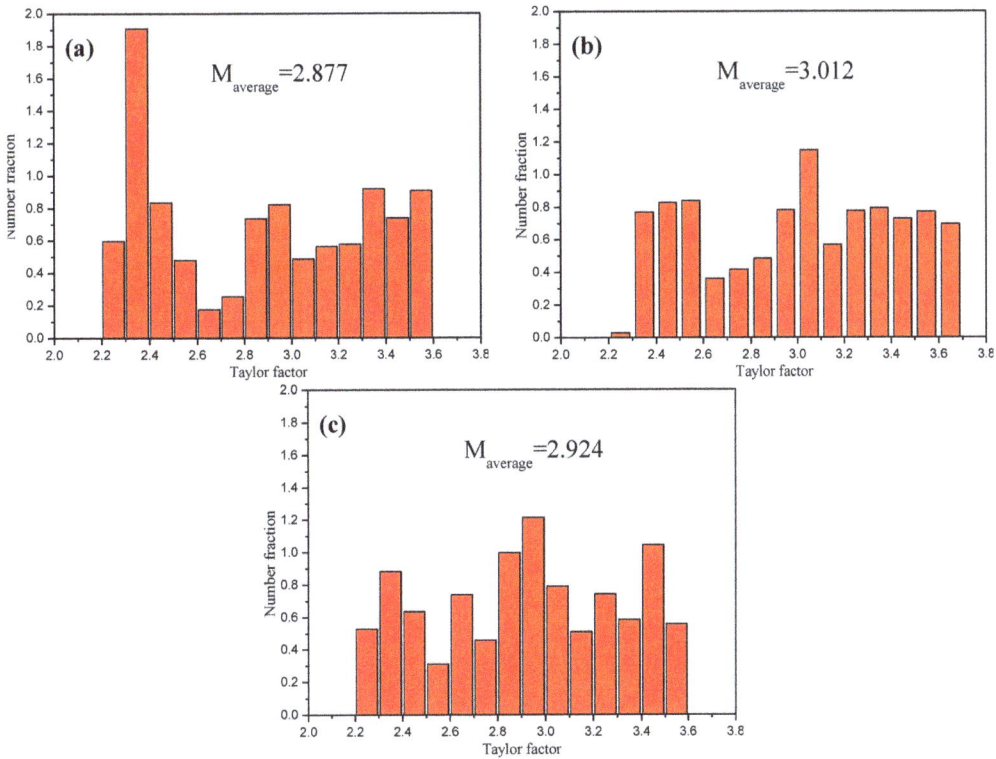

Figure 8. Taylor factors after solid solution (a) $\varepsilon = 0\%$ (b) $\varepsilon = 30\%$ (c) $\varepsilon = 70\%$.

According to Equations (2)–(4), the effects of texture and work hardening after solid solution on the YS are shown in Table 2. When rolling reduction increased from 0% to 30%, the contribution of texture and work hardening on the YS was approximately 146.2 MPa; whereas, the contribution decreased by 131.4 MPa, when the rolling reduction reaches to 50%. Even though, we might consider the improvement of grain refinement on the YS, there is still a big difference between the calculated and experimental values, as $\Delta\sigma_c$ shown in Table 2. It might be considered that c the precipitation of δ phase has an effect on the YS of SSS.

Table 2. Effects of texture and work hardening on the YS of solid solutioned specimen (SSS).

$\beta \times 10^{-3}$/rad			τ_i/MPa			$\Delta\sigma_T$/MPa		$\Delta\sigma_c$/MPa	
0%	30%	50%	0%	30%	50%	0% → 30%	30% → 50%	0% → 30%	30% → 50%
7.3	9.7	7.5	134.3	176.8	137.2	146.2	−131.4	166.8	−30.5

In Figures 3b and 4a, there was 1.82% (wt. %) of δ phase precipitated from un-deformed specimen during solid solution. Meanwhile, the corresponding YS was increased by 77.9 MPa.

Considering the fraction of phase and the changes of YS [23], there is a liner relationship between the fraction of δ phase and yield strength's improvement, expressed as

$$\sigma_\delta = 15.9W_\delta + 59.7 \tag{5}$$

where σ_δ is the contribution of δ phase to yield strength's improvement of GH4169 superalloy, W_δ is the fraction (wt. %) of δ phase.

According to Equation (5), when the fraction of δ phase increased from 1.82% to 3.5%, the contribution of δ phase on the YS of SSS increased from 88.6 MPa to 115.7 MPa, and 27.1 MPa, respectively. When the rolling reduction reaches to 50% and the fraction of δ phase increases to 6.93%, the contribution was also increased to 169.9 MPa. It increases by 54.2 MPa compared to 30% rolling reduction is 30%; however, YS still reveals a decreasing trend. When the rolling reduction reaches <50%, the texture and work hardening were completely disappeared, due to static recrystallization. It leads to YS decrease by 131.4 MPa according to Equations (2)–(4), as shown in Table 2. Besides, according to Ref. [1], it shows that when the second phase precipitated from the matrix, the contribution of solid solution strengthening decreased approximately to 20 MPa due to consumption of the strengthening elements. Therefore, considering the influence of δ phase's precipitation on the solid solution strengthening, the reason of the decreasing yield strength when the rolling reduction is larger than 50% are grain refinement, precipitation of δ phase, disappear of texture and work hardening. Moreover, the influence degrees are in the following order: texture and work hardening > grain refinement > precipitation of δ phase.

4. Conclusions

(1)   Following the increase of cold rolling reduction, the yield strength and ultimate tensile strength of GH4169 superalloy sheet were substantially increased, however, they were increased then decreased following solid solution.

(2)   The main reasons for yield strength changes of GH4169 superalloy are texture and work hardening, followed by the grain refinement and precipitation of δ phase. When the rolling reduction was below 30%, the influence of δ phase was greater than grain refinement and the opposite happen when the rolling reduction was larger than 50%.

(3)   The precipitation of δ phase promotes the improvement of yield strength and the relationship between the fraction of δ phase and the improvement of yield strength satisfies the following equation: $\sigma_\delta = 15.9W_\delta + 59.7$.

Acknowledgments: This work was supported by the fund of Special Inventive Fund of Science and Technology in Shenyang under the Contract Number F15-172-6-00 and the State Key Laboratory of Advanced Processing and Recycling of Non-ferrous Metals, Lanzhou University of Technology under the Contract Number SKLAB02014001.

Author Contributions: During the writing of this paper, we benefited from suggestions and critical insights provided by S.H.Z., M.C. and H.W.S. Valuable comments on a first draft were received from H.W.Z. and P.B.W. All experiments and analysis in this paper were accomplished by N.Y.Y.

Conflicts of Interest: The authors declare no conflict of interest.

References

1.   Ahmadi, M.R.; Povoden-Karadeniz, E.; Whitmore, L.; Stockinger, M.; Falahati, A.; Kozeschnik, E. Yield strength prediction in Ni-base alloy 718Plus based on thermo-kinetic precipitation simulation. *Mater. Sci. Eng. A* **2014**, *608*, 114–122. [CrossRef]

2.   Fisk, M.; Ion, J.C.; Lindgren, L.E. Flow stress model for IN718 accounting for evolution of strengthening precipitates during thermal treatment. *Comput. Mater. Sci.* **2014**, *82*, 531–539. [CrossRef]

3.   Zheng, L.; Zhang, M.C.; Chellali, R.; Bouchikhaoui, H.; Dong, J.X. Oxidation property of powder metallurgy EP741NP Ni-based superalloy at elevated temperatures. *Mater. Technol. Adv. Perform. Mater.* **2013**, *28*, 2–128. [CrossRef]

4.   Lin, Y.C.; Chen, X.M.; Wen, D.X.; Chen, M.S. A physically-based constitutive model for a typical nickel-based superalloy. *Comput. Mater. Sci.* **2014**, *83*, 282–289. [CrossRef]

5.   Zheng, L.; Zhang, M.C.; Chellali, R.; Dong, J.X. Investigations on the growing, cracking and spalling of oxides scales of powder metallurgy Rene 95 nickel-based superalloy. *Appl. Surf. Sci.* **2011**, *257*, 9762–9767. [CrossRef]

6.   Zheng, L.; Zhang, M.C.; Chellali, R.; Dong, J.X. Hot corrosion behavior of powder metallurgy Rene 95 nickel-based superalloy in molten $NaCl-Na_2SO_4$ salts. *Mater. Des.* **2011**, *32*, 1981–1986. [CrossRef]

7.   Sundararaman, M.; Mukhopadhay, P.; Banerjee, S. Precipitation of the δ-Ni_3Nb phase in two Ni base superalloys. *Metall. Mater. Trans. A* **1992**, *23*, 2015–2028. [CrossRef]

8.   Li, H.Y.; Kong, Y.H.; Chen, G.S.; Xie, L.X.; Zhu, S.G.; Sheng, X. Effect of different processing technologies and heat treatments on the microstructure and creep behavior of GH4169 superalloy. *Mater. Sci. Eng. A* **2013**, *582*, 368–373. [CrossRef]

9.   Yeh, A.C.; Lu, K.W.; Kuo, C.M.; Bor, H.Y.; Wei, C.N. Effect of serrated grain boundaries on the creep property of Inconel 718 Superalloy. *Mater. Sci. Eng. A* **2011**, *530*, 525–529. [CrossRef]

10. Kong, Y.H.; Liu, R.Y.; Chen, G.S.; Xie, L.X.; Zhu, S.G. Effects of different heat treatments on the microstructures and creep properties of GH4169 superalloy. *J. Mater. Eng. Perform.* **2013**, *22*, 1371–1377. [CrossRef]

11. Du, J.H.; Lv, X.D.; Deng, Q. Effect of heat treatment on microstructure and mechanical properties of GH4169 superalloy. *Rare Metal Mater. Eng.* **2014**, *43*, 1830–1834. [CrossRef]

12. Zhou, N.; Lv, D.C.; Zhang, H.L.; McAllister, D.; Zhang, F.; Mills, M.J.; Wang, Y. Computer simulation of phase transformation and plastic deformation in IN718 superalloy: Microstructural evolution during precipitation. *Acta Mater.* **2014**, *65*, 270–286. [CrossRef]

13. Wang, K.; Li, M.Q.; Luo, J.; Li, C. Effect of the δ phase on the deformation behavior in isothermal compression of superalloy GH4169. *Mater. Sci. Eng. A* **2011**, *528*, 4723–4731. [CrossRef]

14. Zhang, H.Y.; Zhang, S.H.; Cheng, M.; Li, Z.X. Deformation characteristics of δ phase in the delta-processed Inconel 718 alloy. *Mater. Charact.* **2010**, *61*, 49–53. [CrossRef]

15. Liu, W.C.; Chen, Z.L.; Xiao, F.R.; Yao, M.; Jiang, Z.Q.; Wang, S.G. Effect of cold rolling on the precipitation behavior of δ phase and γ'' phase in Inconel 718. *Acta Aeronaut. Astronaut. Sin.* **1999**, *20*, 279–282. [CrossRef]

16. Hansen, N. Polycrystalline strengthening. *Metall. Trans. A* **1985**, *16A*, 2167–2190. [CrossRef]

17. Liu, W.C.; Xiao, F.R.; Yao, M. Quantitative phases analysis of GH4169 by X-ray diffraction. *J. Mater. Sci.* **1997**, *16*, 769–771.

18. Liu, W.C.; Xiao, F.R.; Yao, M. Relationship between the lattice constant of γ phase and the content of δ phase, γ'' and γ' phases in Inconel 718. *Scr. Mater.* **1997**, *37*, 59–64. [CrossRef]

19. Li, R.B.; Yao, M.; Liu, W.C.; He, X.C. Isolation and determination for δ, γ'' and γ' phases in Inconel 718 alloy. *Scr. Mater.* **2002**, *46*, 635–638. [CrossRef]

20. Cai, D.Y.; Liu, W.C.; Li, R.B.; Zhang, W.H.; Yao, M. On the accurancy of the X-ray diffraction quantitative phases analysis method in GH4169. *J. Mater. Sci.* **2004**, *39*, 719–721.

21. Rauch, E.F.; Gracio, J.J.; Barlet, F. Work-hardening model for polycrystalline metals under strain reversal at large strain. *Acta Materialia* **2007**, *55*, 2030–2948. [CrossRef]

22. Wang, X.Q.; Cui, F.K.; Yan, G.P.; Li, Y.X. Study on the dislocation density change during cold roll-beating of 40Cr. *China Mech. Eng.* **2013**, *24*, 2248–2256.

23. Ye, N.Y.; Cheng, M.; Zhang, S.H.; Song, H.W.; Zhou, H.W.; Wang, P.B. Effect of δ Phase on Mechanical Properties of GH4169 Alloy at Room Temperature. *J. Iron Steel Res. Int.* **2015**, *22*, 752–756. [CrossRef]

Establishment of Heat Treatment Process for Modified 440A Martensitic Stainless Steel Using Differential Scanning Calorimetry and Thermo-Calc Calculation

Huei-Sen Wang * and Pei-Ju Hsieh

Academic Editor: Hugo F. Lopez

Deptartment of Materials Science and Engineering, I-Shou University, Kaohsiung 84001, Taiwan; pjhsieh@isu.edu.tw
*   Correspondce: huei@isu.edu.tw

Abstract: To provide a suitable microstructure and mechanical properties for modified Grade 440A martensitic stainless steel (MSS), which could facilitate the further cold deformation process (e.g., cold rolling), this work used differential scanning calorimetry (DSC) and Thermo-Calc software to determine three soaking temperatures for annealing heat treatment processes (HT1, HT2 and HT3). To verify the feasibility of the proposed annealing heat treatment processes, the as-received samples were initially heated to 1050 °C (similar to the on-line working temperature) for 30 min and air quenched to form a martensitic structure. The air-quenched samples were then subjected to three developed annealing heat treatment conditions. The microstructure and mechanical properties of the heat-treated samples were then investigated. Test results showed that considering the effects of the microstructure and the hardness, the HT1, the HT2 or the soaking temperatures between the HT1 and HT2 were the most recommended processes to modified Grade 440A MSS. When using the recommended processes, their carbides were fine and more evenly distributed, and the microhardness was as low as 210 Hv, which can be applied to the actual production process.

Keywords: heat treatment; martensitic stainless steel (MSS); differential scanning calorimetry (DSC); Thermo-Calc; microstructure; mechanical properties

1. Introduction

Grade 440A martensitic stainless steel (MSS) [1–3] is a high carbon (around 0.6 wt. %) and high chromium (around 16–18 wt. %) MSS. Given the advantages of high strength, moderate corrosion resistance, and good hardness and wear resistance, it is designed to be used for wear components, such as stainless steel knives, bearings, valves, nozzles, precision slides, *etc*. With a high content of C and Cr, a large sized carbide particle may precipitate during cooling due to solidification or high temperature operation. Large sized carbide particles not only can induce stress concentration when this material is utilized but also have a negative impact on the surface roughness of the material [4]. To eliminate the above problem by reducing the content of Cr, recently, a modified Grade 440A MSS (containing 0.6 wt. % of C and 12.7 wt. % of Cr) was developed. The amount and type of carbide particles in the new type of Grade 440A MSS have important effects on hardness, resistance to corrosion, wear, deformation process (e.g., rolling), *etc*., and are significantly influenced by associated heat treatments. Although the heat treatment process for the new type of MSS could be conducted by using the empirical approach or referred to the typical Grade 440A heat treatment process, optimal properties of the new type of MSS may not be obtained. Many recent studies [5–7] used the combination of differential scanning calorimetry (DSC) and Thermo-Calc calculation to obtain the phase transformation temperatures of materials, which were used to design suitable parameters for heat treatment in the deformation process of materials.

In this study, the combination of DSC and Thermo-Calc calculation was used to determine various annealing heat treatment parameters for air-cooled samples. After heat treatment, the microstructure and mechanical properties of the heat-treated samples were investigated to develop the most suitable fully annealing heat treatment conditions for modified Grade 440A MSS.

2. Materials and Experimental Procedure

The chemical composition of the modified Grade 440A MSS (Yieh Haing Enterprise Co., LTD, Kaohsiun, Taiwan) samples identified by inductively coupled plasma optical emission spectrometry (ICP-OES) used in this study is shown in Table 1. For comparison purposes, the chemical composition of a typical Grade 440A MSS is also shown in this table. The alloy was cast, hot rolled and cut in the form of 30 mm diameter by 40 mm length bars. To simulate the on-line process, the test was performed at a heating temperature of 1050 °C (a heating rate of 5 °C/min) for 30 min, which is similar to the on-line working temperature. Then, the alloy was air quenched, and the working temperatures for annealing were characterized using differential scanning calorimetry (DSC) and Thermo-Calc to determine the first temperature interval where the α phase transfers to the γ phase. For DCS analysis, 100 mg samples were placed in a NETZSCH 404C simultaneous thermal analyzer (NETZSCH Gerätebau GmbH, Selb, Germany) which was heated to 1500 °C at a heating rate of 5 °C/min. During the tests, all of the samples were conducted in a purged high-purity argon atmosphere using high-purity alumina crucibles. Simulations were made using Thermo-Calc software (Thermo-Calc software, Stockholm, Sweden), excluding inclusions or residual elements. The same compositions of the modified Grad 440A MMS were used, and an equilibrium phase diagram was plotted. Based on the results of the DSC tests and Thermo-Calc calculations, various working temperatures of annealing heat treatment conditions were determined. Annealing heat treatment processes were then carried out on the air-quenched samples. After the heat treatments, the samples were sectioned longitudinally, etched (HNO_3 1 mL + HCL 10 mL + H_2O 10 mL) and prepared for inspection. The microstructures of the samples were characterized using an optical microscope (OM, Olympus BX51M, Olympus, Tokyo, Japan.) and a scanning electron microscope (SEM; Hitachi S-4700, Hitachi High-Technologies Corporation, Tokyo, Japan) equipped with an energy dispersive spectrometer (EDS; HORIBA 7200-H, HORIBA, Ltd., Kyoto, Japan). The volume factions of the large sized carbides were measured using the intercept method, following the ASTM E562-08 standard [8] (standard test method for determining volume fraction by systematic manual point count). The Vickers microhardness was measured using a microhardness tester (Shimadzu HMV-2, Shimadzu Corporation, Kyoto, Japan) under a load of 300 g.

Table 1. Chemical compositions of typical and modified Grade 440A MSS.

Element (wt. %)	C	Mn	P	S	Cr	Mo
Typical 440A	0.60–0.75	1.00	0.35 max	0.030 max	16.0–18.0	0.075
Modified 440A	0.65	0.68	0.02	0.003	12.55	0.05

3. Results and Discussion

3.1. Thermo-Calc Prediction and DSC Analysis

The Fe–Cr–C ternary phase diagram calculated using Thermo-Calc is shown in Figure 1. The carbon content is plotted along the horizontal axis. In this figure, the vertical line indicates phase transformation of the test sample *vs.* heating temperature. Figure 2 shows the DSC heating curve of as-received materials. As seen, four major endothermic peaks in the heating curves of the materials were observed. Referring to Figure 1, the first peak corresponds to the phase transformation of α → γ, the second peak corresponds to the phase transformation of $C1(M_{23}C_6) \rightarrow C2(M_7C_3)$, the third peak corresponds to the dissolution of C2 and the fourth peak corresponds to the melting of the matrix.

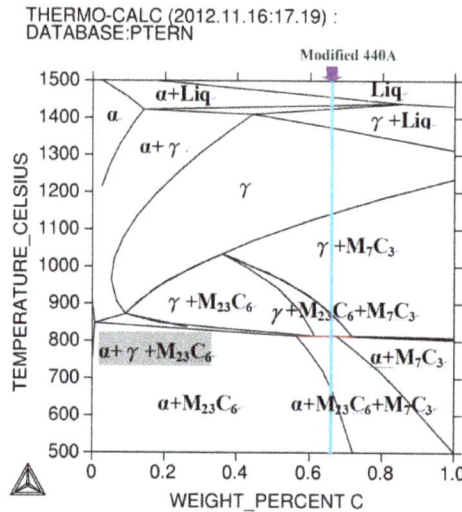

Figure 1. Phase diagram of the test sample calculated using Thermo-Calc.

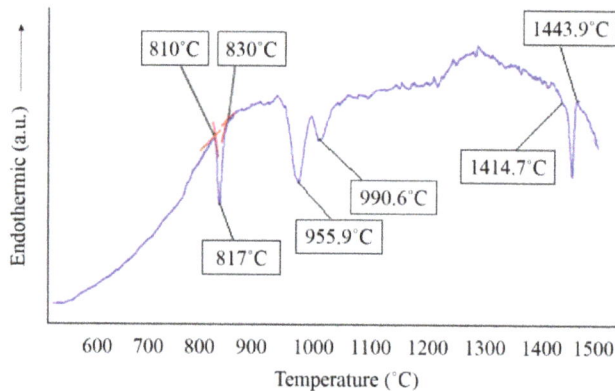

Figure 2. DSC traces measured from the as-received modified Grade 440 MSS at a heating rate of 5 K/min.

3.2. Design of Annealing Heat Treatment Processes

To prove the feasibility of the developed annealing heat treatment processes, the samples were heated to 1050 °C (similar to the on-line working temperature) for 30 min. Referring to Figure 1, at a slow heating rate (5 °C/min = 0.08 °C/s), an approximation γ phase and un-dissolved $C2(M_7C_3)$ carbides were obtained. After a short time of soaking at 1050 °C, the samples were air quenched. It was expected that the lath martensite, the most common phase in the microstructure; fine carbides (could be C1 or C2); retained austenite; and ferrite may be found in the microstructure. To provide a suitable microstructure and mechanical properties for a further cold deformation process, the annealing processes were required. Considering the size of the matrix grain, three soaking temperatures (840, 850 and 865 °C, defined as HT1, HT2 and HT3 processes, respectively), as shown in Figure 3, for annealing were determined in the temperature interval corresponding to the phase transformation of $\alpha \rightarrow \gamma$ (*i.e.*, 810 °C to 870 °C, referring to Figure 1). A slow heating rate (5 °C/min = 0.08 °C/s) was applied for the annealing process. After soaking followed by furnace cooling to 500 °C, the M_7C_3 phase fully transformed to $M_{23}C_6$ (see Figure 1). The test samples were then cooled by air.

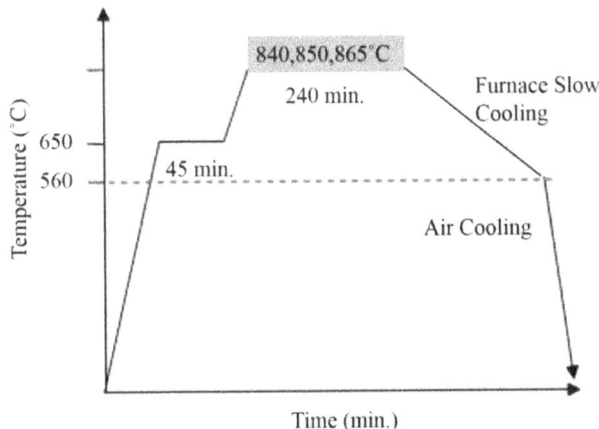

Figure 3. Schematic of various working temperatures of annealing heat treatment processes.

3.3. Microstructure Analysis of As-Received, As-Quenched and Heat-Treated Materials

Figure 4a–c show three micrographs of the as-received modified 440A martensitic stainless steel which were taken from the edge, quarter and center areas, respectively. The as-received sample was cut from the hot-rolled bar. As seen, due to the rapid cooling in this area, the microstructure consists of martensite and different types of carbides (grain size of carbides $\leqslant 1.5$ µm, and averaged volume fraction is 15%) distributed in a matrix of the edge area. In the quarter and center areas, due to the slower cooling rate in these areas, the microstructure mainly consists of perlite and different types of carbides (grain size of carbides $\leqslant 3$ µm, and averaged volume fraction is around 1%). Furthermore, it clearly exhibits larger sized globular carbide particles mainly formed in the quarter and center areas.

Figure 5 shows the microstructure of air-quenched modified 440A MSS from a temperature of 1050 °C to room temperature. As can be seen, a transformation from the austenite to martensitic phase was observed. Furthermore, the size and amount of carbide particles were decreased (grain size $\leqslant 1.5$ µm, and averaged volume fraction $\leqslant 1\%$) According to the phase diagram of the test sample calculated using Thermo-Calc (Figure 1), in addition to the martensitic phase, it is expected that the mixed structure of retained austenite, $M_{23}C_6$ (C1) and M_7C_3 (C2) carbide particles should also exist in the as-quenched samples. However, C2 carbide is generally unwanted due to its hard and brittle nature, which may have deleterious effects on the following cold working processes. To eliminate this phase, three annealing processes (HT1, HT2 and HT3) were conducted in this study. Figure 6 shows the microstructure of modified 440A MSS after HT1, HT2 and HT3 processes. It was found that the matrix structure of the martensitic phase transforms into a ferrite structure after three annealing processes. When compared to the as-quenched sample, more carbide particles were found on those samples. Referring to Figure 1, as the annealing samples were furnace cooled to 500 °C, the M_7C_3 phase in the as-quenched samples tended to fully transform to $M_{23}C_6$. Therefore, the carbide particles in those figures were mainly $M_{23}C_6$. However, as the soaking temperature was increased (e.g., HT3), more volume factions of the large sized carbide particles were observed.

In general, large sized carbide tends to result in a larger crack along the carbide grain during the cold working. This finding was also observed (see Figure 7) in our on-line production work. In this case, larger cracks were observed along the large sized (>5 µm) carbides. However, after HT1, HT2 and HT3 processes, it was observed the volume factions of the large sized (>5 µm) carbides are 0.6%, 0.8% and 2.0%, respectively. Hence, the use of HT1 or HT2 may be the better choice for actual production needs.

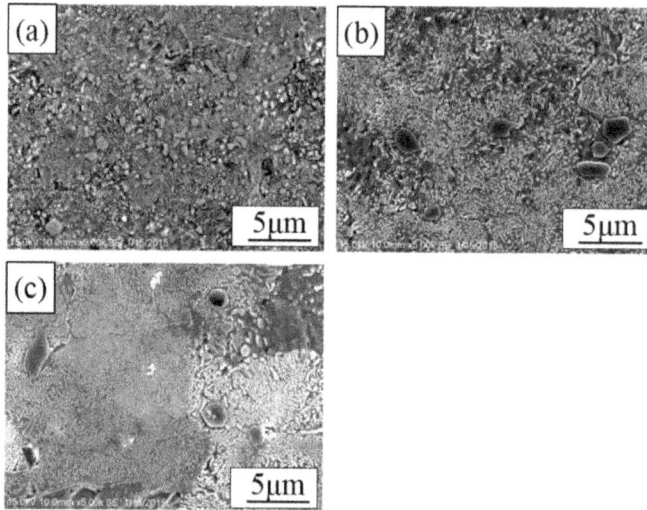

Figure 4. Microstructure of the as-received modified 440A MSS which were taken from (**a**) edge; (**b**) quarter; and (**c**) center areas, respectively.

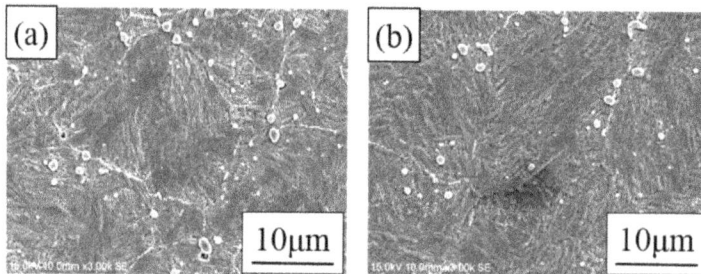

Figure 5. Microstructure of as-quenched modified 440A MSS at a temperature of 1050 °C, taken from the (**a**) quarter and (**b**) center areas, respectively.

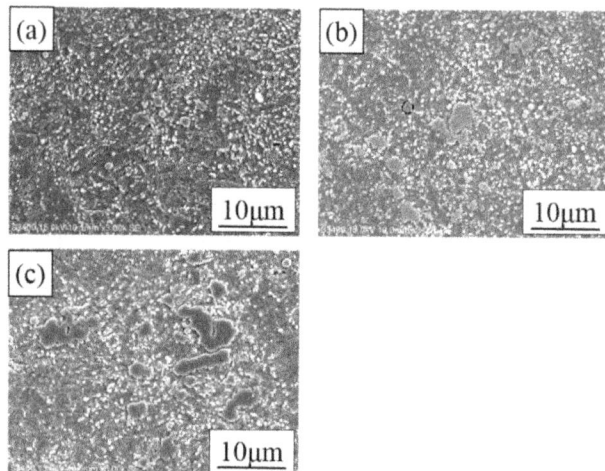

Figure 6. Microstructure of modified 440A MSS after (**a**) HT1; (**b**) HT2 and (**c**) HT3 processes.

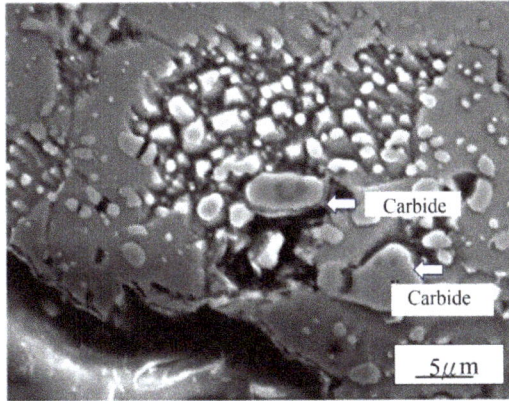

Figure 7. Initial position of crack during cold working.

3.4. Microhardness under Various Heat Treatment Processes

Figure 8 shows the microhardness of modified 440A MSS under various heat treatment processes. As seen, an average hardness value of 723.3 Hv was obtained from the as-quenched samples. After HT1, HT2 and HT3 processes, their hardness dropped significantly, approximately 210–230 Hv. As mentioned in the previous section, the decrease in hardness is attributed to the fact that most martensite transforms to ferrite after annealing heat treatment processes. However, considering the effects of the microstructure and hardness, the HT1, HT2 or the soaking temperature between the HT1 and HT2 are the most recommended approaches in actual production.

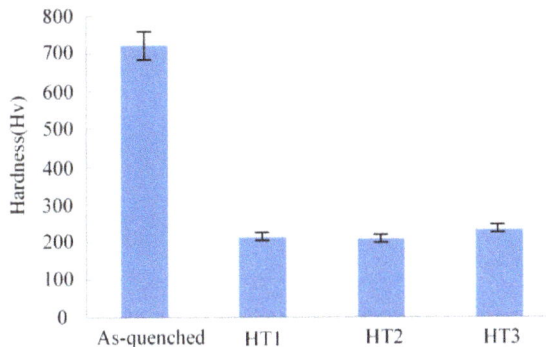

Figure 8. Microhardness of modified 440A MSS under various heat treatment processes.

4. Conclusions

Based on the results, the conclusions are drawn as follows:

1. The combination of DSC and the Thermo-Calc calculation approach was used to determine various annealing heat treatment parameters (HT1, HT2 and HT3) for the air-quenched samples.
2. After the HT1, HT2 and HT3 processes, the volume factions of the larger sized carbide (>5 µm) were 0.6%, 0.8% and 2.0%, respectively.
3. After the HT1, HT2 and HT3 processes, hardness values were approximately 210–230 Hv.
4. Considering the effects of the microstructure and hardness, the HT1, HT2 or soaking temperatures between HT1 and HT2, were the most recommended processes for the modified Grade 440A MSS.

Acknowledgments: This research is financially supported by Yieh Hsing Enterprise Co., LTD, Taiwan, under Project No. ISU101-IND-123.

Author Contributions: H.S.W. designed the experimental procedure; H.S.W. and P.J.H. conducted the experiments and analyzed the data; H.S.W. performed the analysis tools; H.S.W. wrote the paper.

Conflicts of Interest: The authors declare no conflict of interest.

References

1. Fan, R.; Gao, M.; Ma, Y.; Zha, X.; Hao, X.; Liu, K. Effect of heat treatment and nitrogen on microstructure and mechanical properties of 1Cr12NiMo martensitic stainless steel. *J. Mater. Sci. Technol.* **2012**, *28*, 1059–1066. [CrossRef]

2. Aksoy, M.; Yilmaz, O.; Korkut, M.H. The effect of strong carbide-forming elements on the adhesive wear resistance of ferritic stainless steel. *Wear* **2001**, *249*, 639–646. [CrossRef]

3. Andres, C.G.; Caruana, G.; Alvarez, L.F. Control of $M_{23}C_6$ Carbides in 0.45C–13Cr martensitic stainless steel by means of three representative heat treatment parameters. *Mater. Sci. Eng. A* **1998**, *241*, 211–215. [CrossRef]

4. Lin, C.C.; Lin, Y. Microstructure and mechanical properties of 0.63C–12.7Cr martensitic stainless steel. *Taiwan Chung Hua J. Sci. Eng.* **2009**, *7*, 41–46.

5. Andersson, J.O.; Helander, T.; Hdghmd, L.; Shi, P.; Sundman, B. Thermo-Calc & DICTRA computational tools. *Mater. Sci.* **2002**, *26*, 273–312.

6. Zhang, J.; Singer, R.F. Hot tearing of nickel-based superalloys during directional solidification. *Acta Mater.* **2002**, *50*, 1869–1879. [CrossRef]

7. Shi, Z.; Dong, J.; Zhang, M.; Zheng, L. Solidification characteristics and segregation behavior of Ni-based superalloy K418 for auto turbocharger turbine. *J. Alloy. Compd.* **2013**, *571*, 168–177. [CrossRef]

8. ASTM Standard E562–11. *Standard Test Method for Determining Volume Fraction by Systematic Manual Point Count*; ASTM International: West Conshohocken, PA, USA, 2011.

3

Investigation of Elastic Deformation Mechanism in As-Cast and Annealed Eutectic and Hypoeutectic Zr–Cu–Al Metallic Glasses by Multiscale Strain Analysis

Hiroshi Suzuki [1,*], Rui Yamada [2], Shinki Tsubaki [3], Muneyuki Imafuku [3], Shigeo Sato [4], Tetsu Watanuki [5], Akihiko Machida [5] and Junji Saida [2]

Academic Editor: Klaus-Dieter Liss

[1] Quantum Beam Science Center, Japan Atomic Energy Agency, Tokai, Naka, Ibaraki 319-1195, Japan
[2] Frontier Research Institute for Interdisciplinary Sciences, Tohoku University, Sendai, Miyagi 980-8578, Japan; rui-yamada@fris.tohoku.ac.jp (R.Y.); jsaida@fris.tohoku.ac.jp (J.S.)
[3] Faculty of Engineering, Tokyo City University, Setagaya, Tokyo 158-8857, Japan; shinki_tsubaki@yahoo.co.jp (S.T.); imafukum@tcu.ac.jp (M.I.)
[4] Graduate School of Science and Engineering, Ibaraki University, Hitachi, Ibaraki 316-8511, Japan; shigeo.sato.ar@vc.ibaraki.ac.jp
[5] Quantum Beam Science Center, Japan Atomic Energy Agency, Sayo, Hyogo 679-5148, Japan; wata@spring8.or.jp (T.W.); machida@spring8.or.jp (A.M.)
[*] Correspondence: suzuki.hiroshi07@jaea.go.jp

Abstract: Elastic deformation behaviors of as-cast and annealed eutectic and hypoeutectic Zr–Cu–Al bulk metallic glasses (BMG) were investigated on a basis of different strain-scales, determined by X-ray scattering and the strain gauge. The microscopic strains determined by Direct-space method and Reciprocal-space method were compared with the macroscopic strain measured by the strain gauge, and the difference in the deformation mechanism between eutectic and hypoeutectic Zr–Cu–Al BMGs was investigated by their correlation. The eutectic $Zr_{50}Cu_{40}Al_{10}$ BMG obtains more homogeneous microstructure by free-volume annihilation after annealing, improving a resistance to deformation but degrading ductility because of a decrease in the volume fraction of weakly-bonded regions with relatively high mobility. On the other hand, the as-cast hypoeutectic $Zr_{60}Cu_{30}Al_{10}$ BMG originally has homogeneous microstructure but loses its structural and elastic homogeneities because of nanocluster formation after annealing. Such structural changes by annealing might develop unique mechanical properties showing no degradations of ductility and toughness for the structural-relaxed hypoeutectic $Zr_{60}Cu_{30}Al_{10}$ BMGs.

Keywords: hypoeutectic Zr–Cu–Al bulk metallic glass; structural relaxation; X-ray scattering; pair distribution function; elastic modulus

1. Introduction

Bulk metallic glasses (BMG) exhibit interesting mechanical features such as high strength with high ductility (low Young's modulus), which is a different trend from typical metallic materials. On the other hand, structural relaxation has been known as a thermal behavior of metallic glasses that changes various mechanical properties with a few percent volume shrinkage by a heat treatment below the glass transition temperature T_g. Especially, structural relaxation-induced embrittlement would be a factor to degrade unique mechanical properties of the metallic glass. Meanwhile, Yokoyama *et al.* recently found that a hypoeutectic Zr–Cu–Al BMG with a Zr composition of 10% more than the eutectic composition shows no degradations of ductility and toughness after complete structural relaxation [1,2]. Furthermore, the fatigue property on the hypoeutectic BMG is independent of the annealing temperature, while that on a eutectic BMG changes after annealing. In addition, crystal-like ordering and icosahedral-like contrast are partially recognized in the amorphous glassy matrix after annealing in the hypoeutectic BMG, while the annealed eutectic BMG has homogeneous amorphous, glassy structure. Consequently, the microstructural changes after structural relaxation might be a crucial factor to affect the mechanical and physical properties of BMGs after annealing.

The atomic pair distribution function (PDF) obtained by X-ray scattering, which can evaluate a neighbor atomic distance, can quantitatively estimate the local strain of nanostructures with no or less crystal periodicity. The PDF technique has been utilized so far for the deformation analysis of metallic glasses [3–6]. For instance, it was clarified that the local atomic strain of the metallic glass obtained by the PDF technique is smaller than the macroscopic bulk strain, and various deformation models have been suggested, based on their observations. Therefore, the difference in the deformability between eutectic and hypoeutectic BMGs can be accessed by the PDF technique on a basis of the microscopic deformation behavior in an atomic level. In this study, the microscopic deformation behaviors of the eutectic and hypoeutectic Zr–Cu–Al BMGs are evaluated by the PDF technique with synchrotron high energy X-ray scattering, and a change in the mechanical properties, induced by structural relaxation for their BMGs, was discussed on a basis of the correlation between microscopic and macroscopic deformations.

2. Experimental Procedure

Specimens used in this study were as-cast and annealed eutectic $Zr_{50}Cu_{40}Al_{10}$ and hypoeutectic $Zr_{60}Cu_{30}Al_{10}$ BMGs. Conditions of annealing were 697 K for 2 min for the $Zr_{50}Cu_{40}Al_{10}$ BMG ($T_g = 706K$) and 661 K for 2 min for the $Zr_{60}Cu_{30}Al_{10}$ BMG ($T_g = 671K$). Hereafter, the $Zr_{50}Cu_{40}Al_{10}$ BMG and the $Zr_{60}Cu_{30}Al_{10}$ BMG call Z50 and Z60, respectively.

The X-ray scattering experiments were performed using high energy X-rays of 69.8 keV at BL22XU in SPring-8, Hyogo, Japan [7]. Figure 1a shows the schematic layout of the optical system used in this study. The dog-bone shaped specimen with 1.2 mm in thickness (see Figure 1b) was mounted on a load frame and was irradiated by an incident beam with a size of 0.3 mm × 0.3 mm. Diffraction from the specimen was measured by an Imaging Plate (IP) with 400 mm × 400 mm in size. An aluminum plate with 4 mm in thickness was set in front of the IP to reduce the background by fluorescent X-rays. Tensile loadings were applied to the specimen by using the load frame until 500 to 600 MPa with a crosshead speed of 0.1 mm/min, and the diffraction patterns were measured while holding each applied stress at seven different steps. The distances, L from the IP to the specimen were set to be 300 mm and 700 mm, and exposure times were 300 s and 120 s, respectively. Diffraction patterns in the loading and transverse directions were extruded by circumferentially integrating a range of $\pm 5°$ in the corresponding direction of the two-dimensional scattering image using the WinPIP software [8].

Two novel techniques, suggested by Poulsen *et al.* [3], were utilized for the strain analysis of amorphous metallic glasses, *i.e.*, Reciprocal-space (Q-space) method (QSM), which can measure the local strain from the peak shift of the first peak of the intensity function $I(Q)$ or the structure function $S(Q)$, and Direct-space method (DSM), which can measure the local strain directly from a change in the atomic distance obtained from the pair distribution function $G(r)$. The diffraction patterns measured at $L = 700$ mm and 300 mm were provided for QSM and DSM, respectively. The PDF was produced by the PDFgetX3 program [9] with a Q-range of Fourier transform from 1.4 to 17 Å^{-1}.

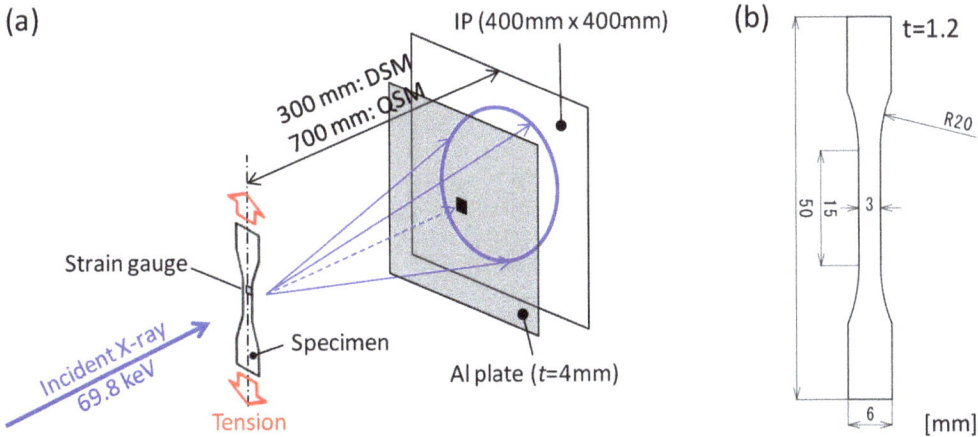

Figure 1. (a) Schematics of the optical layout and (b) the specimen used in this study.

3. Results

3.1. PDF and Microstructure

Figure 2 shows $I(Q)$, $S(Q)$ and $G(r)$ for Z50 and Z60. The radius, r in $G(r)$ indicates the distance from an average atom located at the origin. All functions for Z50 in Figure 2a,c,e seems to be unchanged by annealing. It is known that the free-volume of Z50 decreases due to structural relaxation by annealing [10]; however, any changes cannot be observed in their patterns. On the other hand, the first peak in $I(Q)$ and $S(Q)$ for Z60 sharpens after annealing, as shown in Figure 2b,d, suggesting the development of atomic ordering. Furthermore, some small diffraction peaks are recognizable in the annealed Z60. These diffraction peaks can be clearly observed in the difference curve of $I(Q)$ before and after annealing, and the difference also appears in $G(r)$, as shown in Figure 2f. Comparing with the diffraction patterns of Zr_2Cu (bct) and $ZrCu$ (fcc), calculated by the Rietveld simulation using RIETAN-FP [11], the diffraction peak positions measured approximately correspond to their both diffraction patterns. Considering the precipitation temperature of ZrCu, more than 988K [12], the crystalline phase precipitated in Z60 is expected to be Zr_2Cu since the annealing temperature was less than T_g (697 K). The volume fraction of this crystalline phase is predicted to be a few percent, and, hence, the crystalline phase with such a small volume fraction would not affect the macroscopic deformation behavior.

Figure 3 shows the Transmission Electron Microscope (TEM) images of the annealed Z60. Ambiguous fringe contrast related to nanocrystallization can be seen in the glassy amorphous matrix, as shown in Figure 3a. In addition, many close-range crystal-like orderings could be recognized in Figure 3b, which are typically marked by white circles and shown in the magnified images in Figure 3c,d, although the structure in general remains amorphous. This is a similar feature of the microstructure to the previous works [1,2]. Furthermore, relatively large crystal grains with about a few hundred nanometers were also observed slightly, which may contribute to diffraction peaks shown in Figure 2b.

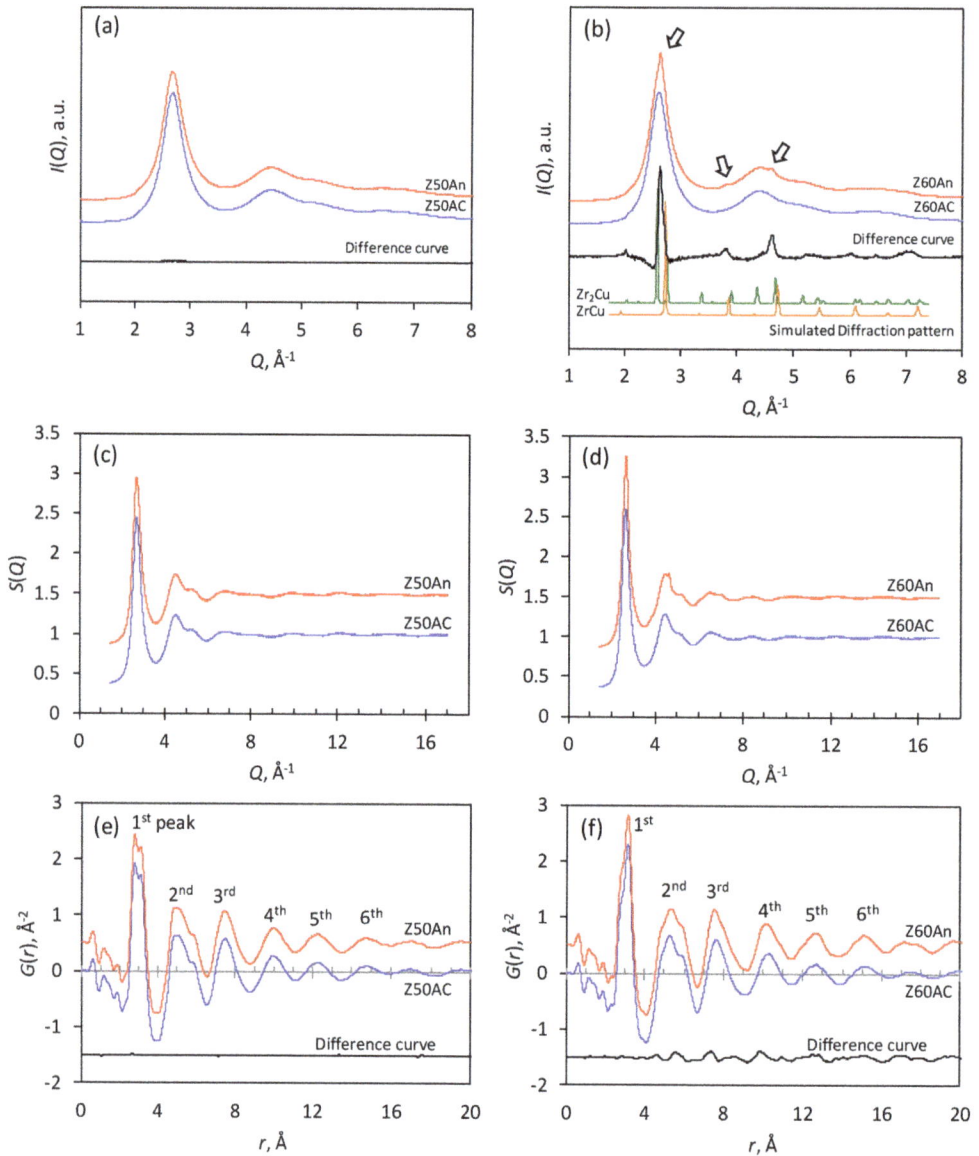

Figure 2. The intensity functions $I(Q)$ of (**a**) $Z_{50}Cu_{40}Al_{10}$ (Z50) and (**b**) $Z_{60}Cu_{30}Al_{10}$ (Z60) BMGs before (blue line) and after (red line) annealing and difference curve between them. "AC" and "An" after sample name denotes "as-cast" and "annealed", respectively. Arrows in (**b**) indicate the distinct diffraction peaks. For comparison, diffraction peak patterns of ZrCu and Zr_2Cu obtained by Rietveld simulation are shown in (**b**). (**c**) and (**d**) show the structure function $S(Q)$ of Z50 and Z60 before (blue line) and after (red line) annealing, respectively. (**e**) and (**f**) show the atomic pair distribution functions $G(r)$ before (blue line) and after (red line) annealing and difference curve between them for Z50 and Z60, respectively. In figures (**c**) to (**f**), $S(Q)$ and $G(r)$ after annealing are intentionally offset by +0.5 for easy comparison.

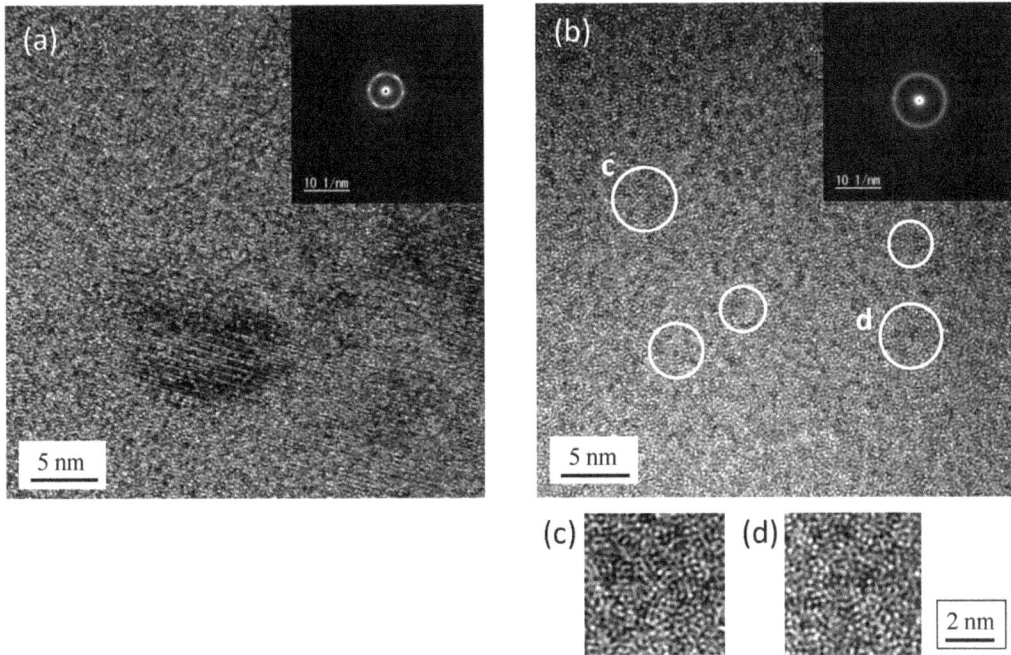

Figure 3. TEM images of the annealed Z60. Ambiguous fringe contrast related to nanocrystallization can be seen in (**a**), and random close-range crystal-like orderings could be recognized as typically shown by white circles in (**b**). Images in the circles, c and d, in (**b**) are magnified in (**c**) and (**d**), respectively. Electron diffraction in the inset of (**a**) shows the diffraction spots and that in the inset of (**b**) indicates the halo ring showing amorphous structure.

3.2. Tensile Deformation Behavior

Figure 4 shows a comparison of the strain changes measured by the strain gauge and QSM for Z50 and Z60. The macroscopic Young's modulus, E_M of Z50 measured by the strain gauge is increased from 87 GPa to 96 GPa by annealing, which is a typical trend caused by structural relaxation [2]. The macroscopic Poisson's ratios, ν_M before and after annealing are 0.34 and 0.36, respectively. A decrease in the Poisson's ratio with a few % to 10% is commonly observed after annealing [13–15], but cannot be found in this result due to less accuracy. In contrast, E_M for Z60 is constant at 86 GPa, regardless of whether annealing was performed. The macroscopic Poisson's ratios ν_M shows slightly decreasing from 0.37 to 0.36 by annealing, which is a typical trend caused by structural relaxation.

The microscopic deformations measured by QSM are derived from a shift in the first peak of $Q(r)$, fitted by the Voigt function for an initial state of the applied loading, providing the microscopic Young's modulus, E_Q and Poisson's ratio, ν_Q. The Young's modulus E_Q of the as-cast Z50 is 103 GPa that is larger than E_M (=87 GPa), and slightly increases to 106 GPa after annealing. The Poisson's ratio ν_Q before annealing is 0.32 that is smaller than the macroscopic Poisson's ratio ν_M (=0.34), and decreases to 0.29 after annealing. On the other hand, E_Q of the as-cast Z60 is 93 GPa that is larger than E_M (=86 GPa), and increases to 100 GPa after annealing, which is a different feature from the macroscopic Young's modulus showing a constant level before and after annealing. The microscopic Poisson's ratio ν_Q of the as-cast Z60 is 0.30 that is smaller than ν_M (=0.37), and slightly decreases to 0.29 after annealing. As described above, the microscopic deformation obtained by QSM exhibits different trends from the macroscopic deformation, which can be also found for the Zr–Al–Ni–Cu BMG reported by Sato *et al.* [5]. Furthermore, the relation between macroscopic and microscopic elastic moduli for the hypoeutectic metallic glass is different from that for the eutectic metallic glass.

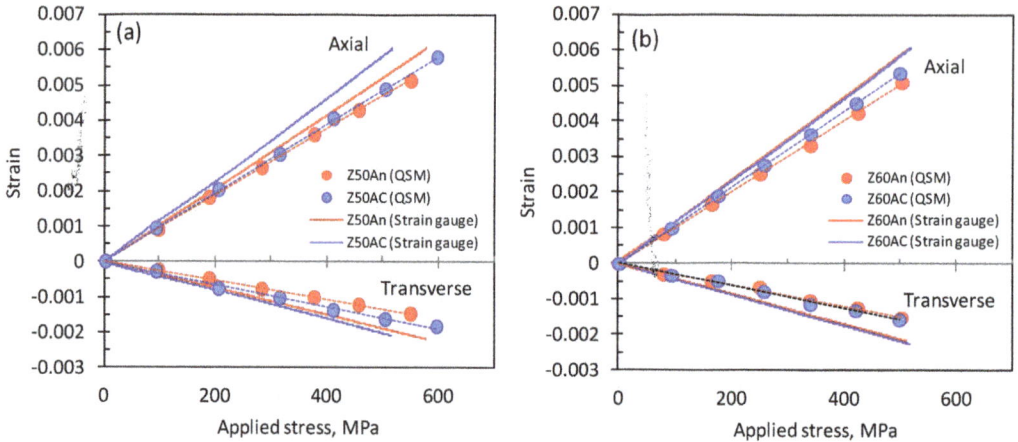

Figure 4. Stress-strain relations of (a) Z50 and (b) Z60 before (blue) and after (red) annealing, derived from macroscopic strains measured by the strain gauge and from microscopic strains determined by QSM. Average error bars are from ± 1.3 to $\pm 2.4 \times 10^{-4}$.

3.3. Comparison of Young's Modulus with Different Scales

To understand the microscopic elastic behavior accurately, the microscopic Young's moduli, E_D were assessed by a shift in each peak of $G(r)$, and they are plotted in Figure 5, as a function of r. The comparison of the Young's moduli and the Poisson's ratios measured by the strain gauge, QSM and DSM, is presented in Table 1. The Poisson's ratios were determined from the stress-strain relations in the axial and transverse directions, measured by each method.

Figure 5a shows the microscopic Young's modulus E_D as a function of r for Z50, compared with E_Q and E_M. The Young's modulus of the first peak at the lowest value of r is larger than the Young's moduli determined at larger values of r beyond the second peak. This is a typical trend for the metallic glasses as shown in previous studies [3–5]. The local strain at the lowest value of r represents the local structural deformation between first nearest-neighbor atoms, which depends on the inherent stiffness of atomic bonds. In contrast, the strains at larger values of r exhibit the average structural deformation including a redistribution of the free-volume. Here, the Young's modulus, E_{D_A} derived from the average structural deformation was calculated by averaging values in a flat region of E_D, and those before and after annealing were calculated to be 96 GPa and 100 GPa, respectively. These values are approximately 7 GPa smaller than the microscopic Young's modulus E_Q obtained by QSM. The difference between E_{D_A} and E_Q can be explained by considering E_Q to be increased by an influence of the local structural deformation at the lowest value of r. Moreover, a slight increase in E_{D_A} by annealing would be caused by hardening due to structural relaxation. In contrast, E_{D_A} before annealing is 9 GPa larger than E_M (=87 GPa) measured by the strain gauge, whereas E_M approaches E_{D_A} after annealing, i.e., $E_{D\_A} = 100$ GPa and $E_M = 96$ GPa.

Figure 5b shows a change in the microscopic Young's modulus, E_D as a function of r for Z60, compared with E_Q and E_M. The trend of a change in E_D is similar to that for Z50 shown in Figure 5a. However, a rise in E_D at the lowest value of r is smaller than that for Z50, suggesting the inherent stiffness of atomic bonds for Z60 to be smaller. The Young's modulus of the average structural deformation E_{D_A} before annealing is larger than the macroscopic Young's modulus E_M, i.e., $E_{D\_A} = 91$ GPa and $E_M = 86$ GPa. Furthermore, E_Q is 93 GPa that is slightly larger than E_{D_A}, but the Young's moduli for all scales tend to be comparable for the as-cast Z60. The Young's moduli E_M and E_{D_A} are almost unchanged by annealing, while E_Q is clearly increased. In addition, the annealed Z60 shows a specific trend that the microscopic Young's modulus E_D determined at the second peak of $G(r)$ increases after annealing.

The Poisson's ratio also shows a characteristic trend that the macroscopic Poisson's ratio ν_M is larger than the microscopic Poisson's ratios ν_Q and ν_D, which is independent of the sample condition. Moreover, any drastic changes cannot be observed before and after annealing.

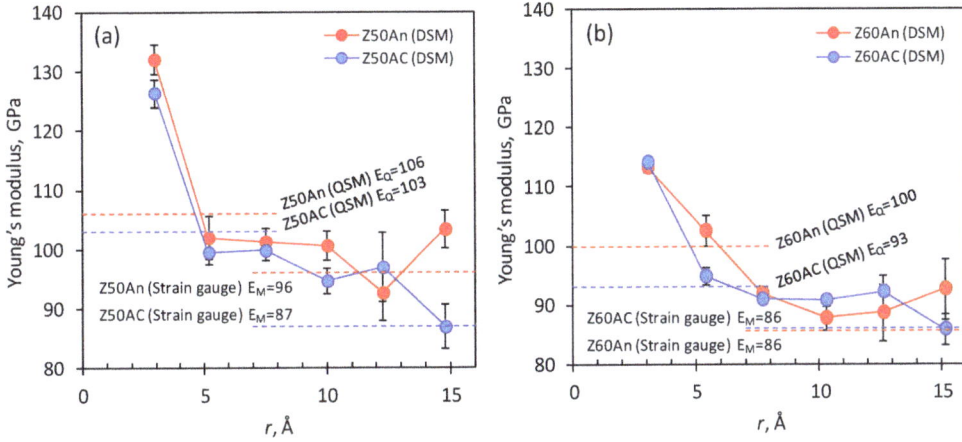

Figure 5. Microscopic Young's moduli of (a) Z50 and (b) Z60 before (blue) and after (red) annealing, which are derived from DSM, compared with Young's moduli evaluated by QSM and the strain gauge.

Table 1. Young's moduli and Poisson's ratios of Z50 and Z60 evaluated by reciprocal-space (Q-space) method (QSM), direct-space method (DSM) and the strain gauge. "AC" and "An" denote "as-cast" and "annealed", respectively.

Elastic Modulus	Z50		Z60	
	AC	An	AC	An
E_Q (QSM), GPa	103 ± 1	106 ± 1	93 ± 0	100 ± 1
ν_Q (QSM)	0.32	0.29	0.30	0.29
E_{D_A} (DSM Average), GPa	96 ± 5	100 ± 4	91 ± 3	90 ± 2
ν_{D_A} (DSM Average)	0.28	0.27	0.27	0.24
E_M (Strain gauge), GPa	87 ± 0	96 ± 0	86 ± 0	86 ± 0
ν_M (Strain gauge)	0.34	0.36	0.37	0.36

4. Discussion

4.1. Strain-Scale Observed by Each Technique

First of all, let us classify strain-scales observed by QSM, DSM and strain gauge. The strain-scale determined by QSM indicates an average deformation of nanoscale structures with less crystal periodicity involving crystal-like orderings and a glassy amorphous structure or none at all. On the other hand, the strain scale determined by DSM provides an average microscopic deformation of all composed structures including crystalline phases. In contrast, the strain-scale determined by the strain gauge is a macroscopic deformation of the specimen.

4.2. Deformation Model of Eutectic Z50-BMG

The microstructural model suggested by Ichitsubo et al. [16] provides an idea of the deformation model for the metallic glass. Figure 6a shows the schematic illustration of the simplified microstructural model of Z50 before and after annealing. The microstructure of a metallic glass is known to be a heterogeneous structure composed of strongly-bonded regions (SBRs) with low mobility and high density (low free-volume fraction), surrounded by weakly-bonded regions (WBRs) with high mobility and low density (high free-volume fraction). It is further suggested that an icosahedral atomic configuration generally exists in SBR in Zr-based metallic glasses [17]. Therefore, $G(r)$ in Figure 2e involves the structural information of both regions, *i.e.*, SBR and WBR, weighted by each volume fraction. Since the volume fraction of SBRs for the Zr-based BMG is typically larger than that of WBRs [18], the microscopic strain determined by DSM should be dominated by SBRs. Consequently, E_{D_A} in Table 1 predominantly represents the microscopic Young's modulus of SBR. On the other hand, the macroscopic deformation measured by the strain gauge must be a total deformation including both WBRs and SBRs. As the elastic constant of WBR is softer than that of SBR, the macroscopic Young's modulus E_M depends on the volume fraction of each region [18]. In particular, E_M can approach the microscopic Young's modulus E_{D_A} as the volume fraction of SBRs increases. In the present study, E_M is significantly smaller than E_{D_A} before annealing, suggesting relatively inhomogeneous microstructure with high volume fraction of WBRs. After annealing, in contrast, reduction in the free-volume in WBRs due to structural relaxation increases the volume fraction of SBRs, constructing relatively homogeneous microstructure. Therefore, E_M approaches E_{D_A} by annealing since the elastic homogeneity is developed with an increase in the volume fraction of SBRs.

The relation between macroscopic and microscopic Poisson's ratios can be explained by the same mechanism of the Young's modulus with different Poisson's ratios between WBR and SBR. Ichitsubo *et al.* suggests that the Poisson's ratio of WBR is higher than that of SBR [18], explaining the present results showing the macroscopic Poisson's ratio to be larger than the microscopic Poisson's ratio. However, we cannot see any changes in the Poisson's ratio by annealing, suggesting that the measurement resolution might be insufficient for observing its changes

The elastic homogeneity inhibits generation of a shear transformation zone, leading to decrease in its toughness or ductility and plasticity. Therefore, degradation of deformability of the eutectic Z50 due to structural relaxation would be originated from homogenization of the glassy nanostructure by reduction in the volume fraction of WBRs.

4.3. Deformation Model of Hypoeutectic Z60-BMG

Figure 6b shows the schematic illustration of the simplified microstructural model of Z60 before and after annealing. Since the crystal-like orderings are partially recognized in the annealed Z60, the as-cast Z60 might originally have some crystal-like ordered SBRs which can be nuclei of nanoclusters after annealing. This is supported by the fact showing the first peaks of $I(Q)$ and $S(Q)$ for the as-cast Z60 in Figure 2b, and d is slightly sharper compared with those for the as-cast Z50. In addition, the as-cast Z60 has relatively homogeneous microstructure with large volume fraction of SBRs since E_M is originally close to E_{D_A}.

It is known that the intensity damping of $G(r)$ is correlated to damping of the structural coherence [19]. Accordingly, a discrete $G(r)$ for the crystal-like ordered regions with the size of a few nanometers or less decreases immediately with an increase of r. Therefore, an increase in E_D at the second peak of $G(r)$ for the annealed Z60 would be affected by nanocluster formation after annealing. A similar trend appears in E_Q to be increased after annealing since E_Q could be affected by the change in the Young's modulus E_D at the second peak of $G(r)$. In contrast, the average deformation determined at larger values of r above the third peak of $G(r)$ represents the microscopic Young's modulus of the average microstructure (E_{D_A} in Table 1), and is almost unchanged by annealing. Therefore, an increase in E_D at the second peak of $G(r)$ after annealing suggests that there are locally harder regions such as nanoclusters in relatively softer amorphous glassy matrix. As described above, the annealed Z60 exhibits structural inhomogeneity owing to nanocluster formation, likewise elastic homogeneity would be decreased after complete structural relaxation. However, the result that E_M and E_{D_A} are almost unchanged by annealing suggests that the elastic properties of the annealed SBRs are almost unchanged from that of the as-cast SBRs, and that WBRs surrounding nanoclusters would play a role of damper to minimize an influence of nanoclusters on the macroscopic elastic constant. In addition, nanocluster formation in SBRs after annealing might improve deformability of glassy structure by branching and pinning the shear band around the nanoclusters [20]. This would be one reason why the hypoeutectic Z60 inhibits degradation of mechanical properties after complete structural relaxation.

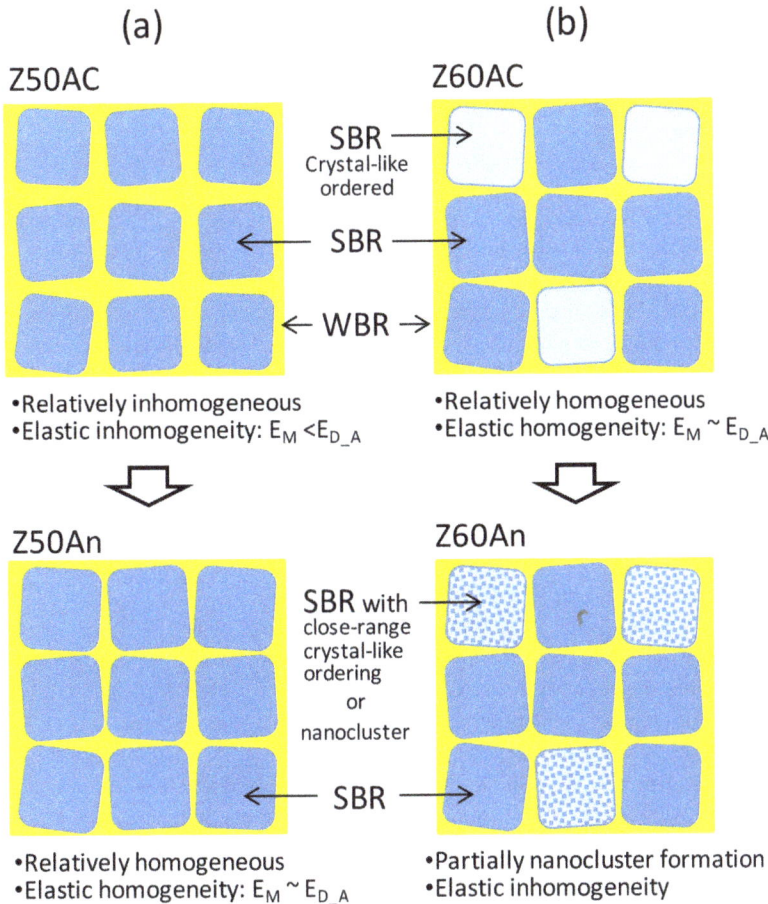

Figure 6. Schematic illustration of the simplified microstructural models of (**a**) Z50 and (**b**) Z60 before and after annealing. The weakly-bonded regions (WBR) shown by the yellow color surround the strongly-bonded regions (SBR) shown by the blue color.

5. Conclusions

In this study, the difference of the elastic deformation behaviors between eutectic $Zr_{50}Cu_{40}Al_{10}$ and hypoeutectic $Zr_{60}Cu_{30}Al_{10}$ BMGs was investigated by comparing strains with different scales obtained by the Direct-space method (DSM), Reciprocal-space (Q-space) method (QSM) and the strain gauge method. The eutectic $Zr_{50}Cu_{40}Al_{10}$ BMG obtains more homogeneous microstructure by free-volume annihilation after annealing, improving resistance to deformation but degrading ductility because of a decrease in the volume fraction of WBRs (weakly-bonded regions) with relatively high mobility. On the other hand, the as-cast hypoeutectic $Zr_{60}Cu_{30}Al_{10}$ BMG originally has homogeneous nanostructure with high volume fraction of SBRs (strongly-bonded regions) but loses structural and elastic homogeneities because of nanocluster formation partially after annealing. Such structural changes by annealing might develop unique mechanical properties showing no degradations of ductility and toughness for the structural-relaxed hypoeutectic $Zr_{60}Cu_{30}Al_{10}$ BMGs.

Acknowledgments: This work has been supported by a Grant-in-Aid of the Ministry of Education, Sports, Culture, Science and Technology, Japan, Scientific Research (A) (No. 23246109). The synchrotron radiation experiment was performed at the SPring-8 with the approval of Japan Synchrotron Radiation Research Institute (JASRI) as Proposal No. 2013B3724. The authors wish to acknowledge the experimental assistance of T. Shobu and A. Shiro at Japan Atomic Energy Agency (JAEA), and K. Shimizu at Tokyo City University. The authors would also like to acknowledge N. Igawa at JAEA for his beneficial assistance.

Author Contributions: H.S., M.I, S.S and J.S. conceived and designed the experiments; T.W. and A.M. contributed experimental instruments; H.S., R.Y., S.T. and J.S. performed the experiments and analyzed the data; all authors contributed to the interpretation of the data; H.S. wrote the paper.

Conflicts of Interest: The authors declare no conflict of interest.

References

1. Yokoyama, Y.; Yamasaki, T.; Nishijima, M.; Inoue, A. Drastic Increase in the Toughness of Structural Relaxed Hypoeutectic $Zr_{59}Cu_{31}Al_{10}$ Bulk Glassy Alloy. *Mater. Trans.* **2007**, *48*, 1276–1281. [CrossRef]

2. Yokoyama, Y.; Yamasaki, T.; Liaw, P.K.; Inoue, A. Study of the structural relaxation-induced embrittlement of hypoeutectic Zr–Cu–Al ternary bulk glassy alloys. *Acta Mater.* **2008**, *56*, 6097–6108. [CrossRef]

3. Poulsen, H.F.; Wert, J.A.; Neuefeind, J.; Honkimaki, V.; Daymond, M. Measuring strain distributions in amorphous materials. *Nat. Mater.* **2005**, *4*, 33–36. [CrossRef]

4. Hufnagel, T.C.; Ott, R.T. Structural aspects of elastic deformation of a metallic glass. *Phys. Rev. B* **2006**. [CrossRef]

5. Sato, S.; Suzuki, H.; Shobu, T.; Imafuku, M.; Tsuchiya, Y.; Wagatsuma, K.; Kato, H.; Setyawan, A.D.; Saida, J. Atomic-scale characterization of elastic deformation of Zr-Based metallic glass under tensile stress. *Mater. Trans.* **2010**, *51*, 1381–1385. [CrossRef]

6. Liss, K.-D.; Qu, D.D.; Yan, K.; Reid, M. Variability of Poisson's ratio and enhanced ductility in amorphous metal. *Adv. Eng. Mater.* **2013**, *15*, 347–351. [CrossRef]

7. Watanuki, T.; Machida, A.; Ikeda, T.; Ohmura, A.; Kaneko, H.; Aoki, K.; Sato, T.J.; Tsai, A.P. Development of a single-crystal X-ray diffraction system for hydrostatic-pressure and low-temperature structural measurement and its application to the phase study of quasicrystals. *Philos. Mag.* **2007**, *87*, 2905–2911. [CrossRef]

8. Fujihisa, H. Recent progress in the power X-ray diffraction image analysis program PIP. *Rev. High Press. Sci. Technol.* **2005**, *15*, 29–35. [CrossRef]

9. Juhás, P.; Davis, T.; Farrow, C.L.; Billinge, S.J.L. PDFgetX3: A rapid and highly automatable program for processing powder diffraction data into total scattering pair distribution functions. *J. Appl. Crystallogr.* **2013**, *46*, 560–566. [CrossRef]

10. Ishii, A.; Hori, F.; Iwase, A.; Fukumoto, Y; Yokoyama, Y.; Konno, T.J. Relaxation of free volume in $Zr_{50}Cu_{40}Al_{10}$ bulk metallic glasses studied by positron annihilation measurements. *Mater. Trans.* **2008**, *49*, 1975–1978. [CrossRef]

11. Izumi, F.; Momma, K. Three-dimensional visualization in powder diffraction. *Solid State Phenom.* **2007**, *130*, 15–20. [CrossRef]

12. Zeng, K.J.; Hamalainen, M. A new thermodynamic description of the Cu–Zr system. *J. Phase Equilib.* **1994**, *15*, 577–586. [CrossRef]

13. Louzguine-Luzgin, D.V.; Fukuhara, M.; Inoue, A. Specific volume and elastic properties of glassy, icosahedral quasicrystalline and crystalline phases in Zr–Ni–Cu–Al–Pd alloy. *Acta Mater.* **2007**, *55*, 1009–1015. [CrossRef]

14. Kumar, G.; Rector, D.; Conner, R.D.; Schroers, J. Embrittlement of Zr-based bulk metallic glasses. *Acta Mater.* **2009**, *57*, 3572–3583. [CrossRef]

15. Ngai, K.L.; Wang, L.-M.; Liu, R.; Wang, W.H. Microscopic dynamics perspective on the relationship between Poisson's ratio and ductility of metallic glass. *J. Chem. Phys.* **2014**. [CrossRef]

16. Ichitsubo, T.; Matsubara, E.; Yamamoto, T.; Chen, H.S.; Nishiyama, N.; Saida, J.; Anazawa, K. Microstructure of fragile metallic glasses inferred from ultrasound-accelerated crystallization in Pd-based metallic glasses. *Phys. Rev. Lett.* **2005**. [CrossRef] [PubMed]

17. Matsubara, E.; Ichitsubo, T.; Itoh, K.; Fukunaga, T.; Saida, J.; Nishiyama, N.; Kato, H.; Inoue, A. Heating rate dependence of T_g and T_x in Zr-based BMGs with characteristic structures. *J. Alloys Compds.* **2009**, *483*, 8–13. [CrossRef]

18. Ichitsubo, T.; Kato, H.; Matsubara, E.; Biwa, S.; Hosokawa, S.; Matsuda, K.; Uchiyama, H.; Baron, A.Q.R. Static heterogeneity in metallic glasses and its correlation to physical properties. *J. Non-Cryst. Solids* **2011**, *357*, 494–500. [CrossRef]

19. Kodama, K.; Iikubo, S.; Taguchi, T.; Shamoto, S. Finite size effects of nanoparticles on the atomic pair distribution functions. *Acta Crystallogr.* **2006**, *A62*, 444–453. [CrossRef] [PubMed]
20. Saida, J.; Setyawan, A.D.; Kato, H.; Inoue, A. Nanoscale multistep shear band formation by deformation-induced nanocrystallization in Zr–Al–Ni–Pd bulk metallic glass. *Appl. Phys. Lett.* **2005**. [CrossRef]

4

Biodegradable Behaviors of Ultrafine-Grained ZE41A Magnesium Alloy in DMEM Solution

Jinghua Jiang [1,2], Fan Zhang [1], Aibin Ma [1,*], Dan Song [1], Jianqing Chen [1], Huan Liu [1] and Mingshan Qiang [1]

Academic Editors: Vineet V. Joshi and Alan Meier

[1] College of Mechanics and Materials, Hohai University, Nanjing 210098, China; jinghua-jiang@hhu.edu.cn (J.J.); zhangf416@gmail.com (F.Z.); songdancharls@hhu.edu.cn (D.S.); chenjq@hhu.edu.cn (J.C.); liuhuanseu@163.com (H.L.); 15751873896@163.com (M.Q.)

[2] Jiangsu Key Laboratory of Advanced Structural Materials and Application Technology, Nanjing 226000, China

* Correspondence: aibin-ma@hhu.edu.cn

Abstract: The main limitation to the clinical application of magnesium alloys is their too-fast degradation rate in the physiological environment. Bio-corrosion behaviors of the ZE41A magnesium alloy processed by multi-pass equal channel angular pressing (ECAP) were investigated in Dulbecco's Modified Eagle Medium (DMEM) solution, in order to tailor the effect of grain ultrafining on the biodegradation rate of the alloy implant. Hydrogen evolution tests indicated that a large number of ECAP passes decreased the stable corrosion rate of the alloy after the initial incubation period. Potentiodynamic polarization curves showed that more ECAP passes made the corrosion potential nobler and the corrosion tendency lower. Corroded surfaces of the ECAPed alloy indicated a higher resistance toward localized corrosion due to the homogeneous redistribution of broken second phases on the ultrafine-grained Mg matrix. It suggests that grain ultrafining can decrease the biodegradable rate of the magnesium alloy-containing rare-earth elements and tailor the lifetime of the biodegradable material.

Keywords: magnesium alloy; biodegradation; ECAP; grain refinement; bio-corrosion

---

1. Introduction

Several million people suffer from bone fractures annually, which have to be surgically fixed by internal bone implants. Conventional inert metal implants (made of Ti alloys, stainless steels and Co-Cr alloys) have several disadvantages such as stress shielding and additional surgical intervention [1]. To overcome these drawbacks, biodegradable implants are of special interest for clinical applications. Most of the current biodegradable implants are made of polymers, which have relatively low strength and unpredictable degradation rates [2]. Therefore, magnesium and its alloys have garnered more and more attention for their excellent biocompatibility, elastic modulus and compressive yield strength, similar to those of human bone, and other positive effects on the growth of new bone after implanting in a physiological environment [3–5].

Magnesium alloys as biodegradation implants had already been applied to orthopedic and trauma surgery in the middle of the last century. However, the applications are seriously limited due to the excessively rapid degradation rate in the physiological environment [6,7]. Previous studies have shown that grain refinement can improve the corrosion resistance of some Mg alloys in Cl^--containing solution [8,9]. Equal-channel angular pressing (ECAP) is an effective technique to fabricate bulk ultrafine-grained (UFG) metallic materials with almost the same geometrical shape, and exceptional mechanical or/and physical advantages [8,10,11]. Until now, only a few studies have been carried out to investigate the influence of ECAP on corrosion behaviors of magnesium and its alloys [12,13]. The limited literatures present that these ECAP procedures endow some magnesium alloys with higher corrosion resistance and better mechanical properties [8,9,14]. However, the corrosion results in NaCl solution cannot be used to perfectly predict biodegradation behaviors of magnesium alloys in the physiological environment. Therefore, particular attention should be paid to reveal the effect of ECAP on the bio-corrosion behavior of Mg alloys, in order to reveal the effect of grain ultrafining on the biodegradation rate of the alloy implant.

During the degradation of magnesium alloys, most alloying elements will dissolve into the human body. Therefore, the magnesium alloy containing Al, Cd and heavy metals is not appropriate for clinical application from the medical aspect [15]. The ZE41A magnesium alloy with low rare-earth (RE) content is suitable to act a biodegradable material since some RE elements are proven to be tolerated by the human body [14,16]. Our previous work showed that the ZE41A alloy after multi-pass ECAP consists of UFG Mg matrix and homogeneously dispersed second-phase particles, which contribute to better mechanical properties at room temperature and the higher corrosion resistance in NaCl solutions [17,18]. Thus, the aim of this paper is to study the biodegradation behavior of the ECAPed ZE41A alloy in a simulated body fluid, and then tailor the biodegradation rate of the UFG magnesium alloy containing rare-earth elements.

2. Experimental Section

A commercial as-cast ZE41A (Mg-4.9Zn-1.4RE-0.7Zr) alloy was used to prepare bulk UFG samples through a large number of ECAP passes. The multi-pass ECAP was conducted at 603 K using a rotary-die having a channel angle of $90°$ and through route A [8]. The dimension of the ECAP billet was 20 mm × 20 mm × 40 mm. These billets were continuously ECAP-fabricated from eight to 60 passes to obtain the uniformly distributed and UFG microstructure. Three kinds of ZE41A samples with various ECAP passes (eight, 16 and 60, respectively) were applied to investigate the effect of microstructure evolution on the bio-corrosion behavior. The microstructure was observed by means of Nikon Eclipse ME600 optical microscope(Nikon Instruments, Inc., Tokyo, Japan).

The degradation rate of the alloy during immersion can be monitored by the evolved hydrogen volume, and this method is easy to implement and is not prone to errors inherent in the weight loss method [19]. Herein, the ZE41A samples cut from the ECAP billets were immersed in 300 mL Dulbecco's modified eagle medium (DMEM) solution for 1–10 days at 310 K to study the hydrogen evolution rate variation with the immersion time. The DMEM solution was made using DMEM Cell Culture Media (without sodium bicarbonate, Merck Millipore Beijing Skywing technology Co., Ltd, Beijing, China), sodium bicarbonate (reagent grade) and distilled water. The initial pH value of the solution was 7.4, similar to that of human blood plasma.

Each sample was molded into epoxy resin with an exposed area for hydrogen evolution testing and electrochemical measurement, while the working surface was ground with SiC emery papers up to 2000 grit. The cylindrical test samples for eight and 16 ECAP passes had the diameter of 5 mm, while the exposed area of the plate samples for 60 ECAP passes was 7.6 mm × 1.8 mm. Hydrogen bubbles from each sample were collected into a burette, which detailed procedures have been reported elsewhere [20]. The hydrogen evolution rate of the 60-pass sample was also measured in 0.9 wt. % NaCl solution at room temperature as compared with the degradation performance in DMEM solution. Potentiodynamic polarization curves of the samples immersed for four days were measured in the simulated body fluid at 310 K, by means of an advanced electrochemical system of PARSTAT 2273 (Princeton Applied Research, Oak Ridge, TN, USA),. The three-electrode system has a saturated calomel electrode (SCE) as a reference and a platinum electrode as a counter and the sample as a working electrode. The polarization scan started from −2 V to −1 V at a scan rate of 1 mV/s.

After immersion testing for 2–10 days, the samples were removed from the DMEM solution, washed with distilled water and dried in air. The surfaces of the corroded samples were observed by HIROX KH-7700 Digital Microscope (Hirox Asia Ltd., Hong Kong, China), before and after the corrosion products were removed. The corroded samples were cleaned with chromate acid (200 g/L CrO_3 + 10 g/L $AgNO_3$) for 5 min to remove corrosion products.

3. Results and Discussion

3.1. Microstructure of the ECAP-Fabricated ZE41A Alloy

Figure 1 presents optical micrographs of the as-cast ZE41A alloy with various ECAP passes. Primary microstructure of the as-cast alloy consists of equiaxed α-Mg grains and a small amount of net-like second phases, which are distributed mainly at α-Mg grain boundaries. After various ECAP passes, the microstructure of the alloy showed a continuous change with significant grain refinement of the Mg phase and homogeneous redistribution of broken second phases. Large elongated grains with a few percentages of ultrafine grains are observed for the eight-pass sample, and the second phases begin breaking and dispersing in the Mg matrix. Dynamic recrystallization (DRX) has been proven to be an efficient mechanism of grain refinement in the plastic deformation process [21]. After eight passes of ECAP, only small amounts of ultrafine grains are found due to incomplete recrystallization. The grains of the 16-pass sample are further refined with a more homogeneous dispersion of second phase particles. It shows that the percentage of ultrafine α-Mg grains increases with the ECAP passes. The alloy with a large pass number of ECAP can cause complete DRX to general equiaxed ultrafine grains. Hence, the 60-pass sample obtains UFG α-Mg grains (average size of about 2.5 μm) and fine second-phase particles which are homogeneously dispersed.

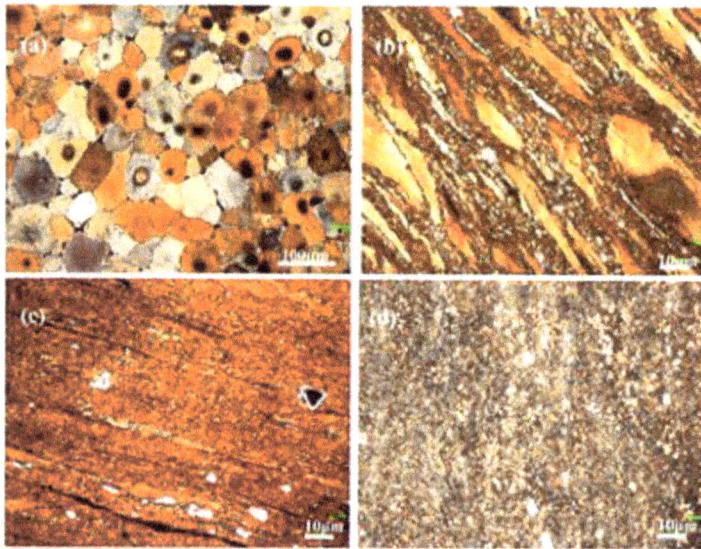

Figure 1. Micrographs of the ZE41A alloy before and after multi-pass ECAP: (**a**) As-cast; (**b**) eight passes; (**c**) 16 passes; (**d**) 60 passes.

Figure 2 presents the results of hydrogen evolution of the ECAPed samples immersed in DMEM solution and the 60-pass sample immersed in 0.9 wt. % NaCl solutions. The degradation process of the Mg alloy after multi-pass ECAP was time-dependent in the simulated body fluid. The hydrogen evolution of the samples immersed in DMEM solution went through three stages: an incubation period, a fast corrosion period and then a steady period. However, the 60-pass sample immersed in 0.9 wt. % NaCl solution basically shows a linear change. For those samples immersed in DMEM solution, a decrement of the incubation period with the increase of the ECAP pass was observed. It is probable due to the higher corrosive tendency of the Mg matrix acquiring lots of energy through the EACP processing. According to the method from Shi and Atrens [22], the corrosion rates of the ECAPed samples in different immersion periods are summarized in Table 1. All three samples immersed in DMEM solution show a close corrosion rate which is less than 1.4 mm/y in the first period. The biggest corrosion rate is observed for the 60-pass sample in the fast corrosion period, which results in the earliest entrance into the steady period. During the steady period, the 60-pass sample shows the best corrosion resistance.

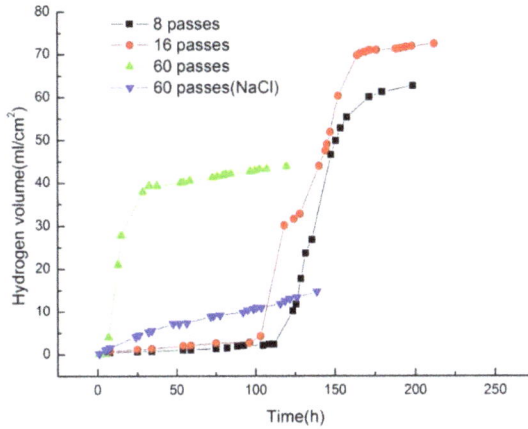

Figure 2. Hydrogen evolution diagrams of the ECAPed samples in DMEM solution.

Table 1. Corrosion rate of ECAPed samples in DMEM solution.

ECAP Pass	Duration of Incubation Period (h)	Hydrogen Evolution Rate (mL/cm^2/h)			Corrosion Rate (mm/y)		
		Incubation Period	Fast Corrosion Period	Steady Period	Incubation Period	Fast Corrosion Period	Steady Period
8	120	0.019	1.38	0.075	1.04	75.48	4.10
16	96	0.022	1.45	0.050	1.20	79.30	2.73
60	5	0.025	2.60	0.045	1.37	142.20	2.46

Figure 3 presents potentiodynamic polarization curves of the ECAPed alloy after four days of immersion in DMEM solution. Three samples with different ECAP passes have similar polarization curves, indicating similar electrochemical corrosion behavior in the DMEM solution. Table 2 lists corresponding parameters including the corrosion potential (E_{corr}), the corrosion current density (I_{corr}) and the corrosion rate (CR_i) calculated from the polarization curves by Tafel extrapolation [22]. It shows that the increase of the ECAP pass makes the corrosion potential higher and the corrosion tendency lower. After immersion for four days, the corrosion current density of the 16-pass sample is the lowest but the 60-pass one is the biggest. Similar to the existing literature [22], the corrosion rates evaluated from polarization curves are lower than those evaluated from hydrogen evolution. Furthermore, the results of polarization measurement are consistent with the curves of hydrogen evolution in Figure 2. It is because the 16-pass sample after four days of immersion is at the incubation period but the 60-pass one is at the steady period. It seems that the anodic polarization plot of the 60-pass sample appears a passive zone.

Table 2. Parameters calculated from the potentiodynamic polarization curves in Figure 3.

ECAP Pass	8	16	60
E_{corr} (mV SCE)	−1358.59	−1352.57	−1337.43
I_{corr} (mA/cm^{-2})	1.71×10^{-2}	1.46×10^{-2}	2.08×10^{-2}
CR_i (mm/y)	0.39	0.33	0.47

Figure 4 presents surface morphologies of the corroded samples after different immersion times with/without corrosion products. It is obvious that pitting corrosion is the main corrosion type for the eight-pass sample and the 16-pass one, and the difference between the two samples is the amount and dimension of the corrosion pits. Several deep and obvious pits are observed for the eight-pass sample, while the surface of the 16-pass sample shows more but smaller pits. A reasonable argument can be deduced that the Mg matrix around second phases was corroded firstly, and the second-phase particles could not be held by the matrix and dropped from the site with the progress of the corrosion, thus leaving the pits after removing the corrosion products. Similar conditions and discussions can be found in other reports [17,18]. Therefore, the smaller pits were formed in the 16-pass sample with finer second-phase particles. It is notable that the 60-pass sample exhibits mainly a uniform corrosion. After 60 ECAP passes, the alloy achieved a large amount of ultrafine grains and homogeneously dispersed second-phase particles. It makes the pits in the 60-pass sample more uniform and smaller.

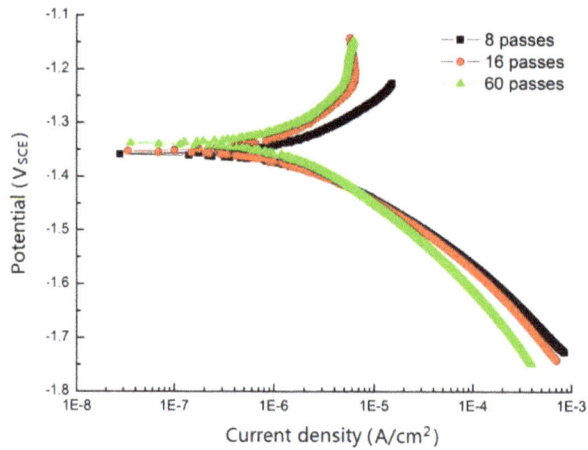

Figure 3. Potentiodynamic polarization curves of the ECAPed ZE41A alloy after immersion for four days in DMEM solution.

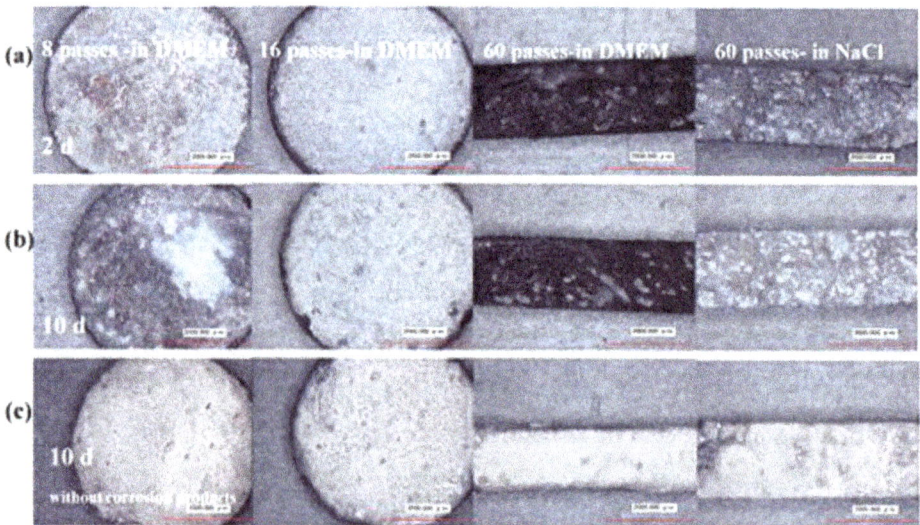

Figure 4. Morphology of the ECAPed samples after immersion for (**a**) two days; (**b**) 10 days; and (**c**) 10 days without corrosion products.

Considering that the degradation rate of Mg alloy in DMEM solution is one order of magnitude higher than that in Hank's solution [23], the corrosion rate of the 60-pass sample is significantly smaller than that of as-cast ZE41A Mg alloy reported by other research [15]. Hence, ECAP is an efficient method to improve the corrosion resistance of ZE41A Mg alloy in the physiological environment and tailor the lifetime of the biodegradable implant. This finding indicates that desirable retardation of the excessively rapid degradation of Mg alloys can be provided by grain ultrafining. It suggests that there is value in further investigating such effects by broadening both the alloy base and the processing techniques in order to control the appropriate biodegradation rate of Mg alloys for the clinical application.

4. Conclusions

The bio-corrosion behavior of a rare-earth magnesium alloy processed by multi-pass ECAP was investigated in a simulated body fluid. The present study shows that the biodegradation rate of the ZE41A alloy in the DMEM solution can be significantly reduced by grain ultrafining. The results open up a new possibility of tailoring the biodegradation of Mg alloys.

(1)   The degradation process of the UFG Mg alloy was time-dependent in the simulated body fluid. The ZE41A alloy via different ECAP passes went through an incubation period, a fast corrosion period and then a steady period in DMEM solution. With the ECAP pass number increasing, the incubation period was shortened and the biodegradation rate at the steady corrosion period was decreased.

(2)   The finer grains and more homogeneous second phases after more ECAP passes can improve the bio-corrosion resistance of the magnesium alloy. It results in a higher resistance toward localized corrosion and a lower biodegradation rate during the steady corrosion period. The UFG alloy for 60 ECAP passes has excellent corrosion resistance in DMEM solution.

(3)   ECAP should be taken into consideration as an efficient technique to control the bio-corrosion rate of Mg alloys in the physiological environment and tailor the lifetime of the biodegradable implant, owing to its significant ability to acquire bulk UFG materials with homogeneous second phases.

Acknowledgments: This work is financially supported by Natural Science Foundation of Jiangsu Province of China (Grant No. BK20131373), Qing Lan Project and the Opening Project of Jiangsu Key Laboratory of Advanced Structural Materials and Application Technology (Grant No. ASMA201404).

Author Contributions: The work presented here was carried out in collaboration between all authors. Aibin Ma and Jinghua Jiang defined the research theme. Jinghua Jiang, Fan Zhang and Dan Song designed methods and experiments, carried out the laboratory experiments, analyzed the data, interpreted the results and wrote the paper. Jianqing Chen, Huan Liu and Mingshan Qiang co-designed experiments, discussed analyses and interpretation. All authors have contributed to, seen and approved the manuscript.
    The author hopes that this paper can make its due contribution to the successful application of the high-performance Mg alloy implant.

Conflicts of Interest: The authors declare no conflict of interest.

References and Notes

1.   Lee, J.Y.; Han, G.S.; Yu, C.K. Effects of impurities on the biodegradation behaviour of pure magnesium. *Met. Mater. Int.* **2009**, *15*, 955–961. [CrossRef]

2.   Levesque, J.; Dube, D.; Fiset, M.; Mantovani, D. Materials and properties for coronary stents. *Adv. Mater. Process.* **2004**, *162*, 45–48.

3.   Mani, G.; Feldman, M.D.; Patel, D.; Agrawal, C.M. Coronary stents: A materials perspective. *Biomaterials* **2007**, *28*, 1689–1710. [CrossRef] [PubMed]

4.   Witte, F.; Hort, N.; Vogt, C.; Cohen, S.; Kainer, K.U.; Willumeit, R.; Feyerabend, F. Degradable biomaterials based on magnesium corrosion. *Curr. Opin. Solid State Mater. Sci.* **2008**, *12*, 63–72. [CrossRef]

5. Witte, F. The history of biodegradable magnesium implants: A review. *Acta Biomater.* **2010**, *6*, 1680–1692. [CrossRef] [PubMed]

6. Mueller, W.D.; Nascimento, M.L.; de Mele, M.F.L. Critical discussion of the results from different corrosion studies of Mg and Mg alloys for biomaterial applications. *Acta Biomater.* **2010**, *6*, 1749–1755. [CrossRef] [PubMed]

7. Staiger, M.P.; Pietak, A.M.; Huadmai, J.; Dias, G.J. Magnesium and its alloys as orthopedic biomaterials: A review. *Biomaterials* **2006**, *27*, 1728–1734. [CrossRef] [PubMed]

8. Valiev, R.Z.; Langdon, T.G. Principles of equal-channel angular pressing as a processing tool for grain refinement. *Prog. Mater.Sci.* **2006**, *51*, 881–981. [CrossRef]

9. Alvarez-Lopez, M.; Pereda, M.D.; del Valle, J.A.; Fernandez-Lorenzo, M.; Garcia-Alonso, M.C.; Ruano, O.A.; Escudero, M.L. Corrosion behaviour of AZ31 magnesium alloy with different grain sizes in simulated biological fluids. *Acta Biomater.* **2010**, *6*, 1763–1771. [CrossRef] [PubMed]

10. Sun, H.Q.; Shi, Y.N.; Zhang, M.X.; Lu, K. Plastic strain-induced grain refinement in the nanometer scale in a Mg alloy. *Acta Mater.* **2007**, *55*, 975–982. [CrossRef]

11. Figueiredo, R.B.; Langdon, T.G. Principles of grain refinement and superplastic flow in magnesium alloys processed by ECAP. *Mater. Sci. Eng.* **2009**, *501*, 105–114. [CrossRef]

12. Feng, X.M.; Ai, T.T. Microstructure evolution and mechanical behavior of AZ31 Mg alloy processed by equal-channel angular pressing. *Trans. Nonferrous Metals Soc. China* **2009**, *19*, 293–298. [CrossRef]

13. Ding, R.; Chung, C.; Chiu, Y. Effect of ECAP on microstructure and mechanical properties of ZE41 magnesium alloy. *Mater. Sci. Eng.* **2010**, *527*, 3777–3784. [CrossRef]

14. Huang, P.; Li, J.; Zhang, S.; Chen, C.; Han, Y.; Liu, N.; Xiao, Y.; Wang, H.; Zhang, M.; Yu, Q.; *et al.* Effects of lanthanum, cerium, and neodymium on the nuclei and mitochondria of hepatocytes: Accumulation and oxidative damage. *Environ. Toxicol. Pharmacol.* **2011**, *31*, 25–32. [CrossRef] [PubMed]

15. Song, G. Control of biodegradation of biocompatable magnesium alloys. *Corros. Sci.* **2007**, *49*, 1696–1701. [CrossRef]

16. Feyerabend, F.; Fischer, J.; Holtz, J.; Witte, F.; Willumeit, R.; Drucker, H.; Vogt, C.; Hort, N. Evaluation of short-term effects of rare earth and other elements used in magnesium alloys on primary cells and cell lines. *Acta Biomater.* **2010**, *6*, 1834–1842. [CrossRef] [PubMed]

17. Zhang, E.; Yang, L. Microstructure, mechanical properties and bio-corrosion properties of Mg-Zn-Mn-Ca alloy for biomedical application. *Mater. Sci. Eng.* **2008**, *497*, 111–118. [CrossRef]

18. Gao, J.H.; Guan, S.K.; Ren, Z.W.; Zhu, S.J.; Wang, B. Homogeneous corrosion of high pressure torsion treated Mg-Zn-Ca alloy in simulated body fluid. *Mater. Lett.* **2011**, *65*, 691–693. [CrossRef]

19. Song, G.; Atrens, A. Understanding magnesium corrosion-a framework for improved alloy performance. *Adv. Eng. Mater.* **2003**, *5*, 837–858. [CrossRef]

20. Abidin, N.I.Z.; Martin, D.; Atrens, A. Corrosion of high purity Mg, AZ91, ZE41 and Mg2Zn0.2Mn in Hank's solution at room temperature. *Corros. Sci.* **2011**, *53*, 862–872. [CrossRef]

21. Galiyev, A.; Kaibyshev, R.; Gottstein, G. Correlation of plastic deformation and dynamic recrystallization in magnesium alloy ZK60. *Acta Materi.* **2001**, *49*, 1199–1207. [CrossRef]

22. Shi, Z.; Liu, M.; Atrens, A. Measurement of the corrosion rate of magnesium alloys using Tafel extrapolation. *Corros. Sci.* **2010**, *52*, 579–588. [CrossRef]

23. Xin, Y.; Hu, T.; Chu, P.K. Influence of test solutions on in vitro studies of biomedical magnesium alloys. *J. Electrochem. Soc.* **2010**, *157*, C238–C243. [CrossRef]

5

The Study of Heat Treatment Effects on Chromium Carbide Precipitation of 35Cr-45Ni-Nb Alloy for Repairing Furnace Tubes

Nakarin Srisuwan [1,*], Krittee Eidhed [2], Nantawat Kreatsereekul [1], Trinet Yingsamphanchareon [3,4] and Attaphon Kaewvilai [3,4]

Academic Editor: Hugo F. Lopez

[1] Thai-French Innovation Institute, King Mongkut's University of Technology North Bangkok, Bangsue, Bangkok 10800, Thailand; natthawat.k@tfii.kmutnb.ac.th

[2] Faculty of Engineering, King Mongkut's University of Technology North Bangkok, Bangsue, Bangkok 10800, Thailand; krittee.eed@gmail.com

[3] Department of Welding Engineering Technology, College of Industrial Technology, King Mongkut's University and Technology North Bangkok, Bangsue, Bangkok 10800, Thailand; trinet2518@hotmail.com (T.Y.); attaphonk@kmutnb.ac.th (A.K.)

[4] Welding Engineering and Metallurgical Inspection, Science and Technology Research Institute, King Mongkut's University and Technology North Bangkok, Bangsue, Bangkok 10800, Thailand

[*] Correspondence: nakrin.s@tfii.kmutnb.ac.th

Abstract: This paper presents a specific kind of failure in ethylene pyrolysis furnace tubes. It considers the case in which the tubes made of 35Cr-45Ni-Nb high temperature alloy failed to carburization, causing creep damage. The investigation found that used tubes became difficult to weld repair due to internal carburized layers of the tube. The microstructure and geochemical component of crystallized carbide at grain boundary of tube specimens were characterized by X-ray diffractometer (XRD), scanning electron microscopy (SEM) with back-scattered electrons mode (BSE), and energy dispersive X-ray spectroscopy (EDS). Micro-hardness tests was performed to determine the hardness of the matrix and the compounds of new and used tube material. The testing result indicated that used tubes exhibited a higher hardness and higher degree of carburization compared to those of new tubes. The microstructure of used tubes also revealed coarse chromium carbide precipitation and a continuous carbide lattice at austenite grain boundaries. However, thermal heat treatment applied for developing tube weld repair could result in dissolving or breaking up chromium carbide with a decrease in hardness value. This procedure is recommended to improve the weldability of the 35Cr-45Ni-Nb used tubes alloy.

Keywords: ethylene pyrolysis; furnace tube; chromium carbide; heat treatment; carbide dissolution; carburization

1. Introduction

1.1. Material for Ethane Pyrolysis Furnaces Tube

Ethylene (C_2H_4) can be generated by the thermal cracking of ethane (C_2H_6), which is passed through a coil of reaction tubes externally heated to a temperature of 1000–1150 °C in pyrolysis furnaces. The decomposition of ethane into ethylene is represented by the reaction below (1) [1]. Carbon residue from the combustion products can be deposited at the internal surface of the tube wall as adherent coke and must be removed repeatedly via decoking in water vapor and air. Consequently, the internal carbide formation can occur after a corrosion phenomenon called "carburization", which reduces the mechanical properties of materials and causes damages [2,3]. Thus, the alloy material must be suitable to accommodate the high process temperature. These alloys are selected for better high creep strength, carburization resistance, thermal shock resistance and weldability.

$$C_2H_6 {}_{(g)} \xrightarrow[\Delta]{} C_2H_4 {}_{(g)} + H_2 {}_{(g)} \tag{1}$$

The Ni-Cr-Fe alloys are known for producing material for tubular coils within an industrial pyrolysis furnace. They have been developed for using at elevated temperature where relatively severe mechanical stresses are encountered and high surface stability is required [4]. The structure of Ni-Cr-Fe alloys consist of the primary phase of γ austenitic FCC matrix (Figure 1a), plus a variety of secondary phases [5], which are the metal carbides denoted only by the compound M_xC such as MC, M_6C and $M_{23}C_6$ (M = Metals, C = carbide), as shown in Figure 1b.

(a)

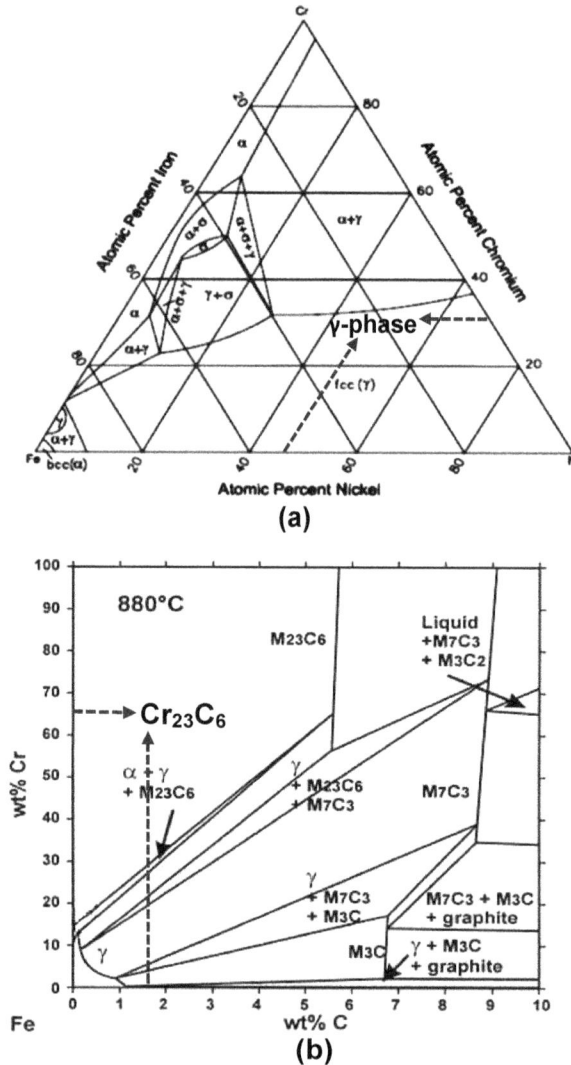

(b)

Figure 1. Phases of 35Cr-45Ni-Fe alloy: (a) primary γ-phase; and (b) secondary carbides phase.

Previously, the small additions of niobium (Nb) to the tube production were able to increase their resistance to thermal shock, while the Nb acts as a carbide stabilizer. Moreover, the nickel and niobium could be combined in the matrix of an alloy to form body-centered tetragonal (BCT) called the metastable γ''-Ni_3Nb phase. This phase provides very high strength at low and intermediate temperatures; however, it is unstable at high temperatures above 650 °C [6,7].

The transition of carbides in the Ni-Cr-Fe alloys can be considered by the Ellingham diagram (relation between temperatures and Gibs free energy; ΔG) [8]. In the range of temperatures over 850 °C, the ΔG of nickel carbide and iron carbide (Ni_3C and Fe_3C) are positive such that the formation of these carbides are a nonspontaneous process. On the other hand, the ΔG of chromium and niobium are negative, which indicates that the chromium and niobium tend to combine with carbon to form chromium carbides ($Cr_{23}C_6$ or Cr_7C_3) and niobium carbide (NbC and Nb_2C), called carbide precipitation or sensitization [9]. The type of formed carbide depends on the contents of the metal and carbon in the alloy. Moreover, the precipitation of silicon carbide (SiC) has also occurred in the structure of Ni-Cr-Fe alloys, but temperature is a very important factor in the determination of which polytype of SiC is formed [6].

From the carbide precipitation, the corrosion resistance of Ni-Cr-Fe alloy decreased while the hardness and brittle were increased [10]. The change in microstructure and in the properties of the Ni-Cr-Fe alloy is a serious problem in the repairing of tube furnaces by welding.

1.2. Weldability with Thermal Heat Treatment

In general, post-weld heat treatments (PWTH) are usually not required for the non-precipitation-hardenable Ni-Cr-Fe alloy weldments except for an additional agreement between the owner and the welding contractor. However, in dissimilar-metal welding (DMW), the identity and properties of the two metals being joined, and of the filler metal joining them, must be considered. For example, if the Ni-Cr-Fe alloy being joined has various different physical and metallurgical properties after long-term service, preheat should be used to make a dissimilar-metal weld. Another variable that should be considered might be the need for a post-heat treatment for reducing the weldability problems from different materials [11]. Additionally, the formation of chromium carbides is readily reversed by thermal heat treatment [9,12]. Therefore, the tubing repair process could consider a dissimilar-metal weld as well as a pre-heat treatment as an important parameter. The failure of the Ni-Cr-Fe alloy in the petrochemical industry has been studied, and the influence of carburization, carbide formation, thermal treatment conditions and welding repair technology has been investigated by some researchers [1–3,13–19].

However, this study is related to the problems of the weld repair process of used tubes in ethylene pyrolysis furnace from the petrochemical industry in Thailand. The failure analysis was performed via visual inspection and hardness measurements of the used tubes. Chromium carbide precipitation was investigated. The results obtained provide guidance on applications of weld repair procedures when the Ni-Cr-Fe alloy tube has a difference level of carburization.

2. Materials and Methods

2.1. Material Verification

A Ni-based super alloy 35Cr-45Ni-Nb grade (Elemental composition: 45% Ni, 35% Cr, 14.5% Fe, 1.8% C, 1.7% Si, 1% Nb and 1% Mn) was used as a material. This alloy does not belong to the standardized ASME-II Part B code (Non-ferrous metal); however, it was typically used in carburizing applications. The information and visual inspection of experimental material are shown in Table 1.

Table 1. Information of 35Cr-45Ni-Nb tube specimen and visuals inspection.

Specimen	Information and Visual Inspection		Image
Tube No. 1 (New)	Diameter (OD)	63.4 ± 0.2 mm	
	Outside wall	Light gray shade	
	Inside wall	Original smooth surface	
	Visual inspection	No coke deposit, original roundness, no swell	
Tube No. 2 (Used)	Location	9 m on the ground	
	Diameter (OD)	67.8 ± 0.8 mm	
	Outside wall	Black shade & rough surface	
	Inside wall	Thin coke deposit	
	Visual inspection	Blockage of coke inside tube, poor roundness	
Tube No. 3 (Used)	Location	2 m on the ground	
	Diameter (OD)	65.0 ± 0.4 mm	
	Outside wall	Black shade & rough surface	
	Inside wall	Thin coke deposit	
	Visual inspection	A little soot inside tube, Poor roundness	
Tube No. 4 (Used)	Location	1 m on the ground	
	Diameter (OD)	65.0 ± 0.4 mm	
	Outside wall	Black shade & rough surface	
	Inside wall	Thin coke deposit inside tube	
	Visual inspection	Near original roundness	

Specimen No. 1 and used tubes No. 2–4 (after 6 years of service at pressure 32 bar, and a temperature of about 800–1100 °C) were carefully cut into the outside diameter at 6.3–6.8 cm and 1.0 cm of thickness. After that, the new tube and used tubes were analyzed by an X-ray diffractometer, (XRD: Philips X-Pert-MPD, Eindhoven, The Netherlands), a scanning electron microscope (SEM: JEOL JSM-6310F, Peabody, MA, USA) with back-scattered electrons mode (BSE), and an energy-dispersive X-ray spectrometer (EDS: EDAX, New York, NY, USA) with elemental mapping.

2.2. Heat Treatment, Hardness and Microstructure

The heat treatments of specimens were performed in three conditions, as reported in Table 2. The specimens before (No. 1) and after heat treatment (No. 2–4) were tested by the Vickers micro-hardness testing and further analyzed the microstructure by combining techniques of SEM-BSE and EDS mapping.

Table 2. Heat treatment conditions.

Conditions	Heat Treatment Methods		Note
C1	Without heat treatment		New tube (No. 1)
C2	Pre-heat 600 °C, 1 h	Cooled down in air	Used tube (No. 2–4)
C3	Pre-heat 900 °C, 1 h	Cooled down in air	Used tube (No. 2–4)

For hardness testing, each of specimen was divided into four quadrants (area 1–4) for comparing the hardness at different position on the cross sectional areas as shown in Figure 2. During the micro-hardness testing, all specimens corresponded to the NACE Standard TM 0498-98 and ASTM A370, with accepted load settings of 500 kg.

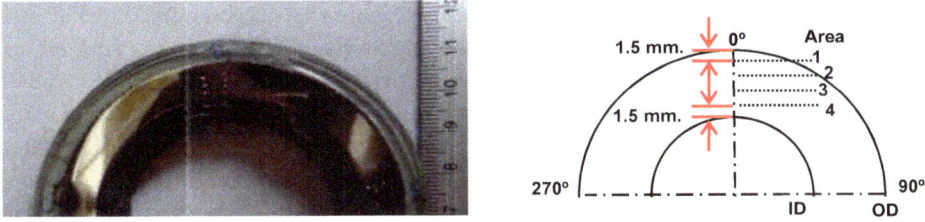

Figure 2. The location on the cross section of tube specimen for micro-hardness testing.

All specimens were cut into smaller pieces and then mounted into Bakelite cylinders. Thereafter, the specimen's surfaces were polished by using abrasive SiC paper (No. 120, 240, 400, 600, 800 and 1000). After every step of polishing, the surfaces were etched in mixed solutions (40 mL glycerol, 20 mL HCl and 20% HNO_3). Additionally, the specimens were cleaned in an ultrasonic bath. Lastly, the SEM-BSE combined with EDS mapping were used to analyze the microstructure and elemental composition.

3. Results and Discussion

3.1. Material Verification

A visual inspection of Specimen No. 1 found that it had no significant internal or external damage. It showed a good quality of structures with no visible pits, no deposit and no signs of carburization. On the other hand, carburization reactions and coke deposits were observed at different configurations in the three height positions studied. Specimen No. 2 had the most coke deposits and signs of carburization, but also a roundness and shape that was significantly more deformed than Specimens No. 3 and No. 4, respectively. However, no cracking in Specimens No. 2, No. 3 or No. 4 were observed (Table 2).

Figure 3 shows the XRD patterns of the Cr-Ni-Nb alloy before and after use in the pyrolysis furnace. Specimen No. 1 exhibited both diffraction peaks corresponding to the γ' austenitic (FCC) phase and the γ'' (BCT) phase of Cr-Ni-Nb alloy and Ni_3Nb, respectively, as shown in Figure 3a. In the case of Specimen No. 2, the XRD pattern (Figure 3b) displayed the same major peaks of FCC and BCT phases as the new tube. Moreover, Specimen No. 2 exhibited the additional peaks which correspond to the carbide structures of NbC (JCPDS No. 00-038-1364), SiC (JCPDS No. 000-029-1129) and $Cr_{23}C_6$ (JCPDS No. 00-035-0783). This result clearly indicated that the precipitated carbide of the tube furnace was produced in the process of ethane pyrolysis.

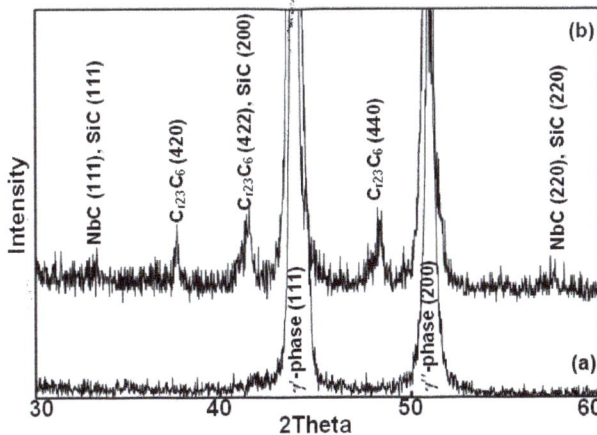

Figure 3. XRD patterns of (a) the new tube and (b) the used tube.

The microstructure of the Cr-Ni-Nb alloy and its carbide was observed by SEM-BSE and EDS mapping, as shown in Figure 4. BSE is the electron signal from elastic scattering interaction between electron beam and atomic specimen. This interaction depends on the atomic weight of the atom. The more heavy an atom is, the more a BSE signal can be generated, giving results in a brighter contrast. Therefore, the BSE is used to detect contrasts between areas with different elemental compositions [19,20]. Figure 4a,b shows the backscattered electrons images (BEI) of Specimen No. 1 (the new tube), in which three different phases (bright, gray and dark) are found. The Nb, a heavy element (high atomic number), exhibited more BSE than those of the lighter elements (low atomic number), and thus appeared brighter in BEI. The gray area was identified as the major part of the matrix alloy, while the dark area in BEI might be the precipitated carbide.

By elemental analysis, the EDS mapping (Figure 4c–h) of Specimen No. 1 confirmed a good dispersion of elementals in the alloy. The EDS revealed that the results agreed to XRD, and that the matrix (Ni-Cr-Fe solid solution) was observed as γ austenitic (FCC), with a variety of secondary phases of γ''-Ni_3Nb (BCT), a strengthener phase comprised of an austenite matrix. The small numbers of coarse chromium carbide precipitates were observed within the matrix and at grain boundaries. Moreover, the mapping of Cr (Figure 4f) was found to be similar to the dark area in BEI (Figure 4b), which indicated that the chromium and carbon appeared as a small chromium carbide precipitation. However, this carbide phase cannot be observed in XRD because of the detection limit of the XRD technique.

In the case of one of the used tubes (Specimen No. 2), the BEI and EDS (Figure 4i–p) showed the microstructure of the austenite matrix and more Cr-rich carbides (Figure 4n). Moreover, a phase separation of mixed NbC and SiC was found, shown in Figure 4l,p. The obtained results of XRD and SEM-BSE with EDS confirmed that the pyrolysis caused the carbide precipitation in the microstructure of the Cr-Ni-Nb alloy. The carbide precipitate decreased the tensile strength, creep resistance and corrosion resistance, and increased the brittle fracture property, the hardness, and the ductility of the material [21].

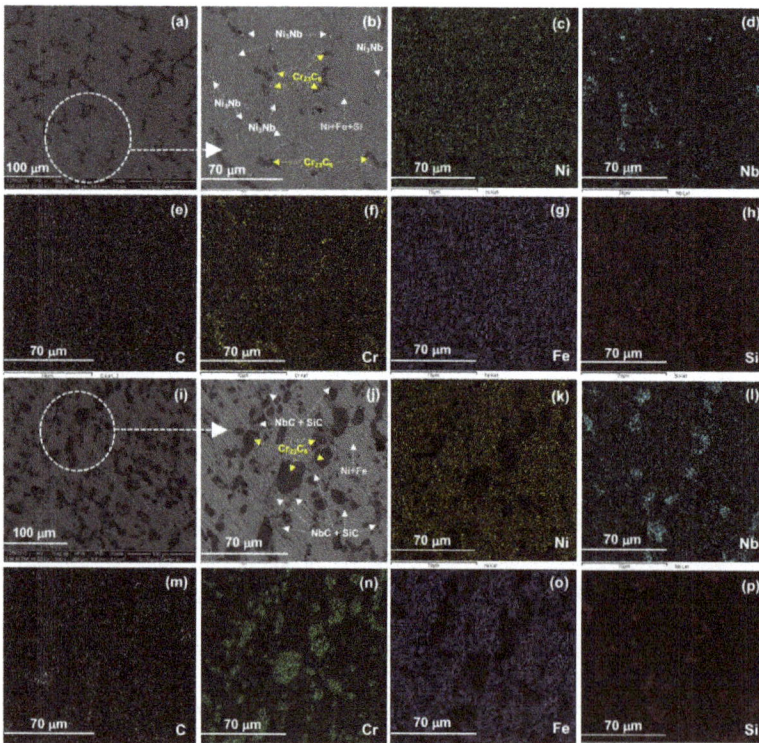

Figure 4. BEI and EDS mapping of (**a–h**) Specimen No. 1 (new); and (**i–p**) Specimen No. 2 (used).

3.2. *Heat Treatment, Hardness and Microstructure*

The hardness of specimens before (No. 1) and after heat treatment (No. 2–4) in four areas were tested and compared, as shown in Figure 5.

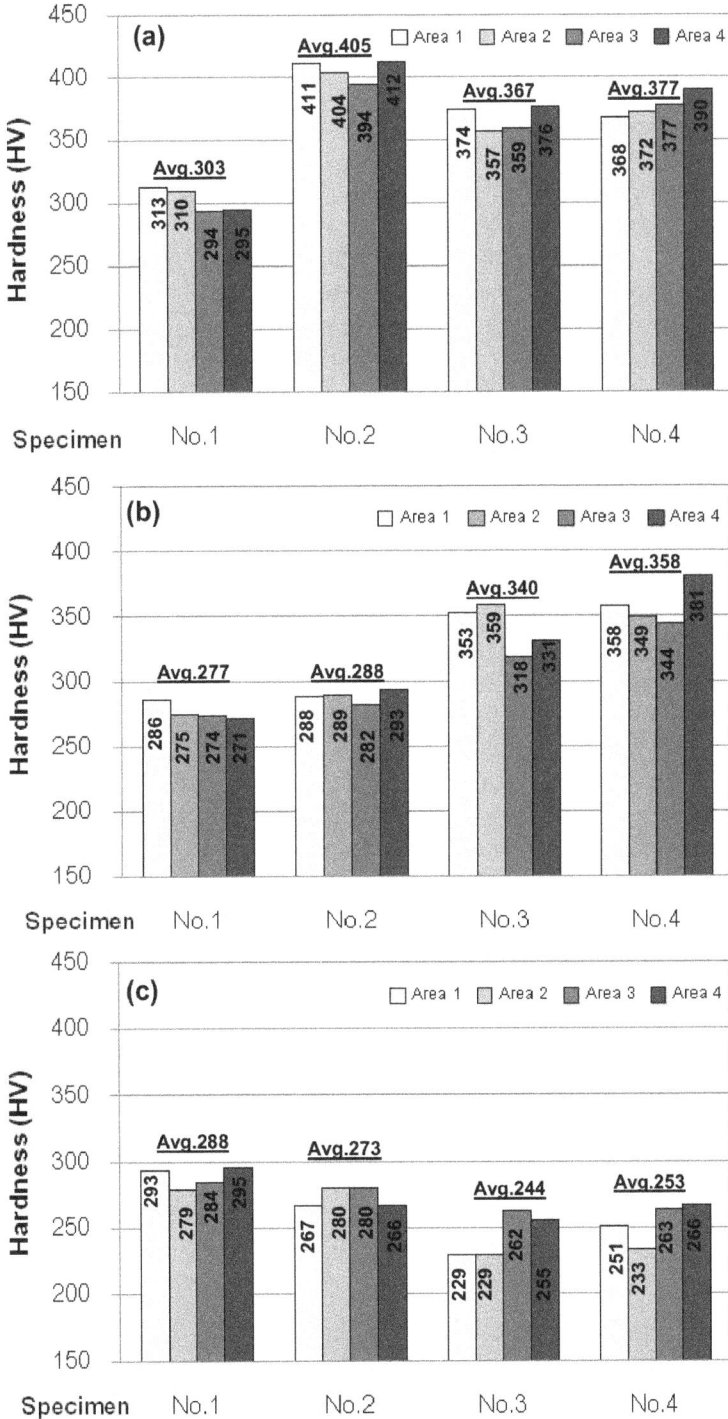

Figure 5. The hardness value of four areas in each specimen with different heat treatments: (**a**) without heat treatment; (**b**) 600 °C/1 h; and (**c**) 900 °C/1 h.

Figure 5 presents the hardness of each specimen with various heat treatment conditions. Each bar shows the hardness value averaged from two readings at four different positions on a cross section of specimens. In condition C1 (without heat treatment), the lowest hardness value was observed in Specimen No. 1 with an average value of 303 HV, as shown in Figure 5a. On the other hand, Specimen No. 2 with thick coke deposit inside the wall of a higher position of the pyrolysis furnace had the highest hardness value, average as 405 HV. However, the average hardness was reduced to 377 HV and 367 HV for Specimen No. 4 and No. 3, respectively.

In Figure 5b, the effect of the heat treatment process showed a considerable decrease in the hardness value of all specimens. In the case of condition C2 (annealing 600 °C, 1 h/air cooled), the average hardness values decreased to 288, 340 and 358 HV for Specimens No. 2–4, respectively. However, the annealing temperature of condition C2 was within the range of sensitization, so chromium carbide could be precipitated at grain boundary. Therefore, the hardness values of used tube specimens were still greater and higher than that of Specimen No. 1 (average of 277 HV). In addition, Specimen No. 2 was acquired from the top part of the furnace tube, where the high carbon content was accumulated as coke and tended to form a large amount of carbide phase. Therefore, the significant decrease of hardness in sample 2 after heat treatment at 600 °C might be due to the superior loss of these carbide grains when compared to those of Specimens No. 3–4 .

Relating to condition C3 (annealing 900 °C, 1 h/air cooled), as shown in Figure 5c, the lowest hardness values were determined to be 273, 244 and 253 HV for Specimens 2–4, respectively, when compared to the heat treatment condition of C1 and C3. The hardness measurement indicated that condition C3 was a suitable condition because it could change the hardness values of the used tubes close to the new tube for avoiding cracking damage during the weld repair and can be used to improve the weldability of 35Cr-45Ni-Nb alloy.

Furthermore, SEM/EDS analysis could confirm a significant reason for a using thermal heat treatment to a 35Cr-45Ni-Nb used tube for welding repair. The microstructures of each specimen at heat treatment conditions of C2 and C3 by using SEM/EDS mapping are shown in Figures 6 and 7 respectively.

In Figure 6a, after heat treatment, Specimen No. 1 showed that the white precipitate adjacent to the grain boundaries was Nb-rich in the metastable γ''-Ni$_3$Nb phase as shown by EDS mapping. Whereas, a small number of chromium carbide precipitates were observed within the matrix and at the grain boundaries. Thus, it could be suggested that the good mechanical properties were exhibited because the phase transformation could not be observed in the microstructure of Specimen No. 1.

Figure 6b–d shows the microstructures in the used tube specimen. Specimens No. 2 and No. 3 also revealed coarse chromium carbide precipitation and a continuous carbide lattice at austenite grain boundaries, while the metastable γ''-Ni$_3$Nb phase could not be observed in the matrix austenite. Additionally, the SiC and NbC were separated from the matrix austenite, which corresponded to the degree of hardness values. The microstructure of Specimen No. 4 showed small amounts of SiC and NbC within the area of chromium carbide precipitation at grain boundaries. This could be explained by the laws of thermodynamic and the Gibbs free energy such that the significant alloying elements (such as Nb and Si) were quickly activated with carbon atoms and reduced the degree of sensitization [15,21–23]. Therefore, it could be recommended that condition C2 (annealing 1 h (600 °C)/air cooled) could not give a beneficial effect to ensure that the precipitation of Cr$_{23}$C$_6$ was dissolved properly in the used tube for repair welding procedure.

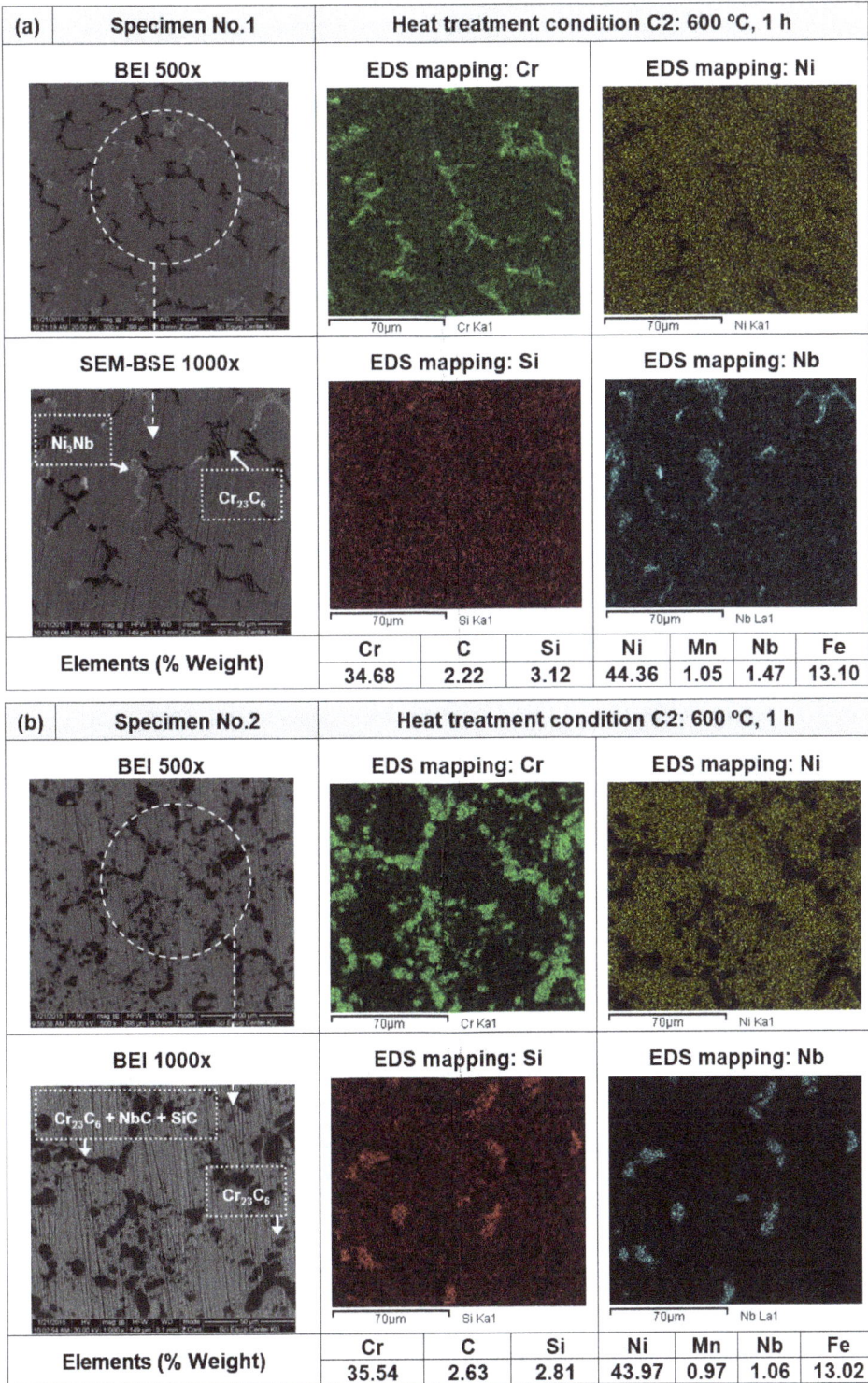

(a)	Specimen No.1	Heat treatment condition C2: 600 °C, 1 h

BEI 500x | EDS mapping: Cr | EDS mapping: Ni

SEM-BSE 1000x | EDS mapping: Si | EDS mapping: Nb

Elements (% Weight)	Cr	C	Si	Ni	Mn	Nb	Fe
	34.68	2.22	3.12	44.36	1.05	1.47	13.10

(b)	Specimen No.2	Heat treatment condition C2: 600 °C, 1 h

BEI 500x | EDS mapping: Cr | EDS mapping: Ni

BEI 1000x | EDS mapping: Si | EDS mapping: Nb

Elements (% Weight)	Cr	C	Si	Ni	Mn	Nb	Fe
	35.54	2.63	2.81	43.97	0.97	1.06	13.02

Figure 6. *Cont.*

(c)	**Specimen No.3**	**Heat treatment condition C2: 600 °C, 1 h**	
	BEI 500x	**EDS mapping: Cr**	**EDS mapping: Ni**
	BEI 1000x	**EDS mapping: Si**	**EDS mapping: Nb**
	$Cr_{23}C_6$ NbC + SiC		

Elements (% Weight)	Cr	C	Si	Ni	Mn	Nb	Fe
	36.6	2.67	2.9	42.72	1.05	0.81	13.25

(d)	**Specimen No.4**	**Heat treatment condition C2: 600 °C, 1 h**	
	BEI 500x	**EDS mapping: Cr**	**EDS mapping: Ni**
	BEI 1000x	**EDS mapping: Si**	**EDS mapping: Nb**
	$Cr_{23}C$ $Cr_{23}C_6$ + NbC + SiC		

Elements (% Weight)	Cr	C	Si	Ni	Mn	Nb	Fe
	35.45	2.59	2.94	44.28	0.63	0.94	13.17

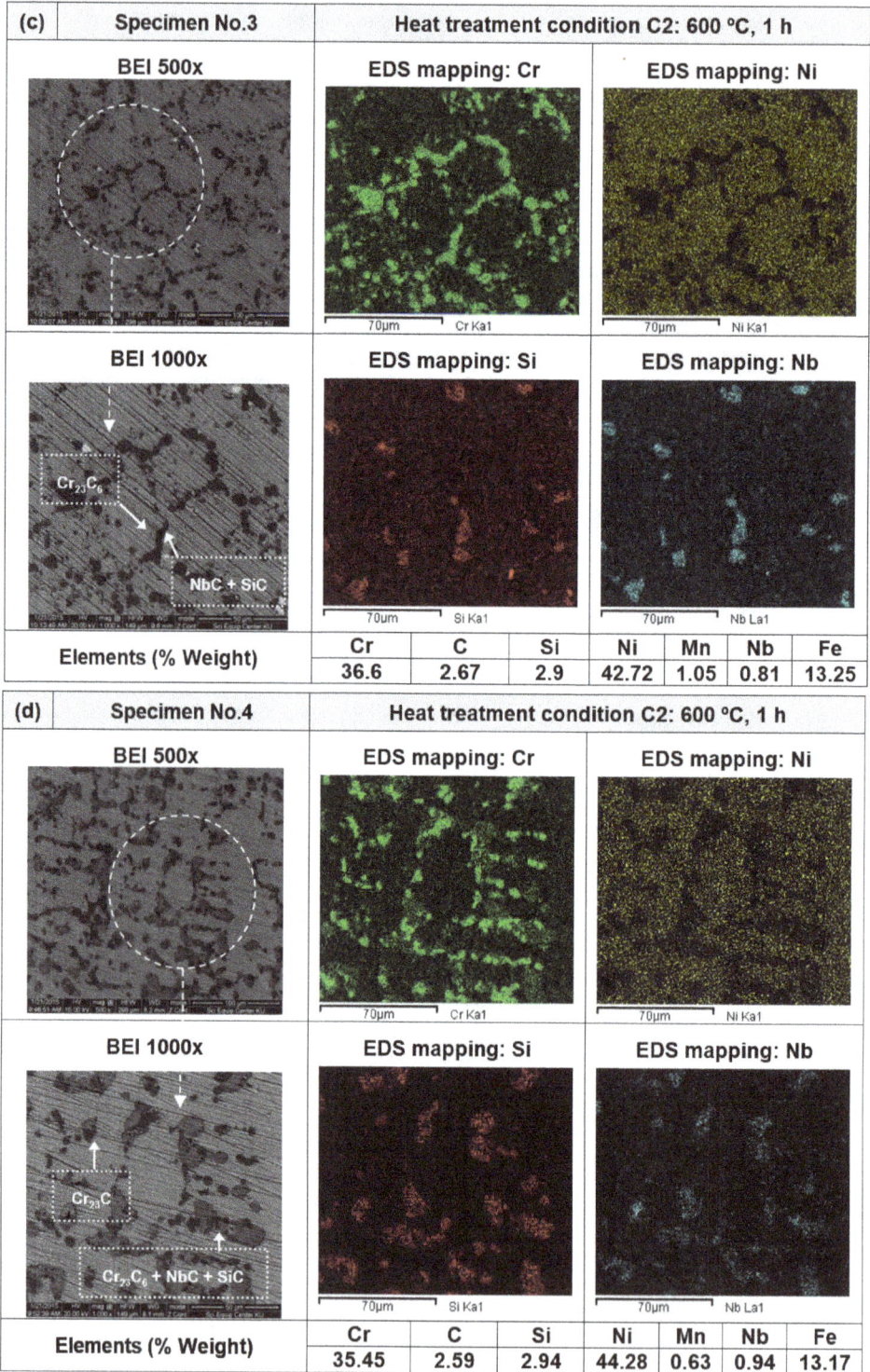

Figure 6. BEI and EDS mapping of the 35Cr-45Ni-Nb specimens after 1 h of thermal heat treatment at 600 °C: (**a**) Specimen No. 1; (**b**) Specimen No. 2; (**c**) Specimen No. 3; (**d**) Specimen No. 4.

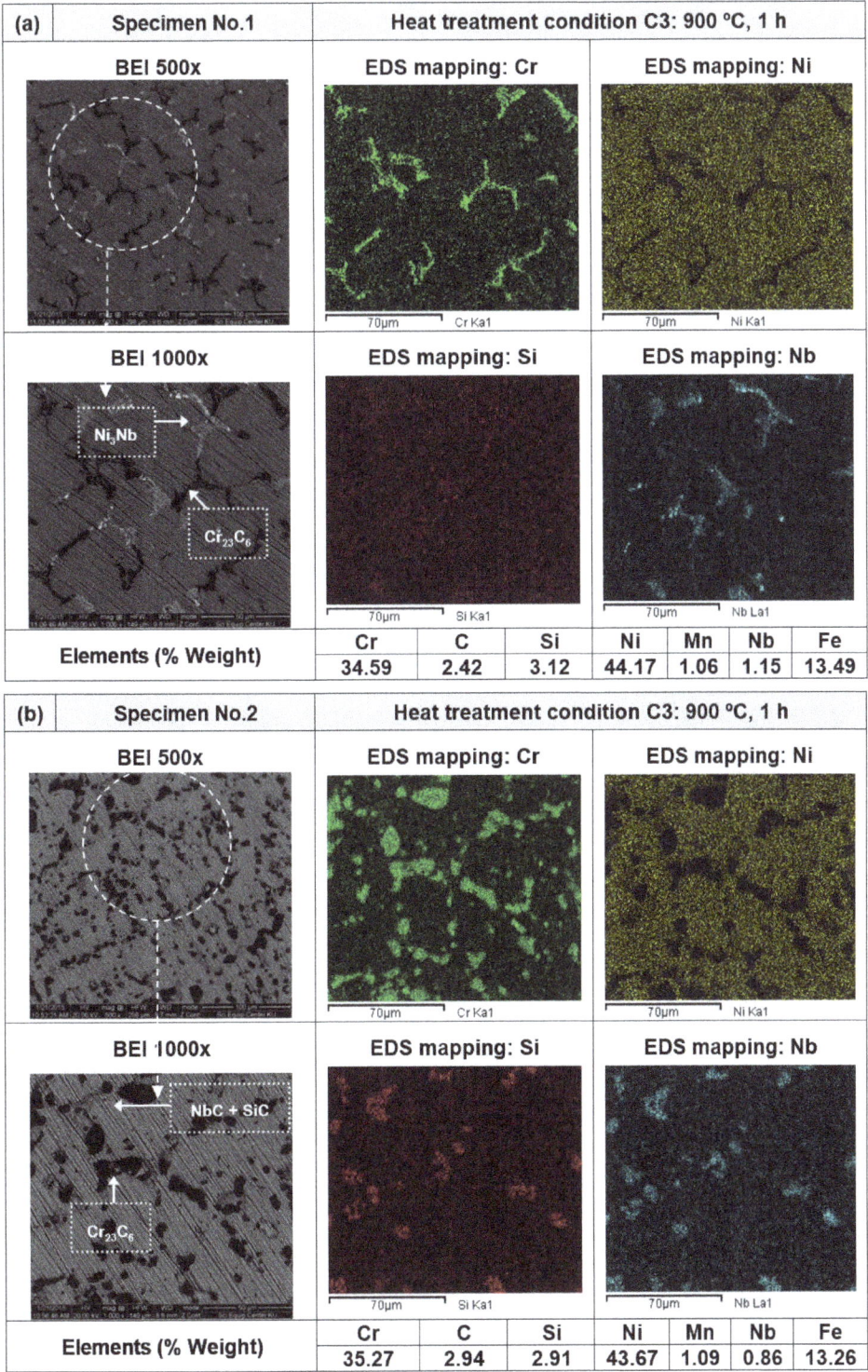

(a)	Specimen No.1	Heat treatment condition C3: 900 °C, 1 h				
Elements (% Weight)		Cr	C	Si	Ni	Mn	Nb	Fe
		34.59	2.42	3.12	44.17	1.06	1.15	13.49

(b)	Specimen No.2	Heat treatment condition C3: 900 °C, 1 h				
Elements (% Weight)		Cr	C	Si	Ni	Mn	Nb	Fe
		35.27	2.94	2.91	43.67	1.09	0.86	13.26

Figure 7. *Cont.*

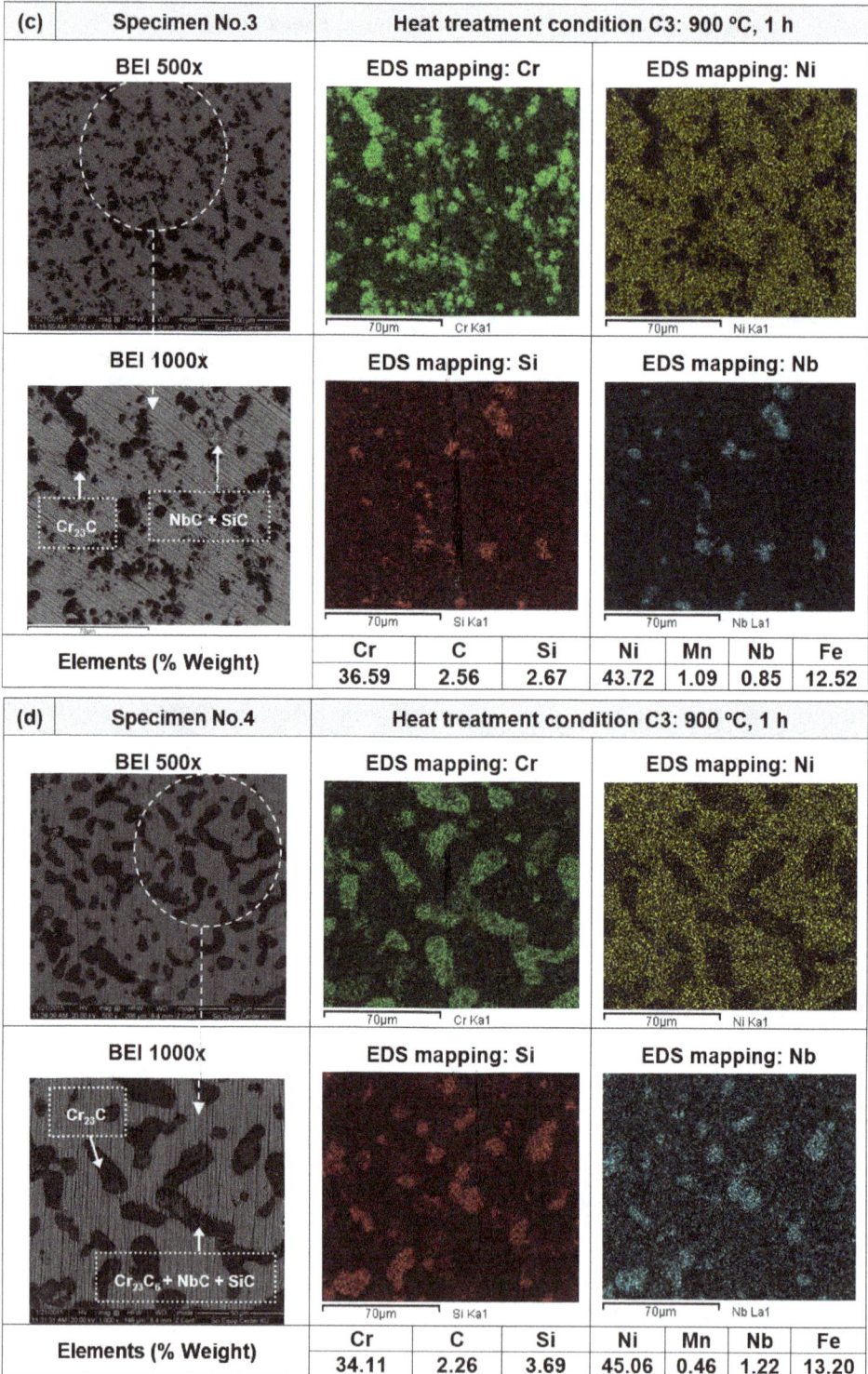

| (c) | Specimen No.3 | Heat treatment condition C3: 900 °C, 1 h | | | | | |

Elements (% Weight)	Cr	C	Si	Ni	Mn	Nb	Fe
	36.59	2.56	2.67	43.72	1.09	0.85	12.52

| (d) | Specimen No.4 | Heat treatment condition C3: 900 °C, 1 h | | | | | |

Elements (% Weight)	Cr	C	Si	Ni	Mn	Nb	Fe
	34.11	2.26	3.69	45.06	0.46	1.22	13.20

Figure 7. BEI and EDS mapping of the 35Cr-45Ni-Nb specimens after 1 h of thermal heat treatment at 900 °C: (**a**) Specimen No. 1; (**b**) Specimen No. 2; (**c**) Specimen No. 3; (**d**) Specimen No. 4.

In condition C3, the formation of the metastable γ''-Ni_3Nb phase occurred in the austenite matrix, as shown in Figure 7a. However, the degree of sensitization at grain boundaries was enlarged when compared to Specimen No. 1 in the condition C2. This was due to the fact that, after annealing at high temperature of 900 °C, the higher cooling rate could increase the reverse transformation of carbide particles [24].

Figure 7b–d shows the microstructure of carbide precipitates. It was found that the particles of chromium carbide were dispersed and discontinued in the matrix austenite, and the small particles of SiC and NbC occurred in the area of chromium carbides precipitation. In addition, there might be other factors affecting the hardness, such as the redistribution of the phase component in the alloy, particularly Ni, Si and Nb phases. Therefore, the mechanical properties and weldability of used tube specimens were enhanced.

As a result, it is suggested that condition C3 (annealing 900 °C 1 h/air cooled) was a suitable method for chromium carbides dissolution which help decreased the hardness values of the used tuebes and might improved the weldability of furnace tube repair.

4. Conclusions

The aim of this work was to find an suitable heat treatment process to decrease the hardness value and/or reduce the chromium carbide precipitation at grain boundary of material due to the major problem in welding repair of ethylene pyrolysis tubes, a dissimilar material between used tubes and new tubes. This experiment confirmed the carburization in the 35Cr-45Ni-Nb alloy tube after long-term service at high temperature in an ethylene pyrolysis furnace. The carbide precipitation at grain boundary of the austenite matrix was found to increase the hardness values of the specimen. However, after thermal heat treatment (annealing 900 °C 1 h/air cooled), a decrease in the carbide precipitate in the austenite matrix and in the hardness value of each specimen was observed when compared to those before heat treatment. The temperature range and cooling rate were important influences for the dissolution of carbides and reverse transformation. Furthermore, the next study will investigate parameters which are significant for the successful joints of furnace tube repair, such as wire electrode, welding process and post-weld heat treatment in the welding procedures.

Acknowledgments: This research was funded by King Mongkut's University of Technology North Bangkok (Contract No. KMUTNB-GEN-57-55). The authors are very grateful to IRPC Public Company Limited for supporting information and materials that used in this experiment. Additionally, thanks to Chockchai Singhatham, in support of sample preparation for hardness testing.

Author Contributions: N. Srisuwan performed reseach and wrote the article. K. Eidhed and N. Kreatsereekul helped in the experimental part. T. Yingsamphancharoen and A. Kaewvilai assisted in the material characterization, data analysis and revised manuscript.

Conflicts of Interest: The authors declare no conflict of interest.

References

1. Ul-Hamid, A.; Tawancy, H.M.; Al-Jaroudi, S.S.; Mohammed, A.I.; Abbas, N.M. Carburisation of Fe-Ni-Cr alloys at High Temperatures. *Mater. Sci. Pol.* **2006**, *24*, 219–331.

2. Grabke, H.J. Carburization, Carbide Formation, Metal Dusting, Coking. *J. Mater. Technol.* **2002**, *36*, 297–305.

3. Babakr, A.; Habiby, F. Lengthening, Cracking and Weld ability Problems of Fe-Ni-Cr Alloy Tube. *J. Miner. Mater. Charact. Eng.* **2009**, *8*, 133–148.

4. Verdier, G.; Carpentier, F. Consider New Materials for Ethylene Furnace Applications. *Hydrocarb. Process.* **2011**, *90*, 61–62.

5. Raghavan, V. Cr-Fe-Ni (Chromium-Iron-Nickel), Section II: Phase Diagram Evaluations. *J. Phase Equilib. Diffus.* **2009**, *30*, 94–95. [CrossRef]

6. Davis, J.R. *ASTM Specially Handbook: Heat-Resistant Materials*; ASM International: Novelty, OH, USA, 1997; p. 224.

7.   Yang, Z.G.; Paxton, D.M.; Weil, K.S.; Stevenson, J.W.; Singh, P. *Materials Properties Database for Selection of High-Temperature Alloys and Concepts of Alloy Design for SOFC Applications*; Pacific Northwest National Laboratory: Richland, WA, USA, 2002.

8.   Shatynski, S.R. The Thermochemistry of Transition Metal Carbides. *Oxid. Met.* **1979**, *13*, 105–118. [CrossRef]

9.   Rashid, M.W.A.; Gakim, M.; Rosli, Z.M.; Azam, M.A. Formation of $Cr_{23}C_6$ during the Sensitization of AISI 304 Stainless Steel and its Effect to Pitting Corrosion. *Int. J. Electrochem. Sci.* **2012**, *7*, 9465–9477.

10. Durand-Charre, M. *Microstructure of Steels and Cast Irons*; E-Book: New York, NY, USA, 2004.

11. Avery, R.E.; Tuthill, A.H. *Guidelines for the Welded Fabrication of Nickel Alloys for Corrosion-Resistance Service*; The Nickel Development Institute: Toronto, ON, Canada, 1994; p. 22.

12. Punburi, P.; Tareelap, N. Post Weld Heat Treatment to Reduce Intergranular Corrosion Susceptibility of Dissimilar Welds Between Austenitic 304 and Ferritic 430 Stainless Steels. In Proceedings of the 1st Mae Fah Luang University International Conference, Bangkok, Thailand, 29 November–1 December 2012; pp. 1–9.

13. Singhatham, C.; Srisuwan, N.; Intho, R.; Theerawatanachai, N.; Pitiariyanan, K.; Eidhed, K. Root Cause Analysis of the Carburized 35Cr-45Ni-Nb Tube in the Ethylene Furnace. *J. Appl. Sci. Res.* **2003**, *9*, 5999–6009.

14. Chasselin, H. Carburization in Ethylene Radiant Coils. AIChE Paper No. 295746. In Proceedings of the 2013 Spring National Meeting, San Antonio, TX, USA, 28 April–2 May 2013.

15. Loto, C.A. Microstructural Analysis of Ethylene Furnace Steel Alloy Tubes. *Mater. Perform. NACE Int.* **2011**, *50*, 2–8.

16. Takcuchi, Y.; Kato, Y.; Yokota, N.; Tsuchiya, M.; Shimizu, T.; Tanaka, I. Multi-Layered Anti-Coking Heat Resistant Metal Tube and Method for Manufacture Thereof. U.S. Patent 6,337,459 B1, 8 January 2002.

17. Kennedy, R. Life Improvement of Alloys in Heat Treatment Furnaces and Fixtures. In *Degree of Bachelor of Science in Mechanical Engineering*; Worcester Polytechnic Institute: Worcester, MA, USA, 2013.

18. Dossett, J.L.; Boyer, H.E. *Practical Heat Treating*, 2nd ed.; ASM International: Novelty, OH, USA, 2006.

19. Wells, O.C.; Gordon, M.S.; Gignac, L.M. Past, Present, and Future of Backscatter Electron (BSE) Imaging. *Proc. SPIE* **2012**. [CrossRef]

20. Santos, M.; Guedes, M.; Baptista, R.; Infante, V.; Cláudio, R.A. Effect of severe operation conditions on the degradation state of radiant coils in pyrolysis furnaces. *Eng. Fail. Anal.* **2015**, *56*, 194–203. [CrossRef]

21. Shao, G. Thermodynamic Modeling of the Cr-Nb-Si System. *J. Intermet.* **2005**, *13*, 69–78. [CrossRef]

22. Serna, A.; Rapp, R.A. Carburization of Austenitic and Ferritic Alloys in Hydrocarbon Environments at High Temperature. *J. Rev. Metal. Madrid* **2003**, *39*, 162–166. [CrossRef]

23. Marinkovic, B.A.; Avillez, R.R.; Barros, S.K.; Assunção, F.C.R. Thermodynamic Evaluation of Carbide Precipitates in 2.25Cr-1.0Mo Steel for Determination of Service Degradation. *J. Mater. Res.* **2002**, *5*, 491–495. [CrossRef]

24. Xu, M.L. Secondary carbide dissolution and coarsening in 13% Cr martensitic stainless steel during austenitizing. In *Mechanical Engineering Dissertations*; Northeastern University: Boston, MA, USA, 2012.

6

Deformation in Metallic Glasses Studied by Synchrotron X-Ray Diffraction

Takeshi Egami [1,2,3,*], Yang Tong [1,4,†] and Wojciech Dmowski [1,†]

Academic Editor: Klaus-Dieter Liss

[1]   Department of Materials Science and Engineering, Joint-Institute for Neutron Sciences, University of Tennessee, Knoxville, TN 37996, USA; yangtong@um.cityu.edu.hk (Y.T.); wdmowski@utk.edu (W.D.)
[2]   Department of Physics and Astronomy, University of Tennessee, Knoxville, TN 37996, USA
[3]   Oak Ridge National Laboratory, Materials Science and Technology Division, Oak Ridge, TN 37831, USA
[4]   Department of Mechanical and Biomedical Engineering, City University of Hong Kong, Hong Kong, China
[*]   Correspondence: egami@utk.edu
[†]   These authors contributed equally to this work.

Abstract: High mechanical strength is one of the superior properties of metallic glasses which render them promising as a structural material. However, understanding the process of mechanical deformation in strongly disordered matter, such as metallic glass, is exceedingly difficult because even an effort to describe the structure qualitatively is hampered by the absence of crystalline periodicity. In spite of such challenges, we demonstrate that high-energy synchrotron X-ray diffraction measurement under stress, using a two-dimensional detector coupled with the anisotropic pair-density function (PDF) analysis, has greatly facilitated the effort of unraveling complex atomic rearrangements involved in the elastic, anelastic, and plastic deformation of metallic glasses. Even though PDF only provides information on the correlation between two atoms and not on many-body correlations, which are often necessary in elucidating various properties, by using stress as means of exciting the system we can garner rich information on the nature of the atomic structure and local atomic rearrangements during deformation in glasses.

Keywords: metallic glasses; mechanical deformation; anisotropic PDF analysis; high-energy X-ray diffraction

1. Introduction

Glass generally is a symbol of something extremely fragile. Indeed, conventional glasses, mostly oxide glasses, shatter helplessly upon impact. However, a relatively new family member of glasses, metallic glass, is stronger and tougher than oxide glasses, and even compares favorably to most of crystalline metallic materials in their mechanical properties. For this reason they are promising as structural materials [1], and are beginning to be used in watches and mobile phones. However, it is not easy to understand why they are strong and how they mechanically fail. Crystalline materials lattice defects, such as dislocations, can be readily defined as deviations from lattice periodicity, and crystalline materials fail because of the motion of these defects. In glasses, defects cannot be uniquely defined because of the extensive disorder in their structures. Nevertheless, phenomenologically structural defects, such as the shear-transformation-zones (STZs), have been postulated and have facilitated elucidation of mechanical properties [2,3]. However, atomistic details of STZs remain elusive.

Diffraction measurements, by their nature, provide only the information on two-atom positional correlation. However, most physical properties depend on more collective atomic correlations, which diffraction measurements cannot directly assess. Luckily, in crystalline materials, because of lattice symmetry and periodicity, it is possible to construct an accurate three-dimensional model out of two-body correlation alone. However, we are unable to do so with liquids and glasses, and it is not possible because of the extensive structural disorder. The structures of liquids and glasses are usually expressed in terms of the atomic pair-distribution function (PDF), $g(r)$, which describe the probability of finding two atoms separated by the distance, r. PDF has an advantage that it can be determined through the Fourier-transformation of the structure function, $S(Q)$, which can be directly measured by diffraction experiments [4,5]. However, all PDFs, more or less, look alike, and they are not strongly discriminatory in determining the structure, allowing high degeneracy of similar structures.

On the other hand, if the system is perturbed by a field, such as a stress field, the response often reveals the nature of the structure more informatively. Usually, the symmetry is broken by the field, so that response can be accurately detected by symmetry discrimination. In this article we discuss how the measurement of response of metallic glasses to the applied stress contribute to the understanding of the subtle nature of the glassy system. In our view, it constitutes one of the triumphs of synchrotron radiation research in interrogating the nature of metallic materials.

2. Synchrotron X-Ray Diffraction Measurement under Stress

2.1. Anisotropic PDF Analysis

The structure of a macroscopically isotropic glass or liquid is usually described by the atomic pair-density function (PDF) defined by:

$$\rho_0 g(r) = \frac{1}{4\pi r^2 N} \sum_{i,j} \delta\left(r - |r_{ij}|\right) \tag{1}$$

where ρ_0 is the atomic number density, N the number of atoms in the system, and $\delta(x)$ is the delta function [4,5]. PDF is related to the structure function $S(Q)$ by:

$$\rho_0 g(r) = \frac{1}{2\pi^2 r} \int_0^\infty \left[S(Q) - 1\right] \sin(Qr) Q dQ \tag{2}$$

where Q is the scattering vector ($= 4\pi\sin\Theta/\lambda$, Θ is the diffraction angle and λ is the wavelength of the probe). $S(Q)$ can be determined by X-ray, neutron, or electron diffraction measurement after correction for geometry, absorption and background [4,5]. Theoretically, the integration in Equation (2) should be carried out to infinity. However, in reality, because $\sin\Theta$ is equal or less than unity the maximum value of Q that can be attained, Q_{Max}, is less than $4\pi/\lambda$. Therefore we have to use high-energy X-ray to access a wide range of Q. Synchrotron radiation is an ideal probe for such a purpose, and X-rays with energy higher than 100 keV are routinely used for PDF measurement. Furthermore, by using a two-dimensional X-ray detector $S(Q)$ can be measured over a wide range of Q at once, making the diffraction measurement very fast [6].

The PDF analysis usually assumes that the sample is isotropic, so $S(Q)$ depends only on the magnitude of Q, not the direction. However, when a solid is subjected to shear or uniaxial stress the sample is no longer isotropic, and both $g(r)$ and $S(Q)$ depend on the direction of r and Q. In this case, we use the spherical harmonics expansion:

$$g(r) = \sum_{\ell,m} g_\ell^m(r) Y_\ell^m(r/r) \tag{3}$$

$$S(Q) = \sum_{\ell,m} S_\ell^m(Q) Y_\ell^m(Q/Q) \tag{4}$$

which are connected through:

$$\rho_0 g_\ell^m(r) = \frac{i^\ell}{2\pi^2} \int_0^\infty S_\ell^m(Q)\, J_\ell(Qr)\, Q^2 dQ \tag{5}$$

where $J_\ell(x)$ is the spherical Bessel function [5,7]. For the isotropic ($\ell = 0$) component, $S_0^0(Q)$, Equation (5) is reduced to Equation (1), because $J_0(x) = \sin x / x$. The anisotropic structure function, $S_\ell^m(Q)$ can be determined from $S(Q)$ by:

$$S_\ell^m(Q) = \iint S(Q)\, Y_\ell^m(Q/Q)\, d\Omega \tag{6}$$

where $d\Omega = d(\cos\theta)d\varphi$, making use of the orthonormal properties of the spherical harmonics. In the case of a sample deformed under a uniaxial stress applied along the z-axis, with the X-ray beam along the y-axis, the data from the two-dimensional detector takes the form of $S(Q, \theta)$. In most cases, the only relevant terms in Equation (4) are $S_0^0(Q)$ (isotropic structure function) and the anisotropic term, $S_2^0(Q)$, which are obtained by:

$$S_0^0(Q) = \int_{-1}^{1} S(Q,\theta)\, d\cos\theta \tag{7}$$

$$S_2^0(Q) = \frac{1}{2}\int_{-1}^{1} S(Q,\theta)\left[3\cos^2\theta - 1\right] d\cos\theta \tag{8}$$

If the deformation is affine, that is, the strain is uniform, and the displacement of the i-th atom is given by a single strain tensor:

$$\Delta r_i^\alpha = \varepsilon^{\alpha\beta} r_i^\beta \tag{9}$$

Then, it can be shown that $S_2^0(Q)$ s given by:

$$S_{2,affine}^0(Q) = -\varepsilon \left(\frac{1}{5}\right)^{1/2} \frac{2(1+v)}{3} Q\frac{d}{dQ} S_0^0(Q) \tag{10}$$

In the same way the anisotropic PDF for affine deformation is given by:

$$g_{2,affine}^0(r) = -\varepsilon \left(\frac{1}{5}\right)^{1/2} \frac{2(1+v)}{3} r\frac{d}{dr} g_0^0(r) \tag{11}$$

Therefore, the measurement of $S_2^0(Q)$ does not provide any new information. In general, however, deformation in glasses is heterogeneous at the atomic level, and the real anisotropic $S(Q)$ and PDF deviate from Equations (10) and (11). In such a case, the measurement does provide new and unique information. Below we describe how the determination of the anisotropic PDF of a metallic glass sample deformed by thermomechanical creep by a diffraction experiment using high-energy X-ray from a synchrotron radiation source facilitated understanding of the microscopic process of deformation.

2.2. Procedure of Determining Anisotropic PDF

The anisotropy in $S(Q)$ is often small, and care has to be exercised in order to measure it accurately. In particular, the sensitivity of a two-dimensional (2D) detector is always slightly inhomogeneous and anisotropic. If we did not compensate for this anisotropy, the data could become very distorted. We make it a standard practice to repeat measurements at two sample orientations, rotated by 90 degrees to each other. In this section, we discuss the procedure of such a measurement in the case of samples processed by thermomechanical (creep) deformation.

2.2.1. Sample Preparation

1) Prepare a cylindrical or rectangular sample of bulk metallic glass of chosen composition. Make sure that the two end surfaces are smooth and parallel to each other.

2) Subject the sample to uniaxial stress, and raise the temperature to T_a, which has to be below the glass transition temperature, T_g. One has to be careful to keep the stress level below the yield stress at that temperature, because the yield stress is strongly temperature dependent, particularly near T_g.

3) Cool the sample down to room temperature with the stress still applied.

4) Remove the stress, and cut a sample thin enough for X-ray diffraction (~0.5 mm) from the crept sample. The cut should be made parallel to the stress axis so that, in transmission geometry, the scattering vector Q probes the direction parallel and normal to the applied stress.

5) The reference sample annealed in the same conditions (same temperature and time) should be prepared with a similar thickness. In general, the thickness is chosen in reference to absorption of X-rays ($t\mu < 0.2$, t is the thickness and μ is the absorption coefficient) and samples containing heavy elements should be thinner.

2.2.2. Diffraction Measurement

6) Set-up the beam line for a high-energy PDF experiment with a 2D detector. The detector should be placed at a distance that covers enough Q space for reliable normalization and Fourier transformation, up to 18–25 Å^{-1}.

7) Place the sample with the plane normal to the beam and the stress axis aligned vertically to the ground.

8) Turn on the X-ray and take data for a sufficiently long time, typically 5–15 min (Data Set #1). The data should be in the form of $I(Q, \theta)$, where θ is the azimuthal angle. The counting time and number of exposures depend on specifics of the detector and X-ray beam intensity. Ideally we want to have ~10^6 counts per inverse Å after integration over θ, at high Q.

9) Turn the sample by 90° around the beam so that the stress axis is horizontal, and repeat the measurement (Data Set #2). This step is useful because the sensitivity and background of the 2D detector are usually not isotropic. It is helpful to have a sample holder that can rotate the sample by 90° without changing its distance to the detector.

10) Rotate Data Set #2 by 90° and subtract it from Data Set #1, to remove the effect of the efficiency, and the flat field anisotropy of the 2D detector.

11) Determine $S(Q, \theta)$ from the raw data, $I(Q, \theta)$, by correcting for absorption, background, Compton intensity and the atomic scattering factor, $<f(Q)>^2$, and normalizing to unity at $Q \to \infty$.

2.2.3. Data Processing

12) Ideally the isotropic and anisotropic $S(Q)$ is obtained by:

$$S_0^0(Q) = \frac{1}{2}\int_0^\pi [S(Q,\theta) + S(Q,-\theta)]\sin\theta d\theta \tag{12}$$

$$S_2^0(Q) = \frac{1}{4}\int_0^\pi [S(Q,\theta) + S(Q,-\theta)]\left[3\cos^2\theta - 1\right]\sin\theta d\theta \tag{13}$$

However, as we noted, we make measurements with two sample orientations to compensate for the anisotropy of detector efficiency, and obtain two sets of data, Data Sets #1 and #2. The method of obtaining $S_2^0(Q)$ from the two sets of data is as follows.

13) The anisotropic part of Data Set #1 has the form:

$$S_{aniso1}(Q,\theta) = \frac{S_2^0(Q)}{2}\left(3\cos^2\theta - 1\right) \tag{14}$$

On the other hand, for Data Set #2, the stress is along the x-axis, so that:

$$S_{aniso2}(Q,\theta) = \frac{S_2^0(Q)}{2}\left(3\sin^2\theta - 1\right) \tag{15}$$

Thus, the difference is:

$$\begin{aligned}\Delta S(Q,\theta) &= S_{aniso1}(Q,\theta) - S_{aniso2}(Q,\theta)\\ &= \frac{3S_2^0(Q)}{2}\left(\cos^2\theta - \sin^2\theta\right) = 2S_{aniso1}(Q,\theta) - \frac{S_2^0(Q)}{2}\end{aligned} \tag{16}$$

Therefore:

$$S_{aniso1}(Q,\theta) = \frac{\Delta S(Q,\theta)}{2} + \frac{S_2^0(Q)}{4} \tag{17}$$

Then:

$$S_2^0(Q) = \frac{1}{8}\int_0^{\pi}\left[\Delta S(Q,\theta) + \Delta S(Q,-\theta)\right]\left[3\cos^2\theta - 1\right]\sin\theta\, d\theta \tag{18}$$

because the second term in Equation (17) integrates out to zero using this integration. Usually, $S_2^0(Q)$ is much smaller than $S_0^0(Q)$, and thus requires a higher precision to determine. By taking the difference in Equation (16), the background and the Compton scattering intensity are automatically removed, making the result more accurate.

14) Carry out the Bessel transformation to obtain the isotropic and anisotropic PDF:

$$\rho_0 g_0^0(r) = \frac{1}{2\pi^2}\int_0^{\infty} S_0^0(Q)\frac{\sin(Qr)}{Qr}Q^2 dQ \tag{19}$$

$$\rho_0 g_2^0(r) = -\frac{1}{2\pi^2}\int_0^{\infty} S_2^0(Q)\left[\left(3 - Q^2 r^2\right)\frac{\sin(Qr)}{Q^3 r^3} - \frac{3\cos(Qr)}{Q^2 r^2}\right]J_\ell(Qr)Q^2 dQ \tag{20}$$

Note that the range of Q is limited to $Q_{max} = 4\pi/\lambda$. Therefore, one has to be careful not to introduce termination errors [4,5].

15) Calculate the anisotropic PDF for affine deformation using Equation (11).

16) Sometimes it is not possible to rotate the sample by $90°$, for instance when we carry out an *in situ* measurement using a heavy mechanical testing machine. In such a case, we use an isotropic sample, such as a metallic glass, well annealed and relaxed at a temperature close to T_g, to determine the $S_2^0(Q)$ of the detector using the step 12), and subtract the detector $S_2^0(Q)$ from the result.

3. Results

Here, we show typical results obtained for the sample treated with thermomechanical creep to demonstrate how the anisotropic PDF can be determined. Metallic glass $Zr_{55}Cu_{30}Ni_5Al_{10}$ ($T_g = 707$ K [8]) with the dimension of $2.5 \times 2.5 \times 5$ (mm) was subjected to compressive stress of 1.0 GPa for 60 min at $T = 623$ K, which is substantially below the glass transition temperature. Using this process the sample undergoes thermomechanical creep deformation and becomes anelastically distorted. The structure is no longer isotropic, but retains the memory of anelastic deformation [7,9]. After cooling the sample to room temperature, a piece with the dimension of $2 \times 0.5 \times 2$ (mm) was cut out of the sample. The X-ray diffraction measurement was carried out at the 6-ID and 1-ID beam lines of the Advanced Photon Source (APS), Argonne National Laboratory. The incident X-ray energy was set to 100 keV. The beam size was 300×300 (µm) and a Perkin-Elmer detector or a GE 2D detector with the pixel size 200×200 (µm) and 2048×2048 pixels was used. The distance between the sample and the detector was 35 cm, which allowed $S(Q)$ to be determined up to $Q = 22$ Å^{-1}.

Figure 1a shows the raw data from the 2D detector. In this figure the diffraction ring is actually distorted elliptically, but it is difficult to recognize it by eye. If one takes the difference between Data

Set #1 and Data Set #2 the distortion is more clearly seen, as shown in Figure 1b. Both data sets were corrected for absorption, Compton scattering, and sample-independent background from the set-up to obtain $S(Q, \theta)$. Then the isotropic and anisotropic $S(Q)$ were determined using Equations (17) and (18). The step size for integration was $\Delta Q = 0.023$ Å^{-1}, $\Delta\theta = 0.052$ radian.

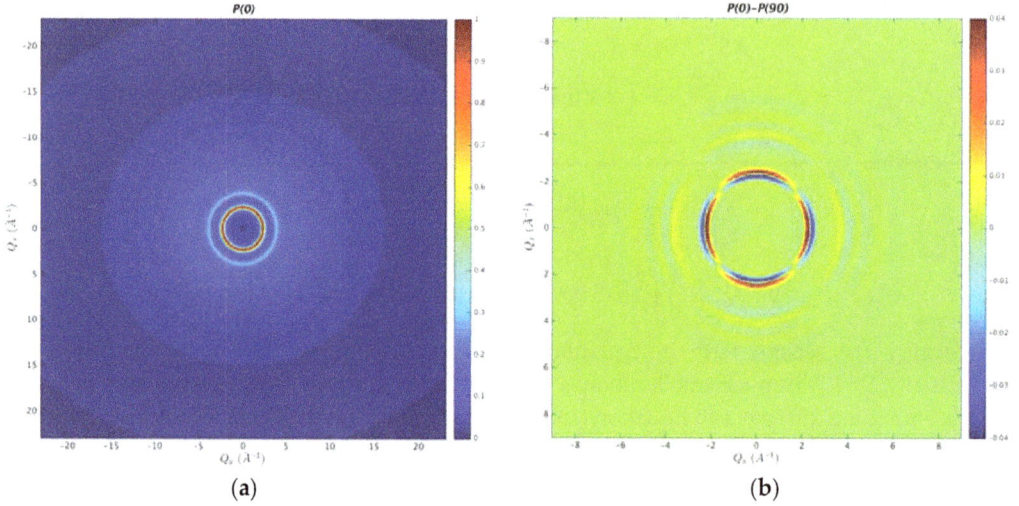

Figure 1. (a) High-energy X-ray diffraction pattern from metallic glass $Zr_{55}Cu_{30}Ni_5Al_{10}$, colors indicating intensity; (b) Difference between Data Set #1 (the stress axis vertical) and Data Set #2 (stress axis horizontal).

$S_0^0(Q)$ and $S_2^0(Q)$ thus obtained are shown in Figures 2 and 3. In Figure 3, $S_2^0(Q)$ is compared to $S_{2,affine}^0(Q)$. They are clearly different due to extensive non-affine deformation and atomic rearrangement as discussed below. The isotropic and anisotropic PDFs, obtained by Equations (19) and (20), are shown in Figures 4 and 5. Again, the anisotropic PDF is compared to the affine anisotropic PDF in Figure 5.

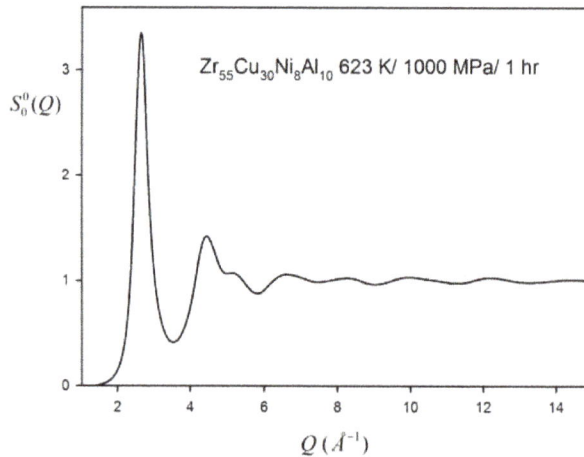

Figure 2. Isotropic structure function $S_0^0(Q)$ for a sample creep deformed at 623 K under 1 GPa for 1 h.

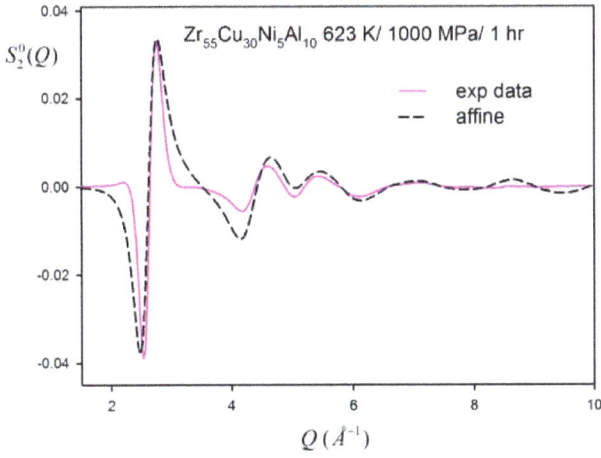

Figure 3. Anisotropic structure function $S_2^0(Q)$. Line in pink shows $S_{2,affine}^0(Q)$.

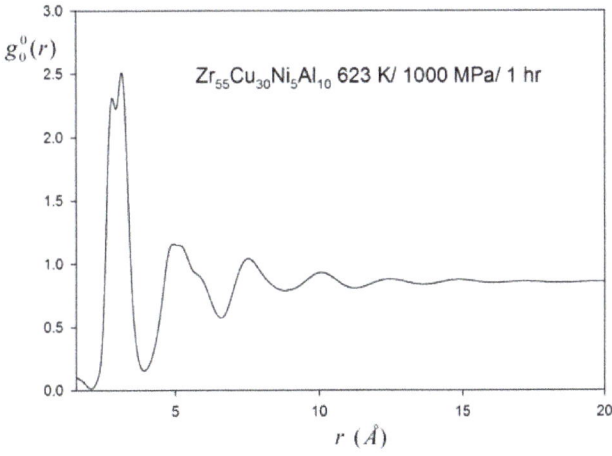

Figure 4. Isotropic PDF, $g_0^0(r)$.

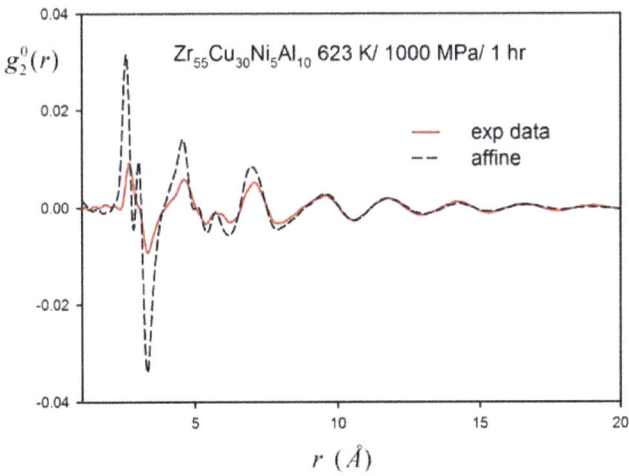

Figure 5. Anisotropic PDF $g_2^0(r)$ compared to the affine PDF (red curve).

It is constructive to compare the results with those obtained for a sample elastically deformed at room temperature with the applied stress well below the yield stress. In this case, in order to deform the sample, heavy mechanical testing equipment is involved, which makes it impossible to rotate the sample by 90°. Thus, we followed the procedure (16) to eliminate the effect of anisotropy in the detector. To test this procedure, we prepared two samples out of the metallic glass sample processed by creep treatment. One was a thin plate piece cut out of the sample with the plane parallel to the applied uniaxial stress direction, and the other was cut perpendicular to the stress. As shown in Figure 6, the one parallel to the stress shows a strong $S_2^0(Q)$ component, whereas the one perpendicular to the stress does not.

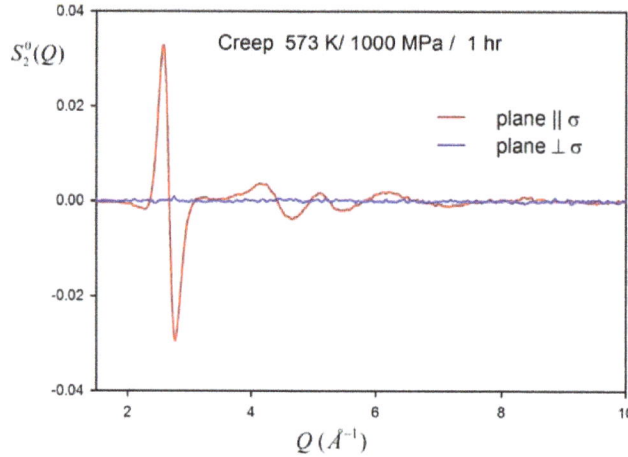

Figure 6. The $S_2^0(Q)$ determined for two cross-sections of the sample after the creep treatment: With the plane parallel to axial stress and plane perpendicular to the uniaxial stress.

During RT deformation , because the applied stress is below the yield stress, metallic glasses respond to stress elastically following Hook's law. However, at the atomic level, the structure changes by breaking and forming of atomic bonds, even below the yield stress [10,11]. Compared to the case of creep deformation shown in Figure 5, the anisotropic PDF $g_2^0(r)$ under applied stress is much closer to the affine anisotropic PDF because the deformation is mostly elastic as shown in Figure 7. However, there are significant differences in the first peak area (2–4 Å) of $g_2^0(r)$, which reflect the intrinsically anelastic nature of a glass [11,12]. Figures 5 and 7 cannot be directly compared, because the result in Figure 7 was obtained while the sample was still under stress, whereas that in Figure 5 was obtained after releasing the stress. To allow closer examination of the deviations from the affine PDF, the difference between the affine PDF and the data PDF, $\Delta g_2^0(r)$, is shown in Figure 8 for the applied stress of 400 and 1000 MPa. The difference is not limited to the first peak area, but extends to 6 Å or beyond. A part of this relaxation reflects the non-affine elastic strain because of the disorder in the structure [13,14]. However, such an effect is rather small [12], and much of it originates from bond cutting and forming [7], which occurs even in the nominally elastic regime [10–12].

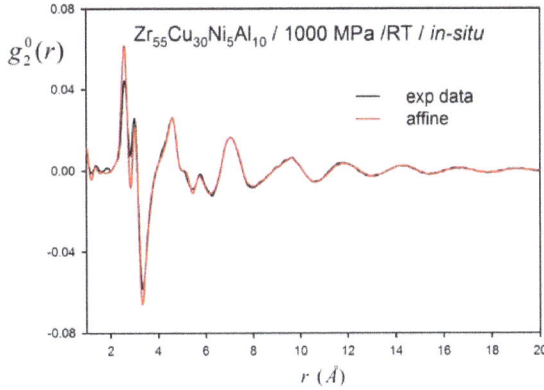

Figure 7. The anisotropic PDF, $g_2{}^0(r)$, determined during *in situ* experiment at 1000 MPa.

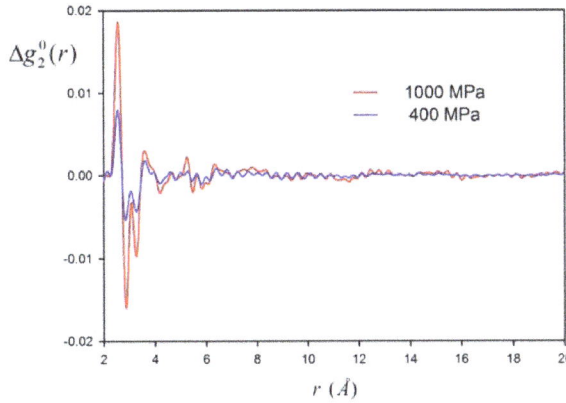

Figure 8. The difference PDF, $\Delta g_2{}^0(r) = g_2{}^0(\text{affine}) - g_2{}^0(\text{data})$, for the applied stress of 400 and 1000 MPa.

4. Discussion

As shown in Figures 3 and 5 the actual anisotropic $S(Q)$ and PDF are significantly deviated from the one for affine deformation. One way to interpret this deviation is to introduce local strain, which is dependent on r, by fitting the following equation to the data:

$$g_2^0(r) = -\varepsilon(r) \left(\frac{1}{5}\right)^{1/2} \frac{2(1+\nu)}{3} r \frac{d}{dr} g_0^0(r) \tag{21}$$

Figure 9 shows the r-dependent local strain, $\varepsilon(r)$, normalized by $\varepsilon(\infty)$ for creep deformation (blue) and elastic deformation (red). A similar analysis was made earlier using the $S(Q)$ data with Q parallel ($\theta = 0$) and perpendicular ($\theta = 90°$) to the stress axis [15–17]. However, in [15–17] the strain was determined from the PDF obtained using Equation (19) rather than Equation (20), which introduced some errors. Note that Equation (19) is only valid for isotropic systems, and once the system becomes anisotropic, we have to use the spherical harmonics expansion shown above. The correct procedure with Equation (20) is now widely used [18–20].

For creep deformation (blue symbols in Figure 9), the local strain $\varepsilon(r)$ is small at short distances and increases with r, saturating to a constant, $\varepsilon(\infty)$, which is equal to the macroscopic recoverable strain. The total creep strain includes both anelastic strain (recoverable) and plastic strain (unrecoverable). Plastic strain leaves no signature on the structure other than some rejuvenation, and $\varepsilon(\infty)$ corresponds only to anelastic strain. The difference, $\Delta\varepsilon(r) = \varepsilon(\infty) - \varepsilon(r)$, represents screening, or the local strain modification, through atomic rearrangements. The sample that underwent creep was treated thermomechanically at a temperature (623 K), which is below T_g, but high enough to activate anelastic deformation. While the sample was held at this temperature under the stress, anelastic local atomic rearrangements occurred to relax the applied stress. When the sample was cooled down and the stress was removed, the regions in which local atomic rearrangements took place were strained in the opposite direction. The stress produced by these strained regions is balanced by the long-range stress, which corresponds to $G\varepsilon(\infty)$, where G is the shear elastic modulus. Consequently, the actual strain at the nearest neighbor is only 30% of the total strain at a long range $\varepsilon(\infty)$ as seen here. The strain is reduced compared to $\varepsilon(\infty)$ up to 10 Å, indicating that the region of atomic rearrangements extend as far as the fourth neighbor shell.

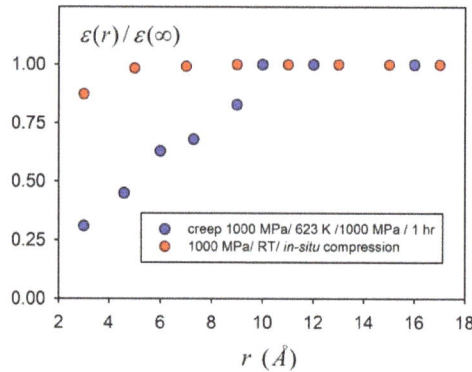

Figure 9. r-dependent local strain determined from Equation (21) for the sample creep deformed at 623 K under 1 GPa for 1 h (blue), and for the sample under load of 1 GPa at room temperature (red).

In comparison, for elastic deformation (red symbols in Figure 9) the strain is nearly 100%. This is mainly because, in this case, the diffraction measurement was made with the applied stress on, so that much of the strain is truly elastic. Nevertheless, $\varepsilon(r)/\varepsilon(\infty)$ is less than unity at the nearest neighbor distance, due to non-collinear atomic displacement [13] and anelasticity [10,11]. It is noteworthy that the local strain modification, $\Delta\varepsilon(r) = \varepsilon(\infty) - \varepsilon(r)$, is much more short-range (only up to 4 Å or so, including only the nearest neighbor shell) for elastic deformation than for creep (up to 10 Å).

In crystals, the topology of atomic connectivity is well defined because of lattice periodicity. Lattice defects, such as dislocations and vacancies, are defined as local deviations from the periodic lattice. In glasses, on the other hand, the topology of atomic connectivity is open, in a sense that it is easily changed by thermal excitation or applied stress, through local bond breaking and forming. Much of the deviations from the affine deformation observed for the anisotropic PDF reflect such local bond rearrangements. The local stresses these rearrangements create are balanced by the long-range stress produced by the long-range strain $\varepsilon(\infty)$, to make the total stress zero.

These local atomic rearrangements are phenomenologically described in terms of the shear-transformation-zones (STZs), which facilitated the elucidation of mechanical properties [2,3]. Whereas the atomic level details of STZ still remain elusive, a variety of computer simulations, such as References [3] and [24], made the outline of STZs clearer. The reported size of STZ varies greatly, from 20 to few hundred atoms [21–24]. However, according to our recent work [25,26], the size of STZ is actually much smaller, involving only about five atoms at the saddle point of the potential energy landscape and about 17 atoms after relaxation from the saddle point to the nearest minimum. Atoms involved in STZ form a small cluster with the size extending only to 2–3 atomic distances.

Such a small size of STZ is more consistent with the spatial extension of the atomic rearrangement for the apparently elastic deformation seen in Figures 8 and 9. Therefore, it appears that in the elastic regime the applied stress activates only individual STZs, so that atomic rearrangements induced by stress are limited to immediate locality, not much beyond the nearest neighbors. When the stress is removed, atomic rearrangements in the opposite direction take place. Thus, macroscopically the system appears elastic, and the original dimension is restored when the stress is removed. However, microscopically it is anelastic, in a sense that local atomic rearrangements take place and the local topology of atomic connectivity is altered during deformation. A proof of such an anelastic nature of deformation below the yield stress is the fact that the system can be rejuvenated even during apparently elastic deformation [27].

On the other hand, it was shown that activation of STZ can induce a cascade or avalanche of other STZ actions [26]. The large extension of the zone of atomic rearrangement during creep, as shown in Figures 5 and 9 indicates such avalanche of STZ activity is taking place during thermomechanical creep. Whereas, at room temperature, STZ avalanche occurs only for rapidly quenched unstable samples [26]. Figure 9 suggests that at elevated temperatures it may happen even in well-annealed stable sample. The sample is then cooled down to room temperature for diffraction measurement. When the stress is removed, the original sample dimension is not recovered because some local deformations are anelastically frozen-in. Much of the frozen-in strain is recovered when the sample is relaxed without stress, but a part of deformation becomes plastic by percolation and is never recovered.

5. Conclusions

By their nature, diffraction measurements provide only the information regarding two-atom positional correlation. However, most physical properties depend on more collective atomic correlations, which diffraction measurements cannot directly assess. This problem is exacerbated in liquids and glasses because of strong disorder. Nevertheless, by perturbing the system and looking at the response, it is possible to learn more about the nature of the system. As an example, we discussed the structural response of metallic glasses to applied stress. Applied stress breaks the symmetry of the system, so it becomes possible to carry out measurements with high accuracy by focusing on the emergent symmetry component induced by stress.

Our analysis indicates that anelastic atomic rearrangements at room temperature induced by applied stress below the yield stress are limited to immediate neighborhood of atoms, most likely representing activation of a single STZ. However, when the stress is applied at a temperature close to T_g atomic rearrangements are extensive and spatially more extended, and the sample shows creep deformation. Such deformation must be a consequence of a cascade or avalanche of multiple STZ actions, resulting in macroscopically anelastic behavior. As illustrated in the results above, the study of the structural change due to applied stress using synchrotron X-ray diffraction leads to revelation of atomistic details of the deformation mechanism in metallic glasses when the result are analyzed using the anisotropic PDF method.

Acknowledgments: This work was supported by the US Department of Energy, Office of Science, Basic Energy Sciences, Materials Science and Engineering Division.

Author Contributions: This paper was written by T.E. with the aid by T.Y. and W.D. who provided data and information on experimental details.

Conflicts of Interest: The authors declare no conflict of interest.

References

1. Greer, A.L. Metallic glasses. *Science* **1995**, *267*, 1947–1953. [CrossRef] [PubMed]
2. Argon, A.S. Mechanisms of inelastic deformation in metallic glasses. *J. Phys. Chem. Solids* **1982**, *43*, 945–961. [CrossRef]
3. Falk, M.L.; Langer, J.S. Dynamics of viscoplastic deformation in amorphous solids. *Phys. Rev. E* **1998**, *57*, 7192–7205. [CrossRef]
4. Warren, B.E. *X-ray Diffraction*; Dover Publications: New York, NY, USA, 1969.

5. Egami, T.; Billinge, S.J.L. *Underneath the Bragg Peaks: Structural Analysis of Complex Materials*, 2nd ed.; Pergamon Materials Series, Elsevier: Amsterdam, The Netherland, 2012.

6. Chupas, P.J.; Qiu, X.; Hanson, J.C.; Lee, P.L.; Grey, C.P.; Billinge, S.J.L. Rapid-acquisition pair distribution function (RA-PDF) analysis. *J. Appl. Cryst.* **2003**, *36*, 1342–1347. [CrossRef]

7. Suzuki, Y.; Haimovich, J.; Egami, T. Bond-orientational anisotropy in metallic glasses observed by X-ray diffraction. *Phys. Rev. B* **1987**, *35*, 2162–2168. [CrossRef]

8. Yokoyama, Y.; Inoue, K.; Fukaura, K. Pseudo Float melting state in ladle arc-melt-type furnace for preparing crystalline inclusion-free bulk amorphous alloy. *Mater. Trans.* **2002**, *43*, 2316–2320. [CrossRef]

9. Dmowski, W.; Egami, T. Observation of structural anisotropy in metallic glasses induced by mechanical deformation. *J. Mater. Res.* **2007**, *22*, 412–418. [CrossRef]

10. Suzuki, Y.; Egami, T. Shear deformation of glassy metals: Breakdown of Cauchy relationship and anelasticity. *J. Non-Cryst. Solids* **1985**, *75*, 361–366. [CrossRef]

11. Dmowski, W.; Iwashita, T.; Chuang, C.-P.; Almer, J.; Egami, T. Elastic heterogeneity in metallic glasses. *Phys. Rev. Lett.* **2010**. [CrossRef] [PubMed]

12. Egami, T.; Iwashita, T.; Dmowski, W. Mechanical properties of metallic glasses. *Metals* **2013**, *3*, 77–113. [CrossRef]

13. Waire, D.; Ashby, M.F.; Logan, J.; Weis, J. On the use of pair potentials to calculate the properties of amorphous metals. *Acta Metall.* **1971**, *19*, 779–788. [CrossRef]

14. Maloney, C.E.; Lemaître, A. Amorphous systems in athermal, quasistatic shear. *Phys. Rev. E* **2006**. [CrossRef] [PubMed]

15. Poulsen, H.F.; Wert, J.A.; Neuefeind, J.; Honkimäki, V.; Daymond, M. Measuring strain distributions in amorphous materials. *Nat. Mater.* **2005**, *4*, 33–36. [CrossRef]

16. Hufnagel, T.C.; Ott, R.T.; Almer, J. Structural aspects of elastic deformation of a metallic glass. *Phys. Rev. B* **2006**. [CrossRef]

17. Wang, X.D.; Bednarcik, J.; Saksl, K.; Franz, H.; Cao, Q.P.; Jiang, J.Z. Tensile behavior of bulk metallic glasses by *in situ* X-ray diffraction. *Appl. Phys. Lett.* **2007**. [CrossRef]

18. Mattern, N.; Bednarcik, J.; Pauly, S.; Wang, G.; Das, J.; Eckert, J. Structural evolution of Cu-Zr metallic glasses under tension. *Acta Mater.* **2009**, *57*, 4133–4139. [CrossRef]

19. Vempati, U.K.; Valavala, P.K.; Falk, M.L.; Almer, J.; Hufnagel, T.C. Length-scale dependence of elastic strain from scattering measurements in metallic glasses. *Phys. Rev. B* **2012**. [CrossRef]

20. Qu, D.D.; Liss, K.-D.; Sun, Y.J.; Reid, M.; Almer, J.D.; Yan, K.; Wang, Y.B.; Liao, X.Z.; Shen, J. Structural origins for the high plasticity of a Zr-Cu-Ni-Al bulk metallic glass. *Acta Mater.* **2013**, *61*, 321–330. [CrossRef]

21. Schuh, C.A.; Lund, A.C.; Nieh, T.G. New regime of homogeneous flow in the deformation map of metallic glasses: Elevated temperature nanoindentation experiments and mechanistic modeling. *Acta Mater.* **2004**, *52*, 5879–5891. [CrossRef]

22. Pan, D.; Inoue, A.; Sakurai, T.; Chen, M.W. Experimental characterization of shear transformation zones for plastic flow of bulk metallic glasses. *Proc. Natl. Acad. Sci. USA* **2008**, *105*, 14769–14772. [CrossRef] [PubMed]

23. Ju, J.D.; Jang, D.; Nwankpa, A.; Atzmon, M. An atomically quantized hierarchy of shear transformation zones in a metallic glass. *J. Appl. Phys.* **2011**. [CrossRef]

24. Choi, I.-C.; Zhao, Y.; Yoo, B.-G.; Kim, Y.-J.; Suh, J.-Y.; Ramamurty, U.; Jiang, J. Estimation of the shear transformation zone size in a bulk metallic glass through statistical analysis of the first pop-in stresses during spherical nanoindentation. *Scr. Mater.* **2012**, *66*, 923–926. [CrossRef]

25. Fan, Y.; Iwashita, T.; Egami, T. How thermally activated deformation starts in metallic glass. *Nat. Commun.* **2014**. [CrossRef] [PubMed]

26. Fan, Y.; Iwashita, T.; Egami, T. Crossover from Localized to Cascade Relaxations in Metallic Glasses. *Phys. Rev. Lett.* **2015**. [CrossRef] [PubMed]

27. Tong, Y.; Iwashita, T.; Dmowski, W.; Bei, H.; Yokoyama, Y.; Egami, T. Structural Rejuvenation in Bulk Metallic Glasses. *Acta Mater.* **2015**, *86*, 240–248. [CrossRef]

7

Influences of Restaurant Waste Fats and Oils (RWFO) from Grease Trap as Binder on Rheological and Solvent Extraction Behavior in SS316L Metal Injection Molding

Mohd Halim Irwan Ibrahim *, Azriszul Mohd Amin †, Rosli Asmawi † and Najwa Mustafa

Advanced Manufacturing and Materials Center, University Tun Hussein Onn Malaysia (UTHM), Parit Raja, Batu Pahat, Johor 86400, Malaysia; azriszul@uthm.edu.my (A.M.A.); roslias@uthm.edu.my (R.A.); nj.wawa@gmail.com (N.M.)

* Correspondence: mdhalim@uthm.edu.my

† These authors contributed equally to this work.

Academic Editor: Hugo F. Lopez

Abstract: This article deals with rheological and solvent extraction behavior of stainless steel 316L feedstocks using Restaurant Waste Fats and Oils (RWFO) from grease traps as binder components along with Polypropylene (PP) copolymer as a backbone binder. Optimal binder formulation and effect of solvent extraction variables on green compacts are being analyzed. Four binder formulations based on volumetric ratio/weight fraction between PP and RWFO being mixed with 60% volumetric powder loading of SS316L powder each as feedstock. The rheological analysis are based on viscosity, shear rate, temperature, activation energy, flow behavior index, and moldability index. The optimal feedstock formulation will be injected to form green compact to undergo the solvent extraction process. Solvent extraction variables are based on solvent temperature which are 40 °C, 50 °C, and 60 °C with different organic solvents of *n*-hexane and *n*-heptane. Analysis of the weight loss percentage and diffusion coefficient is done on the green compact during the solvent extraction process. Differential Scanning Calorimeter (DSC) is used to confirm the extraction of the RWFO in green compacts. It is found that all binder fractions exhibit pseudoplastic behavior or shear thinning where the viscosity decreases with increasing shear rate. After considering the factors that affect the rheological characteristic of the binder formulation, feedstock with binder formulation of 20/20 volumetric ratio between PP and RWFO rise as the optimal binder. It is found that the n-hexane solvent requires less time for extracting the RWFO at the temperature of 60 °C as proved by its diffusion coefficient.

Keywords: restaurant waste fats and oils (RWFO); binder formulation; rheologial behavior; solvent debinding variables

1. Introduction

Metal injection molding (MIM) is a manufacturing process with an advantage of producing intricate and small parts in high volume production with a few shot as compare to other fabrication processes [1–4]. The process involves in developing the feedstock from metal powder and multi components of binder which are then injected by injection molding process to form desired shapes. The components then undergo a debinding process to remove the binder and, finally, a sintering process.

316L stainless steel is one of the most widely used materials for industrial applications. Since 316L stainless steel is a highly alloyed material with good mechanical and corrosion properties it is widely recognized as material for implants [5].

Binder in feedstock is crucial since it will influence the formability of the metal powder into the desired shape [6]. Usually binders are constituted of polymer as backbone binder and other additives for improving flow, wettability, and reducing agglomeration of powder particles. Due to these reasons, good rheological behavior is important since it influences the shape retention of the green compacts during injection molding process [7]. Binders ranging from commercial polymer/wax to sustainable binder from food grade materials have long been successfully implemented as binders in MIM. Although variety of binders have been applied in the MIM process, a sustainable binder is preferable as stated in Brandtland report [8]. Many sustainable binders from food grade materials have been explored in MIM such as carnauba wax [9–11], bees wax [12], palm stearin [13], and palm kernel [14,15]. Although these sources of binders have several advantages, they have some conflict issues between being a human source of foods and industrial sector needs which in turn could rise its price and will disrupt the foods chain of the human population [16].

Debinding is a crucial process of removing binder from green compacts since improper selection of debinding variables could lead to component defects such as cracks and swells [17]. Various techniques can be implemented for the debinding process ranging from thermal, wicking, and solvent. With the shortest time and least impact on the part, the solvent extraction process is preferable since it could result in time and cost savings for both solvent and thermal debinding processes [18]. Solvent is used to remove low molecular weight binder components, which results in porosity inside the components. The existence of porosity inside the components is connected to forming channels that minimize part distortion due to pressure built up inside the components and allow the degradation of the backbone binder to diffuse on the surface easily during a thermal debinding process [19]. Many factors can influence the time taken from solvent extraction process such as the thickness of a part, solvent temperature, types of solvents, and the solvent's flow rate.

Restaurant waste fat and oil (RWFO) derivatives from grease traps have long been analyzed as a potential feedstock for biodiesel [16,20] due to its numerous amounts of fatty acids (Oleic, Stearic. Palmitic and *etc.* [21,22]) which come from animal waste fat and cooking oil [23,24]. Furthermore, its properties of non-toxicity, biodegradability, and renewability [16] make it more interesting for use in biodiesel, cosmetics products, and even as a soap products. With rapids growth of human population, urbanisation, lifestyle changes and nutrition transition has made the RWFO become a more interesting topic in sustainable development [25–27]. Although it is well known for its properties of various fatty acids, its implementation in MIM has not being explored extensively yet [28] and therefore RWFO derivatives as binder components for SS316L feedstock are tested and analyzed in terms of rheological behavior and solvent extraction process.

2. Experimental Section

Water atomized 316L stainless steel powder with chemical composition and particle size distribution as shown in Tables 1 and 2 respectively has irregular shape as revealed in Figure 1b. Particle size distribution is obtained using laser scattering particle analyser FRITSCH Analysette 22 (Universiti Tun Hussein Onn Malaysia (UTHM), Johor, Malaysia) since it is one of the important factors in MIM which influenced the different stages of the MIM process [29]. The selected stainless steel powder is found to have 64.8% Critical Powder Volume Concentration (CPVC) (Figure 1a). CPVC is the point where all the particles are tightly packed and all voids between the particles are filled with binder. The determination of CPVC is important in order to establish the optimum amount of binder in the feedstock. CPVC was experimentally determined by means of oil absorption ASTM D-281-31 method using Thermo Haake Rheomix mixer (Universiti Kebangsaan Malaysia (UKM), Selangor, Malaysia) where 0.5 mL volume of oleic acid was added to 200 g of SS316L powder for every 5 min [30]. If the addition of oil continues, the torque value decreases abruptly indicating that the critical powder volume concentration is reached [31]. Equation (1) is employed to obtain the CPVC using powder volume (V_p) and acid oleic volume (V_o) at the point where the torque is a maximum.

$$\text{CPVC} = \frac{V_P}{V_P - V_o} \times 100\% \tag{1}$$

For MIM feedstock, 2% to 5% volumetric powder loading below the CPVC value is suitable [1]. Volumetric powder loading plays a significant role in preventing parts from slumping or distorting which, in this case 60%, volumetric powder loading (approximately 5% below the CPVC value) was used as SS316L feedstock.

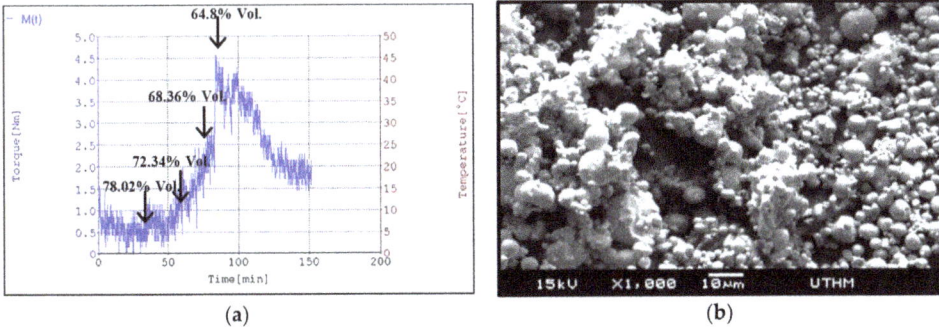

(a)   (b)

Figure 1. (a) CPVC of water atomized SS316L powder [30]; (b) Powder morphology.

Binder consisting of Polypropylene (PP) supplied by Lotte Chemical Titan (Johor, Malaysia) Sdn Bhd is used as a backbone binder and RWFO from a grease trap supplied by Perniagaan Seri Gunung (Selangor, Malaysia) Enterprises is used as primary binder for lubrication and wettability. Thermal characteristics of binder components are shown in Table 3 by using Differential Scanning Calorimeter (DSC, UTHM, Johor, Malaysia) and Thermalgravimetric Analysis (TGA, UTHM, Johor, Malaysia) for melting and degradation temperature, respectively.

Table 1. Chemical composition of water atomized SS316L powders supplied by Epson Atmic Corp., Aomori Prefecture, Japan (wt. %).

				Composition wt. %					
Fe	C	Si	Mn	P	S	Ni	Cr	Mo	Cu
Balance	0.027	0.84	0.19	0.016	0.012	12.2	16.4	2.1	0.03

Table 2. Characteristic of SS316L powders supplied by Epson Atmic Corp. Japan.

D_{10} (μm)	D_{50} (μm)	D_{90} (μm)	Pycnometric Density (g/cm^3)	Tap Density (g/cm^3)
2.72	6.70	15.74	8.0471	4.06

Fourier Transform Infrared (FTIR, UTHM, Johor, Malaysia) was used as a tool to quantify the RWFO received and analysis was done on five samples with different lot of RWFO with the same provider for characterizing the chemical bonds corresponding to various functional groups (Figure 2) [32,33].

Powder and binder volumetric or weight percentage employed in this work are shown in Table 4. Differential Scanning Calorimeter (DSC) analysis was conducted to determine the melting temperature of the binder components and these experiments were performed using DSC TA Instruments Q20 where the samples were heated from room temperature up to 200 °C at 5 °C/min (Figure 3) and then cooled at the same rate under 50 mL/min Nitrogen atmosphere. Thermalgravimetric analysis (TGA) were performed on a LINSEIS Thermobalance from room temperature up to 600 °C at a heating rate of 10 °C/min (Figure 4) in air atmosphere in order to determine decomposition temperature of the binder components.

Powder binder mixtures were carried out in a Brabendeur Plastograph EC (UTHM, Johor, Malaysia) at 175 °C temperature, rotor speed of 30 r.p.m and for 90 min duration. Mixing temperature was selected based on DSC and TGA experiments that was below the lowest degradation temperature of the binder components and had to be higher than the highest melting temperature of the binder components. The feedstock paste was then crushed into small pellets using a granule machine for easy feeding into the capillary rheometer and injection machine. TGA was performed on every feedstock formulation for monitoring the existence of the binder components.

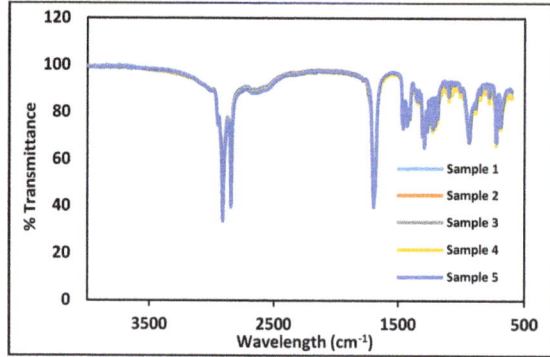

Figure 2. FTIR profile of 5 samples RWFO.

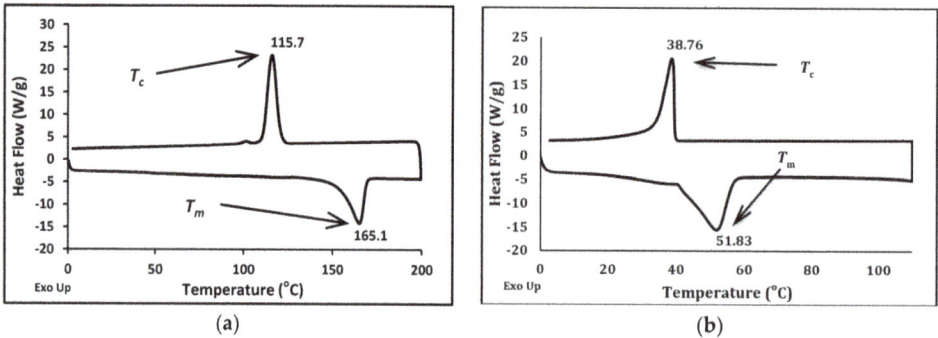

Figure 3. DSC analysis of (a) polypropylene (b) RWFO in nitrogen atmosphere with heating and cooling rate of 5 °C/min to determine the melting and crystalline temperature (T_m and T_c respectively).

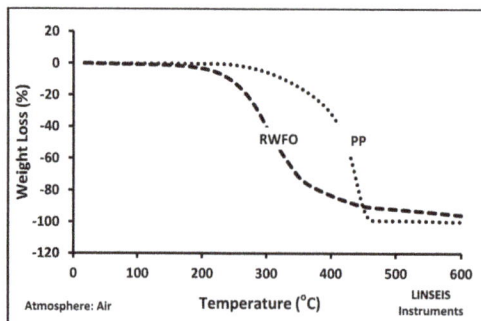

Figure 4. TGA analysis of Polypropylene and RWFO in air atmosphere and heating rate of 10 °C/min to determine degradation temperature of binder constituents.

Rheological behavior of the feedstocks was determined by Instron CEAST SmartRHEO 10 (Universiti Teknikal Malaysia (UTeM), Melaka, Malaysia) with die dimensions of 1 mm in diameter and length 20 mm in order to keep a L/D ratio of 20. The shear rate was chosen in the range of 1000 to 10,000 s^{-1} and a preheat time of 5 min was employed. The effect of temperature on rheological behavior of all feedstock formulations was done with the temperatures of 170 °C, 180 °C, and 190 °C, respectively.

Selected feedstock formulation based on rheological analysis was undergone in the injection process using NISSEI NP 7 Real Mini injection machine (UTHM, Johor, Malaysia). The injection molding parameters were optimized using Taguchi methods (will not be discussed in this paper) as shown in Table 5.

Table 3. Chairacteristic of binder and pure components (T_m and T_d are melting temperature and decomposition temperature respectively).

Material	Density (g/cm^3)	T_m (°C)	T_d (°C)
PP	0.9	165	440.5
RWFO	0.9	52	297.7

Table 4. Components and contents used for 60% volumetric powder loading (93 wt. %) of feedstocks.

Formulation	SS316L Powder (vol. %/wt. %)	Binder Composition and Contents (vol. %/wt. %)
F1	60/93	PP (24/4.2) + RWFO (16/2.8)
F2	60/93	PP (20/3.5) + RWFO (20/3.5)
F3	60/93	PP (16/2.8) + RWFO (24/4.2)
F4	60/93	PP (12/2.1) + RWFO (28/4.9)

Table 5. Injection molding parameter for F2 feedstock.

Injection Temperature (°C)	Injection Pressure (MPa)	Mold Temperature (°C)	Cool Time (s)	Injection Speed (RPM)	Injection Time (s)	Packing Time (s)
165	80.4	40	15	105	1	2

3. Results and Discussion

3.1. FTIR and Thermal Analysis of Binder

FTIR analysis is important since RWFO, which comes from different restaurant location of grease traps are collected and mixed in storage, might influence inconsistency rheological behavior of the feedstock produced. As shown in Figure 2, the results of the FTIR indicates no significant changes in the properties of the RWFO, which confirm the consistency of the constituents in the RWFO.

The melting temperature (T_m) along with crystalline temperature (T_c) can be observed at approximately T_m = 52 °C, T_c = 38.8 °C for RWFO and T_m = 165 °C, T_c =115.7 °C for PP, respectively (Figure 3). Results of TGA curves show the degradation temperature of binder components occurs at the interval of 200 °C to 600 °C and 250 °C to 450 °C, respectively (Figure 4). From these results, mixing temperature of the powder and binder components was selected below than the lowest degradation temperature and higher than the highest melting temperature of the binder components. Data of the degradation and melting temperature of the binder components also will be used as guideline for solvent and thermal debinding process. Information regarding crystalline temperature (Figure 3) of the RWFO and PP is used as the guidelines for mold temperature since crystalline temperature is a transition temperature for substances to form a solid from a liquid state.

The DSC scans obtained for the feedstocks with different binder formulation is presented in Figure 5a. All the peaks are endothermic which show the melting point depression of the PP and RWFO in the different feedstock formulation, respectively. Figure 5a includes the binder components' scans at the top and the feedstock formulation DSC scans at the bottom. It is apparent that the melting peaks of the blends displaced to the left with respect to the peaks of the binder components. For instance, the melting point depression of PP is larger with respect to RWFO for F4 feedstock formulation. The melting point depression was attributed to changes in crystal size because of the presence of the other components in the blends. This reason seems reasonable when considering the way the melting points were obtained [34]. During DSC analysis, the RWFO were first melted followed by PP in succession until the blend was in the liquid state and then quenched, freezing up the melt structure. During the quenching phase, the blend crystallized in a sequence that depended on the rate of crystallization of the polymers. In this regard, the PP crystallized first, then finally the RWFO. In this process, RWFO not yet crystallized could have been "enclosed" in the crystalline structure of a PP already crystallizing reducing the space available for crystallites. This phenomena would explain the reduction of crystallite size and hence, in melting point of RWFO and PP [34].

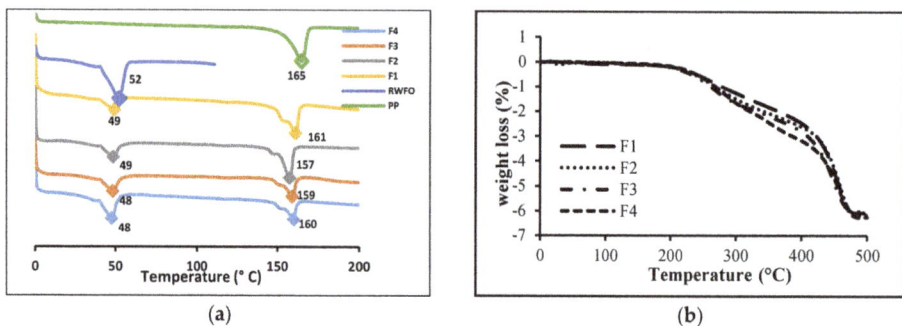

Figure 5. (a) DSC scans of the PP, RWFO and feedstock formulation; (b) TGA profile of different feedstock formulation.

The resulting TGA curves for different binder formulation at a heating rate of 10 °C/min are presented in Figure 5b. The theoretical values of the maximum percentage weight loss possible for these systems are approximately 7%. At the temperature of 500 °C the binder was presumably burned off completely. For F1 feedstock the estimated percentage weight loss of the RWFO is 2.5% and the remaining weight loss is due to the degradation of PP. The evaluation continues with F2 where the estimated of RWFO degradation is 2.9%, followed by F3 which is 3.0%, and finally F4 3.7%. The deviation of TGA profiles between the binder formulations is due to the different weight contents of RWFO and PP in each of the feedstock formulation. The deviation of the total degradation of binder components with respect to the calculated one is due to some binder being left inside the mixer, which reduces its quantity. This also indicates that the miscibility of the binder components with the SS316L powder was good which is crucial in avoiding powder agglomeration [34]. The TGA analysis also suggests that degradation of binder components has not occurred during mixing process. Two degradation slopes were observed which related to RWFO and PP that indicates both components were intact with SS316L powder.

3.2. Rheological Behavior

The rheological behavior of feedstock is crucial to evaluate the ability of mixtures to be injected. The capillary test is the best approach to predict the flow behavior during injection molding. Four different feedstock formulations are shown in Figure 6. It can be observed that the viscosity of all the feedstock decreases with increasing shear rate which indicates pseudoplastic behavior [1] and leads to good wettability of the binder and metal powder [35]. The shear rate range of 100 s^{-1} to 10,000 s^{-1} was selected because it is similar shear rate values obtained during injection stage [31]. For the feedstock to be successfully injected, viscosity values recommended for MIM process must be less than 100 Pa.s for viscosity and 100 s^{-1} to 10,000 s^{-1} in shear rate [31,36].

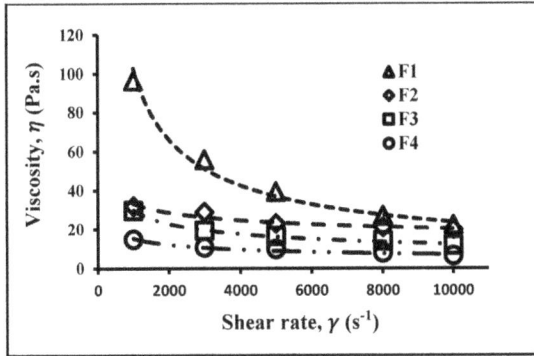

Figure 6. Effect of shear rate on viscosity changes at 170 °C for different binder formulation.

From Figure 6, flow behavior index, n was calculated from the Equation (2).

$$\eta = K\gamma^{n-1} \tag{2}$$

where η is the viscosity, γ is the shear rate, and K is a constant. The value of n indicates the degree of sensitivity of viscosity against shear rate. In the case of MIM process, feedstock must have pseudoplastic behavior where $n < 1$. Higher the ratio of RWFO will aid the viscosity of the blends. When more RWFO being used in the blends, the lower the viscosity of the feedstock. This is due to the RWFO has the lubricating effects which lowering the friction between molecules polymer and hence lower stresses needed in deforming the fluid [37]. Form Figure 6, feedstock with binder formulation of F4 has the lowest viscosity against the shear rate compared to other feedstock formulations.

The sensitivity of viscosity versus shear rate is higher for F1 binder as compared to other binder formulations. This indicates that the F1 formulation highly depends on shear rate and temperature in aiding the flow of the binder due to greater amounts of PP, which is viscous to flow. Other formulations show lower sensitivity due to the amounts of RWFO being higher which indicates that only a small value of shear rate is needed to aid the flow of the feedstock because most of the flow behavior was contributed to by lubrication from RWFO.

Although the n value shows all the binder formulation can be injected, analysis of activation energy base on the Arrhenius equation (Equation (3)) also need to be considered.

$$\ln(\eta_T) - \ln(\eta_0) = E_a/RT \tag{3}$$

From Equation (3), E_a is the flow activation energy, R is the gas constant, T is the temperature, η_0 is the viscosity at reference temperature. If the value of E_a is lower, the viscosity is not so sensitive to temperature variation, which means that any small fluctuation of temperature would not give any sudden viscosity change [37]. The value of E_a/R can be calculated form the gradient of the graph shown in Figure 7. The graph presented in Figures 8 and 9 show the evolution of n, η_0 and E_a rheology parameters with variations of feedstock formulation for the reference conditions of 170 °C and 1000 s^{-1}. A shaded region is the sudden variation of all rheological parameters (n, η_0 and E_a) occurs at F1–F2 feedstock formulation. This region can be considered as critical feedstock formulation based on the 60% volumetric powder loading. A similar region was found by Hidalgo et al. [38] with critical powder loading but in this case for feedstock formulation. Minimum activation energy is obtained for the F2, being the feedstock less sensitive to temperature variations. This minimum marches with a maximum of the η_0 parameter and an inflection point in the value of the flow index.

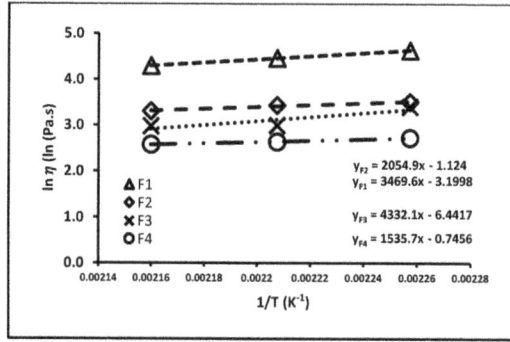

Figure 7. Dependence of shear viscosity with temperature for different feedstock formulations.

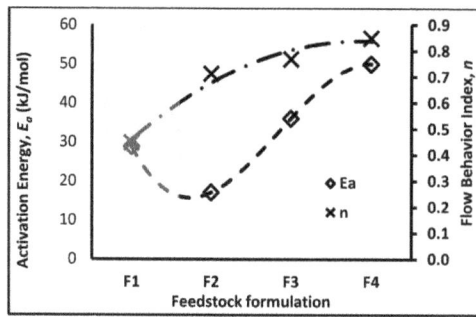

Figure 8. Determination of critical feedstock formulation between E_a and n.

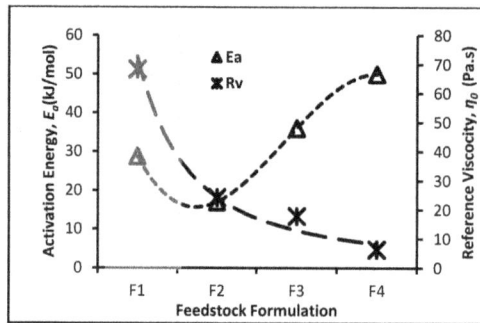

Figure 9. Determination of critical feedstock formulation between Activation energy, E_a and Reference viscosity, η_0.

In the above paragraphs, rheology parameters were independently studied and plotted in dissimilar axes giving unclear or inadequate insights for optimal feedstock formulation. A general moldability index α_{stv} is proposed to summarize and globally describe the rheological behavior of a feedstock taking into account all the parameters [37,38]. The subscripts s, t, and v represent the shear sensitivity, temperature sensitivity, and viscosity, respectively. The general moldability index is defined as in the Equation (4);

$$\alpha_{stv} = \frac{1}{\eta_0} \frac{\left| \dfrac{\partial \log \eta}{\partial \log \gamma} \right|}{\left| \dfrac{\partial \log \eta}{\partial \left(\dfrac{1}{T} \right)} \right|} = \frac{1}{\eta_0} \frac{1-n}{E/R} \tag{4}$$

where η_0 is the reference viscosity at 170 °C and shear rate of 1000 s^{-1}. Higher value of α_{stv} indicates better feedstock formulation. From Figure 10, F2 shows the better general moldability index and this feedstocks will be injected and be analyzed for solvent debinding variables.

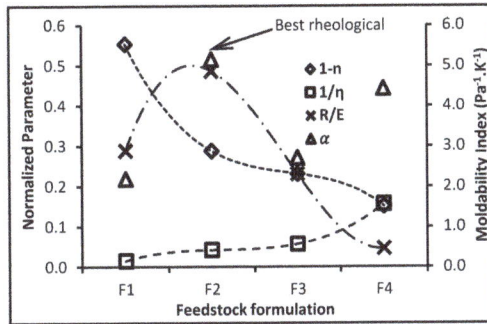

Figure 10. General moldability index *vs.* feedstock formulation.

3.3. Solvents Debinding Behavior

The goal in solvent debinding is to remove the low molecular weight of binder components, which in this case is RWFO, in the shortest time with the least impact on the green molded part. Weight loss of the green compact after solvent extraction was monitored by replication. As can be seen in Figure 11b,c, solubility of the RWFO with hexane solution is good. The colorless hexane solution changes to an orange color, which indicates the existence of extraction process of RWFO.

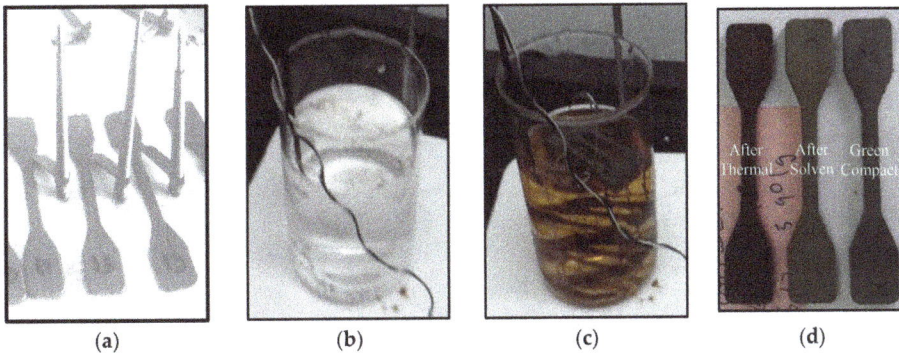

(a)   (b)   (c)   (d)

Figure 11. (a) Injected part (b) Colorless Hexane (c) Color of the hexane solvent changing to orange after green compact being immersed into it during the third hour; (d) comparison of part dimension after thermal and solvent extraction with green compact.

Temperature effect is one of the important variables in the solvent extraction process (Figure 11). There was no indication of swelling or cracks found on the components after solvent extraction process at 60 °C temperature with both solvents. Weight loss analyses were done to monitor any changes in weight after being undergone the solvent extraction process. The diffusion coefficient relation was to determine the rate of diffusion of the RWFO from the green compact. Figure 12 shows result of the solvent extraction process during the first hour with different solvent temperatures and organic solvents.

Weight loss percentage of the green part was calculated base on Equation (5) where W_i is the initial weight of the green part and W_a is the weight of the green part after solvent debound.

$$W_{loss} = \frac{W_i - W_a}{W_i} \times 100\% \tag{5}$$

Using Equation (6), the diffusion coefficient was calculated for RWFO extraction in both Hexane and Heptane solvents where C_a is the average concentration of binder remaining on compressed part, C_i is the initial concentration of binder, C_0 is the boundary condition (zero), t is the debinding time, D is the debinding coefficient and h is the thickness of the compressed part (Nanjo et al., 1993 cited in [18]).

$$\frac{C_a - C_0}{C_i - C_0} = \frac{8}{\pi} e^{\frac{-\pi}{h^2} D t}$$

(6)

Percentage weight loss and diffusion coefficient of green part is increased with temperature from $40\,°C$ to $60\,°C$ using organic solvents of Hexane and Heptane at a fixed solvent to feed ratio (S/F = 14:1) during the first hour. This indicates that RWFO is soluble in both solvents and able to extract RWFO from the green part. During the first hour, Hexane organic solvent shows significant percentage weight loss as compare to Heptane for $40\,°C$ to $60\,°C$ solvent temperature. Diffusion coefficients also increase with temperature that explained the proportional increase in percentage weight loss of the green part. This shows that hexane solvents have better extraction of RWFO as compared to heptane and this might due to the lower carbon number of hexane which increases the solvent diffusion rate into the RWFO [17].

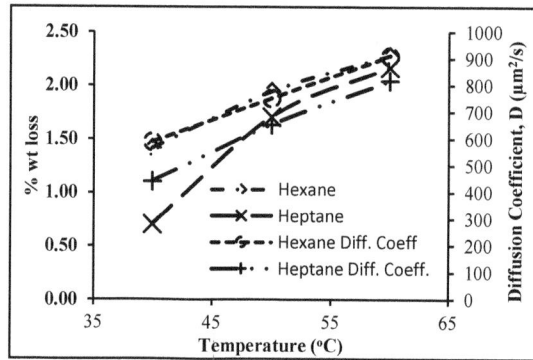

Figure 12. Percentage weight loss and diffusion coefficient at different temperatures during first hour of solvent extraction for F2 feedstock formulation.

From Figure 13, using hexane as the choice of organic solvent, the total amount of RWFO extracted increases by increasing the extraction time from 1 to 3 h, whereas increasing the extraction time from 5 to 6 h does not give any significant change of weight loss percentage on the part. Also as the total percentage weight loss was increase, the diffusion coefficient decreased with time. This is true as stated by Fick's 2nd Law that fluid migration in the matrix occurs mainly through diffusion and concentration drive this migration [39]. During the first hour, concentration of the RWFO was higher in the green compact and reducing after several hours during extraction process. The percentage weight loss is stabilized after 3 h, which reaches about 3 wt. % loss of RWFO. This means that the total amount of binder extracted is not affected by increasing the extraction time over 3 h. The diffusion coefficient of RWFO decreases clearly from 868 to 223.6 $\mu m^2/s$ by increasing the extraction time from 1 to 10 h.

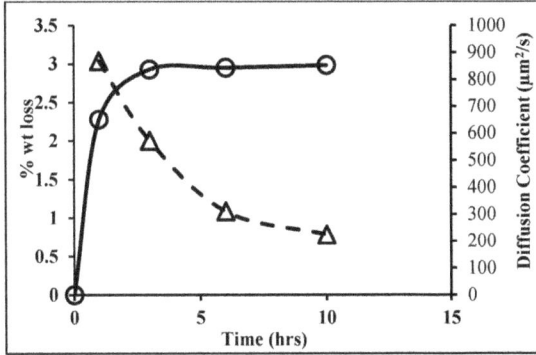

Figure 13. Percentage weight loss and diffusion coefficient of green part using Hexane at temperature of 60 °C.

The green compact after solvent debound for 3 h were then undergoing DSC analysis to monitor the remaining RWFO in the green compact. From Figure 14a, initially two peaks are observed indicating two binder components exist in the green compact. After solvent extraction process, RWFO peaks vanish which leaves only PP melting and crystalline temperature peaks (Figure 14b). This analysis was done on the sample taken from the center of the green compact after solvent extraction had taken place. This indicates that the RWFO were successfully extracted out of the green compact. From Scanning Electron Microscope (SEM) image and Energy dispersive spectroscopy (EDS) analysis (Figure 15a), it can be seen that most of the powder particles are covered by PP and RWFO. Figure 15b shows the results of porosity existence inside the green compact leaving only PP to hold the powder particles after 3 h solvent extraction process. The existence of porosity inside the green compact would create a channel that helps the diffusion of degraded PP during thermal debinding process. Carbon content from the green compact also reduced due to the disappearance of RWFO from the green compact as can be seen on the first peaks of each EDS profile. Figure 11d shows the comparison of the dimensional changes of the green compact before and after the solvent extraction process. No significant dimension changes can be seen on the components. After undergoing thermal debinding process, significant changes in dimension are observed where about 1% shrinkage occurs (will not be discussed in this paper).

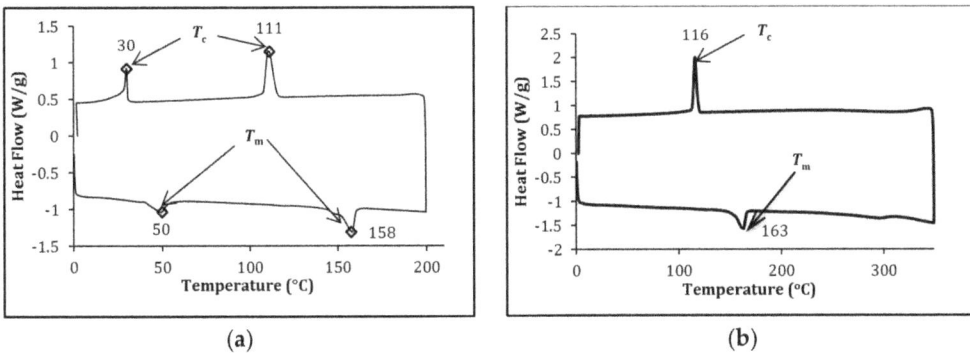

(a)   (b)

Figure 14. DSC scan of the (a) F2 green compact; (b) after solvent extraction under 50 mL/min Nitrogen atmosphere with heating rate of 5 °C/min.

(a)

(b)

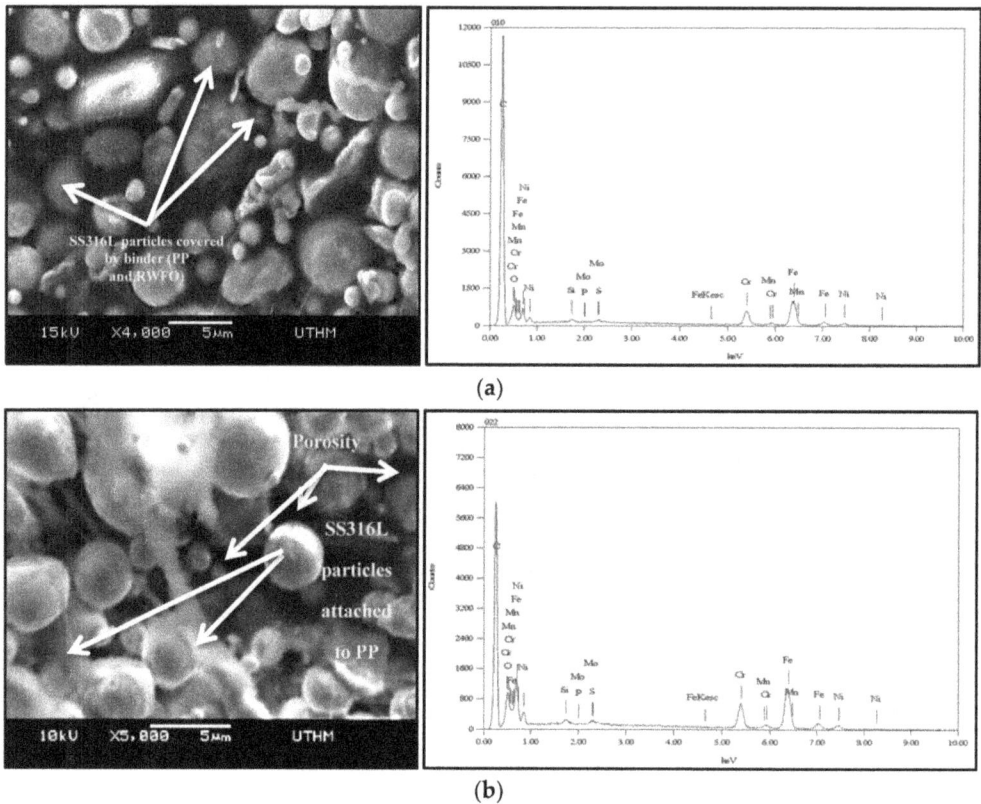

Figure 15. Scanning Electron Microscope (SEM) and Energy Dispersive Spectroscopy (**a**) Green compact morphology (**b**) green compact after 3 h solvent extraction.

4. Conclusions

Rheological behavior of 316L stainless steel feedstock with RWFO as binder components has been analyzed. It is found that pseudoplastic behavior occurs for all binder formulations with viscosity decrease with increasing shear rate. The viscosity decreases with the increasing quantity of RWFO in the binder system showing that lubrication is much influenced by the quantity of RWFO. However the flow activation energy and moldability index shows that binder formulation of F2 is better which strongly indicates that the optimal feedstock formulation between PP and RWFO is a 20/20 volumetric ratio. This might be due to the contents of the stearic and oleic acids components inside the RWFO being less able to strongly influence the wettability of the feedstock [24]. It is also indicative that the RWFO can be used as a sustainable binder system in MIM and further improvement of the binder ratio can be done by additional stearic acids in the binder systems.

F2 feedstock formulation will then been injected to form green compact as shown in Figure 15a and solvent debound was done using heptane and hexane as the organic solvent with temperatures of 40 °C, 50 °C, and 60 °C. From the results above hexane is chooses to be the suitable organic solvent with temperature of 60 °C. This is due to hexane showing a high diffusion rate and percent weight loss of RWFO during extraction process. The results also show no crack formation on the surface of the green compact and quickest extraction can be accomplished within 3 to 4 h.

Acknowledgments: Authors are grateful for the support of this works by the Office of Advanced Manufacturing and Materials Center of Universiti Tun Hussein Onn Malaysia for helpful in preparing samples and analysis.

Author Contributions: Mohd Halim Irwan Ibrahim and Azriszul Mohd Amin performed the experimental work. Mohd Halim Irwan Ibrahim and Azriszul Mohd Amin wrote the paper. Mohd Halim Irwan Ibrahim,

Azriszul Mohd Amin, Rosli Asmawi and Najwa Mustafa contributed to the analyses, interpretation and discussion of results.

Conflicts of Interest: The authors declare no conflict of interest.

References

1. German, R.M.; Bose, A. *Injection Molding of Metals and Ceramics*; Metal Powder Industries Federation: Princeton, NJ, USA, 1997.

2. Abolhasani, H.; Muhamad, N. A new starch-based binder for metal injection molding. *J. Mater. Process. Technol.* **2010**, *210*, 961–968. [CrossRef]

3. Afian, M.; Subuki, I.; Abdullah, N. The influence of palm stearin content on the rheological behavior of 316L stainless steel mim compact. *J. Sci. Technol.* **2012**, *2*, 1–14.

4. Ahn, S.; Jin, S.; Lee, S.; Atre, S.V.; German, R.M. Effect of powders and binders on material properties and molding parameters in iron and stainless steel powder injection molding process. *Powder Technol.* **2009**, *193*, 162–169. [CrossRef]

5. Ibrahim, M.H.I.; Muhamad, N.; Sulong, A.B.; Jamaludin, K.R. Optimization of Micro Metal Injection Molding with Multiple Performance Characteristics using Grey Relational Grade. *Chiang Mai J. Sci.* **2011**, *38*, 231–241.

6. Karatas, C.; Kocer, A.; Ünal, H.I.I.; Saritas, S. Rheological properties of feedstocks prepared with steatite powder and polyethylene-based thermoplastic binders. *J. Mater. Process. Technol.* **2004**, *152*, 77–83. [CrossRef]

7. Huang, B.; Liang, S.; Qu, X. The rheology of metal injection molding. *J. Mater. Process. Technol.* **2003**, *137*, 132–137. [CrossRef]

8. Brandtland, G.H. United Nation Report of the World Commission on Environment and Development: Our Common Future. Available online: http://www.un-documents.net/wced-ocf.htm (accessed on 21 December 2015).

9. Gulsoy, H.O.; German, R.M. Production of micro-porous austenitic stainless steel by powder injection molding. *Mater. Sci.* **2008**, *58*, 295–298. [CrossRef]

10. Kwon, T.H.; Ahn, S.Y. Slip characterization of powder/binder mixtures and its significance in the filling process analysis of powder injection molding. *Powder Technol.* **1995**, *85*, 45–55. [CrossRef]

11. Khakbiz, M.; Simchi, A.; Bagheri, R. Investigation of rheological behavior of 316L stainless steel—3 wt. % TiC powder injection molding feedstock. *Powder Metall.* **2005**, *48*, 144–150. [CrossRef]

12. Supriadi, S.; Baek, E.R.; Choi, C.J.; Lee, B.T. Binder system for STS 316 nanopowder feedstocks in micro-metal injection molding. *J. Mater. Process. Technol.* **2007**, *188*, 270–273. [CrossRef]

13. Omar, M.A.; Subuki, I.; Abdullah, N.S.; Zainon, N.M.; Roslani, N. Processing of Water-Atomised 316L Stainless Steel Powder Using Metal-Injection Processes. *J. Eng. Sci.* **2012**, *8*, 1–13.

14. Salmah, H.; Lim, B.Y.; Teh, P.L. Rheological and thermal properties of palm kernel shell-filled low-density polyethylene composites with acrylic acid. *J. Thermoplast. Compos. Mater.* **2012**. [CrossRef]

15. Mustafa, N.; Ibrahim, M.H.I.; Asmawi, R; Amin, A.M.; Masrol, S.R. Green Strength Optimization in Metal Injection Molding Applicable with a Taguchi Method. *Appl. Mech. Mater.* **2015**, *773–774*, 115–117. [CrossRef]

16. Canakci, M. The potential of restaurant waste lipids as biodiesel feedstocks. *Bioresour. Technol.* **2007**, *98*, 183–190. [CrossRef] [PubMed]

17. Zaky, M.T. Effect of solvent debinding variables on the shape maintenance of green molded bodies. *J. Mater. Sci.* **2004**, *39*, 3397–3402. [CrossRef]

18. Zaky, M.T.T.; Soliman, F.S.S.; Farag, A.S.S. Influence of paraffin wax characteristics on the formulation of wax-based binders and their debinding from green molded parts using two comparative techniques. *J. Mater. Process. Technol.* **2009**, *209*, 5981–5989. [CrossRef]

19. Yang, W.; Yang, K.; Wang, M.; Hon, M. Solvent debinding mechanism for alumina injection molded compacts with water-soluble binders. *Ceram. Int.* **2003**, *29*, 745–756. [CrossRef]

20. Ragauskas, A.M.E.; Pu, Y.; Ragauskas, A.J. Biodiesel from grease interceptor to gas tank. *Energy Sci. Eng.* **2013**, *1*, 42–52. [CrossRef]

21. Williams, J.B.; Clarkson, C.; Mant, C.; Drinkwater, A.; May, E. Fat, oil and grease deposits in sewers: Characterisation of deposits and formation mechanisms. *Water Res.* **2012**, *46*, 6319–6328. [CrossRef] [PubMed]

22. Bart, J.C.J.; Gucciardi, E.; Cavallaro, S. Renewable feedstocks for lubricant production. In *Biolubricants Science and Technology*; Woodhead Publishing Ltd.: Abington, UK, 2013; pp. 1–128.

23. Keener, K.M.; Ducoste, J.J.; Holt, L.M. Properties Influencing Fat, Oil, and Grease Deposit Formation. *Water Environ. Res.* **2008**, *80*, 2241–2246. [CrossRef] [PubMed]

24. He, X.; Francis, L.; Leming, M.L.; Dean, L.O.; Lappi, S.E.; Ducoste, J.J. Mechanisms of Fat, Oil and Grease (FOG) deposit formation in sewer lines. *Water Res.* **2013**, *47*, 4451–4459. [CrossRef] [PubMed]

25. Popkin, B.M. Urbanization, Lifestyle Changes and the Nutrition Transition. *World Dev.* **1999**, *27*, 1905–1916. [CrossRef]

26. Baines, T.; Brown, S.; Benedettini, O.; Ball, P. Examining green production and its role within the competitive strategy of manufacturers. *J. Ind. Eng. Manag.* **2013**, *5*, 53–87. [CrossRef]

27. Chan, H. Removal and recycling of pollutants from Hong Kong restaurant wastewaters. *Bioresour. Technol.* **2010**, *101*, 6859–6867. [CrossRef] [PubMed]

28. Amin, A.M.; Ibrahim, M.H.I.; Asmawi, R.; Mustafa, N. The Influence of Sewage fat Composition on Rheological Bahavior of Metal Injection Molding. *Appl. Mech. Mater.* **2014**, *660*, 38–42. [CrossRef]

29. Sotomayor, M.E.E.; Várez, A.; Levenfeld, B. Influence of powder particle size distribution on rheological properties of 316L powder injection molding feedstocks. *Powder Technol.* **2010**, *200*, 30–36. [CrossRef]

30. Ibrahim, M.H. *Optimization of MicroMetal Injection Molding Parameter by Design of Experiment Method*; Universiti Kebangsaan Malaysia: Selangor, Malaysia, 2011.

31. Sotomayor, M.E.; Levenfeld, B.; Várez, A. Powder injection molding of premixed ferritic and austenitic stainless steel powders. *Mater. Sci. Eng. A* **2011**, *528*, 3480–3488. [CrossRef]

32. Jaggi, H.S.; Kumar, Y.; Satapathy, B.K.; Ray, A.R.; Patnaik, A. Analytical interpretations of structural and mechanical response of high density polyethylene/hydroxyapatite bio-composites. *Mater. Des.* **2012**, *36*, 757–766. [CrossRef]

33. Ghazavi, M.A.; Fallahipanah, M.; Jeshvaghani, H.S. A feasibility study on beef tallow conversion to some esters for biodiesel production. *Int. J. Recycl. Org. Waste Agric.* **2013**, *2*, 1–4. [CrossRef]

34. Adames, J.M. Characterization of Polymeric Binders for Metal Injection Molding (MIM) Process. Ph.D. Thesis, The University of Akron, Akron, OH, USA, December 2007.

35. Li, L.; Liu, X.Q.; Luo, F.H.; Yue, J.L. Effects of surfactant on properties of MIM feedstock. *Trans. Nonferrous Met. Soc. China* **2007**, *17*, 1–8. [CrossRef]

36. Ibrahim, M.H.I.; Muhamad, N.; Sulong, A.B. Rheological Investigation of Water Atomised Stainless Steel Powder for Micro Metal Injection Molding. *Int. J. Mech. Mater. Eng.* **2009**, *4*, 1–8.

37. Aggarwal, G.; Park, S.J.; Smid, I. Development of niobium powder injection molding: Part I. Feedstock and injection molding. *Int. J. Refract. Met. Hard Mater.* **2006**, *24*, 253–262. [CrossRef]

38. Hidalgo, J.; Jiménez-Morales, A.; Torralba, J.M. Torque rheology of zircon feedstocks for powder injection molding. *J. Eur. Ceram. Soc.* **2012**, *32*, 4063–4072. [CrossRef]

39. Liu, H.; Mou, J.; Cheng, Y. Impact of pore structure on gas adsorption and diffusion dynamics for long-flame coal. *Nat. Gas Sci. Eng.* **2015**, *22*, 203–213. [CrossRef]

8

Monitoring of Bainite Transformation Using *in Situ* Neutron Scattering

Hikari Nishijima [1,†], **Yo Tomota** [2,*,‡], **Yuhua Su** [3], **Wu Gong** [3] and **Jun-ichi Suzuki** [4]

Academic Editor: Klaus-Dieter Liss

[1] Department of Applied Beam Science, Graduate school of Science and Enginieering, Ibaraki University, 4-12-1 Naka-narusawa, Hitachi 316-8511, Japan; nishijima.hikari@suzuki-metal.co.jp
[2] Graduate school of Science and Engineering, Ibaraki University, 4-12-1 Naka-narusawa, Hitachi 316-8511, Japan
[3] Japan Atomic Energy Agency, 2-4 Shirane Shirakata Tokai, Ibaraki 319-1195, Japan; yuhua.su@j-parc.jp (Y.S.); gong.wu@jaea.go.jp (W.G.)
[4] Comprehensive Research Organization for Science and Society, 162-1 Shirakata, Tokai, Ibaraki 319-1106, Japan; j_suzuki@cross.or.jp
[*] Correspondence: yo.tomota.22@vc.ibaraki.ac.jp
[†] Current address: Nippon Steel & Sumikin SG Wire Co. Ltd., 7-5-1 Higashinarashino, Narashino, Chiba 275-8511, Japan.
[‡] Current address: National Institute of Materials Science, 1-2-1 Sengen, Tsukuba, Ibaraki 305-0047, Japan.

Abstract: Bainite transformation behavior was monitored using simultaneous measurements of dilatometry and small angle neutron scattering (SANS). The volume fraction of bainitic ferrite was estimated from the SANS intensity, showing good agreement with the results of the dilatometry measurements. We propose a more advanced monitoring technique combining dilatometry, SANS and neutron diffraction.

Keywords: small angle neutron scattering; dilatometry; bainite transformation; *in situ* measurement; neutron diffraction

1. Introduction

Multi-phase steels containing carbon-enriched austenite have been extensively studied because of their excellent combination of strength and ductility/toughness. To obtain multi-phase structures, various material processings have been developed such as intercritical annealing followed by quenching to produce dual-phase steels [1–3], isothermal holding to yield carbon-enriched retained austenite for transformation-induced plasticity (TRIP steels) [4,5], isothermal holding at a low temperature to realize ultra-fine lamellar structure (nano-bainite steel) [6–10], and the partial quenching followed by up-heated isothermal holding (Q&P steels) [11–15]. In particular, nano-bainite steels consisting of nano-scale lamellae of bainitic ferrite and carbon-enriched austenite have attractively exhibited tensile strength greater than 2 GPa and fracture toughness of approximately 30 MPa m$^{1/2}$ [6,7]. The nano-bainite formed by isothermal holding at 300~400 °C shows an extremely slow transformation rate [6,16,17]. This heat treatment is favorable for producing large mechanical components with small residual stresses. However, the acceleration of the transformation must enlarge the application of nano-bainite steels. We have found that a small amount of low temperature ausforming (e.g., at 300 °C) [18,19] or partial quenching below Ms temperature [20] is effective to accelerate nano-bainite transformation. The dislocation structure introduced in austenite at low temperatures is found to assist bainite transformation with strong variant selection where partial dislocations introduced by ausforming play an important role for bainite transformation [19].

In these advanced steels, the processing is complicated enough that it is very important to monitor microstructure evolution quantitatively.

For studying microstructure evolution, *in situ* observations during processing using synchrotron X-ray [21,22] or neutron diffraction (ND) [19,20] have successfully been employed so far. Diffraction profiles provide the insights on volume fractions of constituents, which show good agreement with the results obtained by dilatometry [17], carbon contents in austenite [23], texture [19], and dislocation density [20]. Line broadening analysis for the ND profile provides "coherently diffracting mosaic size" probably related to dislocation cell size in engineering steels. The sizes larger than 1.0 μm such as austenite grain size or ferrite lath size cannot be evaluated by diffraction but hopefully by SANS or Bragg edge (BE) measurements. *In situ* BE measurement during bainite transformation was examined by Huang *et al.* [24]. They reported the changes in austenite volume fraction and carbon concentration with progress of bainite transformation but gave us no data on ferrite lath size. The two populations of austenite with different carbon contents have also been demonstrated by diffraction [19,22], but no information concerning the size of bainite lath has been reported so far. If we employ *in situ* SANS, the insights on the shape and size of the bainite lath would be obtained. In this study, we introduce a dilatometer into SANS-J-II at JRR-3/JAEA. The volume fraction of nano-bainite estimated from SANS data is compared with the results of conventional dilatometry. Then, we measure dilatometry, SANS and ND simultaneously to understand the mechanism of microstructure evolution during heat processing using an industrial neuron diffractometer, iMATERIA, at MLF/J-PARC. The traditional dilatometry provides only the phase fraction estimated from the amount of expansion or contraction of a specimen, whereas neutron experiments provide details in crystallography, chemical compositions, internal stresses, *etc*. This paper reports trials of such a monitoring system combined with complementary multi-methods.

2. Experimental Procedures

2.1. Specimen Preparation

The chemical compositions of the steel used in this study were 0.79C–1.98Mn–1.51Si–0.98Cr–0.24Mo–1.06Al–1.58Co–balanced Fe (mass %). The steel was prepared by vacuum induction melting [8,9]. The ingot was homogenized at 1200 °C for 14.4 ks, followed by hot-rolling in the temperature range 1200–1000 °C to reduce the thickness from 40 mm to 10 mm through 10 successive rolling passes. Plate specimens with $15 \times 15 \times 1$ mm^3 were prepared for SANS measurements at SANS-J-II (JRR-3/Japan Atomic Energy Agency, Tokai, Japan) and iMATERIA (Japan Proton Accelerator Research Complex (J-PARC), Tokai, Japan).

2.2. Small Angle Neutron Scattering Methods

In situ SANS measurements were performed using the SANS-J-II small angle neutron scattering instrument installed at the cold neutron beam line in the JRR-3 research reactor of the Japan Atomic Energy Agency (Tokai, Japan). For the SANS measurement, two two-dimensional (2D) detectors were used to detect neutrons scattered in the 0.005 to 0.199 nm^{-1} scattering vector q-range ($q = (4\pi/\lambda)\sin\theta$, where a half of scattering angle θ, and neutron wavelength λ = 0.656 nm), covering a real microstructure size of 3 to 1000 nm. The detector was positioned 10 m away from the specimen to measure the SANS profiles in the q-range of 0.005 to 0.237 nm^{-1}. Experimental set up is shown in Figure 1. As seen, a dilatometer and a 1.0 T magnet were installed for temperature control and separation of nuclear and magnetic scattering, respectively. A thermo-couple was spot-welded on the specimen surface to control temperature of the specimen. The specimen was heated up to 900 °C with a heating speed of 2 °C/s, held there to obtain an austenite single phase microstructure for starting, and then cooled down to 300 °C with a cooling rate of 5 °C/s, followed by isothermal holding at 300 °C in vacuum under a magnetic field of 1.0 T. The time interval for data acquisition was set to be 10 min (600 s) during the isothermal holding. Using the data in the parallel direction or vertical with respect to the magnetic field direction summing azimuthal sector within 30 degrees, SANS profile, *i.e.*, scattering intensity, was counted *versus q* to obtain profiles of nuclear and magnetic scattering components (for more details see Figure 3).

Concerning the iMATERIA at MLF/J-PARC, by which simultaneous measurements of dilatometry, ND and SANS were examined, the detailed explanation was omitted here because of a preliminary experiment. Brief explanation will be given in Section 3.3 together with some experimental results.

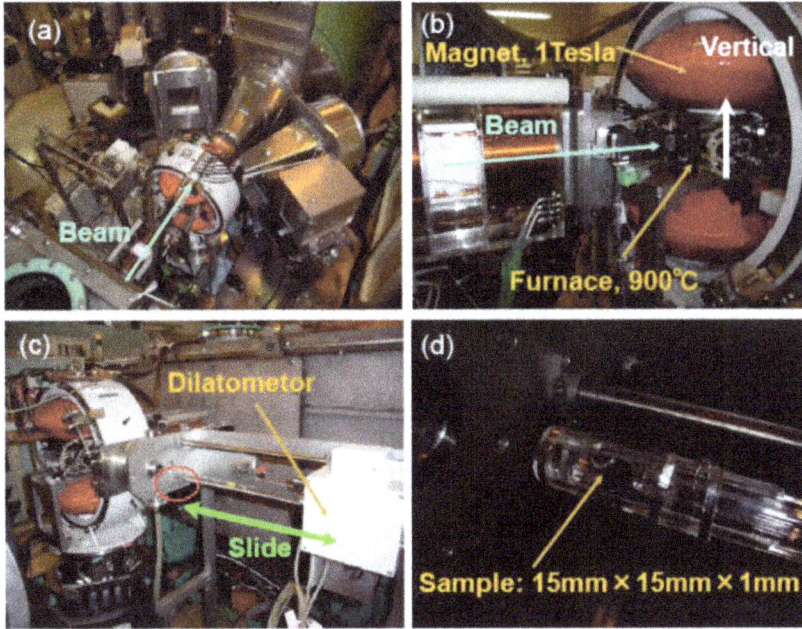

Figure 1. Experiment view of SANS with a 1 T magnet and a dilatometer to monitor the kinetics of bainite transformation at SANS-J-II/JRR-3 (JAEA): (**a**) overall top view; (**b**) magnet; (**c**) dilatometer and (**d**) specimen holder.

2.3. Data Analysis on Small Angle Neutron Scattering

The SANS intensities (I) obtained were plotted as a function of q, where lower q-values correspond to larger size of grain or particle. From the $\ln I(q)$ *versus* $\ln q$ plots, several microstructural parameters can be determined at different q regions, *i.e.*, the radius of gyration (R_g) representing "the effective size of the scattering particle" can be determined by the Guinier plot in the lower q region [25–29].

The q-range for the Guinier approximation depends on the particle size. The particle shape can be recognized from the q-dependence of the scattering intensity, *i.e.*, a slope of "−1" indicates cylinder shape, "−2" disc, and "−4" sphere. On the other hand, the Porod law holds in the high-q region and can be used to calculate other structural parameters. A slope of "−4" suggests that the interface of the particle is smooth and thereby the scattering intensity is proportional to the total interface area.

3. Results and Discussion

3.1. Monitoring of Bainite Transformation by Dilatometry

Figure 2 shows the temperature history of a specimen measured with a thermo-couple and the change in length obtained by dilatometry (DL). As is usually performed, the data obtained by the traditional dilatometry indicate, apparently, dilatation caused by bainite transformation at 300 °C. Though the change in specimen length can be converted to the ferrite volume fraction, *i.e.*, kinetics of bainite transformation, the insights on chemical, crystallographic and microstructural features cannot be found. Therefore, as was described above, *in situ* neutron scattering/diffraction has been performed; *in situ* ND gave us the information on not only the change in the ferrite volume fraction but also the formation of two populations of austenite (the higher carbon concentration region and the lower one) [19,23], texture evolution [19], and dislocation density [20], as was mentioned above. Here, the size of the transformed product is expected to be monitored by SANS measurement.

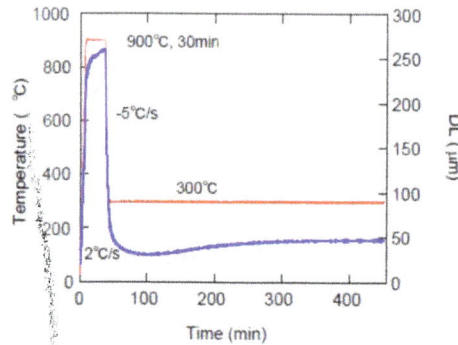

Figure 2. Change in temperature and the length of a specimen (DL) during heat treatment.

3.2. Monitoring of Bainite Transformation with in Situ SANS

Two-dimensional SANS patterns at different holding times are presented in Figure 3. It is found that the scattering intensity increases with the progress of bainite transformation both in the nuclear and magnetic components. These intensities were collected within 30° (see Figure 3c) in the parallel direction or vertical with respect to the magnetic field direction indicated in Figure 3a.

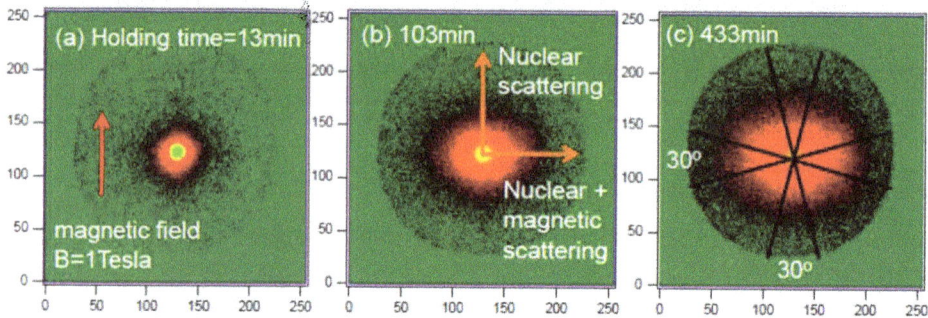

Figure 3. Change in the two-dimensional scattering pattern with holding time at 300 °C: (a) 13 min, (b) 103 min and (c) 433 min.

The nuclear component of the SANS profile obtained with a time interval of 10 min was plotted in Figure 4 as a function of the q-value. After the onset of bainite transformation, the SANS intensity of the nuclear component increased, apparently, with the holding time, *i.e.*, progress of bainite transformation. Here, two interesting features are noticed; one is an increase in scattering intensity in a high q region, the so-called "Porod region" with a slope of -4, and the other is the slope change in a low q region, the so-called "Guinier region". In the Porod region, the scattering intensity is proportional to the total area of the interface of scattering inhomogeneity, the bainitic ferrite lath in the present case, so that the intensity increase means the progress of transformation.

The scattering intensities at lines A and B in Figure 4 were plotted as a function of holding time in Figure 5a,b, respectively. The curves are similar to the dilatation curve shown in Figure 2 and the change in volume fraction determined by *in situ* neutron diffraction in the previous study [18] (the results by dilatometry, SANS and ND obtained by iMATERIA are presented together in Figure 8). As is presented in Figure 6 as an example, the morphology of bainitic ferrite lath would be assumed by a disc shape with a different radius but the same thickness. Hence, the total area of the ferrite/austenite interface is postulated to be proportional to the ferrite volume fraction. This indicates that the scattering intensity in the Porod region is proportional to the ferrite volume fraction showing the bainite transformation kinetics. Because the ferrite phase is magnetic while the austenite phase is non-magnetic at 300 °C, the magnetic scattering component also increases with the increase of the ferrite volume fraction (compare curves labeled N and N + M), suggesting that the ferrite volume fraction could be determined not only by the nuclear scattering component but also by the magnetic scattering component. This means that a magnet is not needed for the evaluation of bainite transformation.

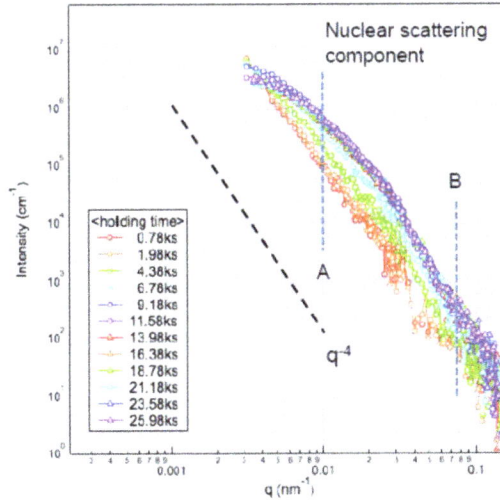

Figure 4. Change in nuclear component of SANS profile with progress in bainite transformation.

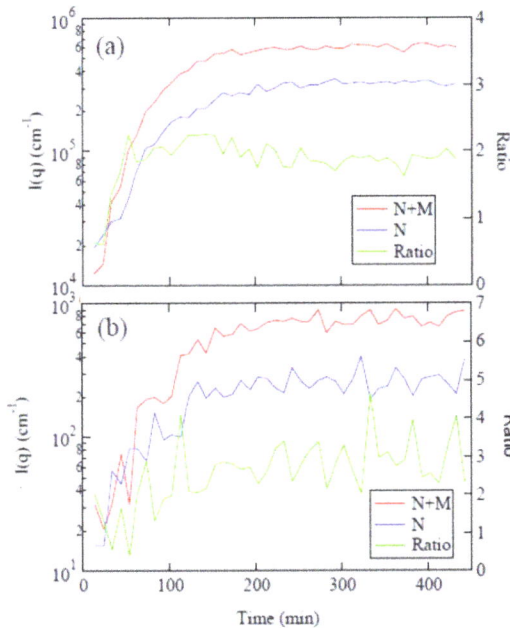

Figure 5. SANS intensity at the Porod region as a function of holding time at 300 °C: (**a**) at $q = 0.01$ nm^{-1} (line A in Figure 4) and (**b**) $q = 0.073$ nm^{-1} (line B in Figure 4).

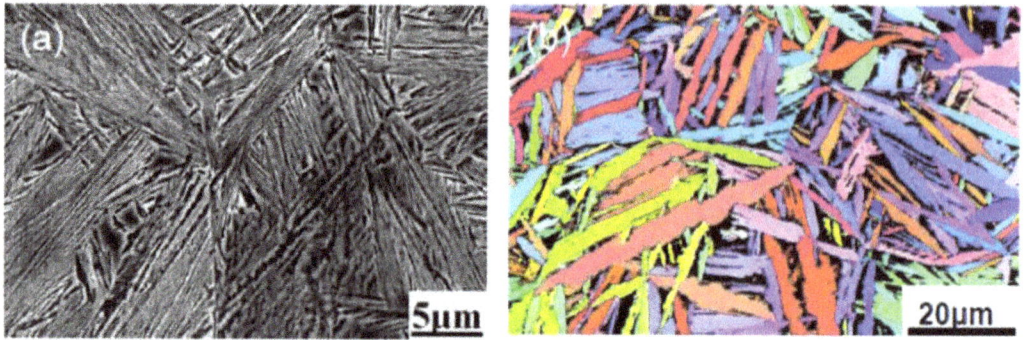

Figure 6. Microstructure of bainite formed at 300 °C: (**a**) SEM image and (**b**) EBSD/IPF map.

In the Guinier region, the slope of the $I-q$ curve would be -2 (disc shape) if it appeared completely. However, the size of the bainite lath is too large to be detected in the present q-range. Hence, as was done previously for non-metallic inclusions in steels [30], we need to expand the measuring q-range to cover smaller values. The influence of multiple scattering [31] is also suspected and, hence, we cannot follow the change in the size of the bainitic lath in this experiment. From microstructure observations, it is very likely that the first lath is large in a scale of prior austenite grain size and that the later-formed laths must be shortened because the pre-formed laths inhibit further growth, although the thickness is nearly constant. Therefore, it is believed that the SANS intensity at the Porod region is proportional to the bainite volume fraction.

3.3. In Situ Measurements of Dilatometry, Small Angle Scattering and Diffraction at iMATERIA

In previous studies, we have successfully used *in situ* neutron diffraction to elucidate the transformation mechanism, particularly the effects of ausforming [18] and partial quenching [19]. The results in the previous Section 3.2 suggest that SANS is of use to monitor the transformation product. Hence, the combined measurements using the conventional dilatometry, ND and SANS were aimed at performing by introducing a new dilatometer into the engineering neutron diffractometer, iMATERIA at MLF/J-PARC, by which back-scatter ND and SANS can be measured simultaneously. A trial was performed using the same steel and some tentative results are presented here to show how to effectively do such an experiment.

Figure 7 shows the results of dilatometry and neutron diffraction obtained at the iMATERIA. As seen, the dilatometry result in Figure 7a is quite similar to that in Figure 2. Changes in austenite 111 and ferrite 110 diffraction profiles are presented in Figure 7b, in which the appearance of two populations of austenite with different amounts of carbon concentration found in the previous studies [19,20] is well confirmed because of the higher resolution of the back-scatter detector at the iMATERIA. The ferrite volume fraction (V_α) was calculated using Equation (1) [32,33] from the *hkl* diffraction intensities determined with the Z-Rietveld software (J-PARC, Tokai, Japan) [34].

$$V\alpha = \frac{\dfrac{1}{m}\sum^m \dfrac{I_{hkl}^\alpha}{R_{hkl}^\alpha}}{\dfrac{1}{m}\sum^m \dfrac{I_{hkl}^\alpha}{R_{hkl}^\alpha} + \dfrac{1}{n}\sum^n \dfrac{I_{hkl}^\gamma}{R_{hkl}^\gamma}} \tag{1}$$

where I_{hkl}^γ, I_{hkl}^α, n and m refer to the measured integrated intensities of austenite, those of ferrite, the number of ferrite peaks and those of austenite, respectively, while R_{hkl}^α and R_{hkl}^γ stand for theoretical values for texture-free material. The obtained results are plotted in Figure 7c showing a good coincidence with the dilatometry result in Figure 7a.

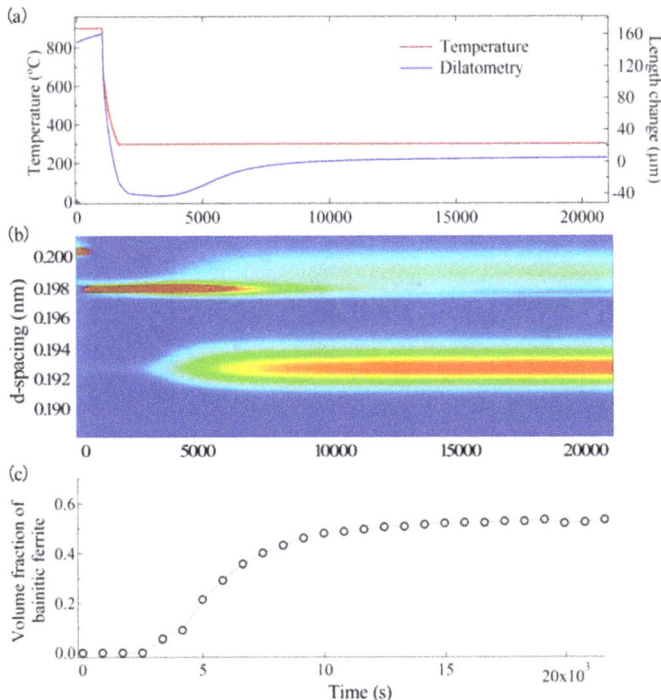

Figure 7. Results of preliminary *in situ* measurements during bainite transformation performed at iMATERIA: (**a**) temperature (red line) and dilatometric change (blue line); (**b**) changes in diffraction profile of austenite 111 and ferrite 110 peaks; and (**c**) volume fraction of ferrite determined from diffraction profiles.

The data analysis for SANS measurements at the iMATERIA is now in progress, so that the scattering intensity at $q = 0.4–0.42$ nm^{-1} was tentatively counted. The scattering intensity obtained was plotted in Figure 8 as a function of holding time together with the results obtained by ND and dilatometry. As can be observed, these three results are in good agreement. This was the first trial to employ SANS for monitoring bainite transformation at the iMATERIA, and, hence, the detailed analysis has not been made yet; nuclear and magnetic components were not separated because a magnetic field was not applied to the specimen in this experiment.

Three methods of neutron scattering and diffraction are applicable to monitor bainite transformation, *i.e.*, ND, SANS and transmission BE [21] measurements. All of these methods can evaluate the transformation kinetics and have the following different possible features:

(1) ND: crystal structure and volume fraction of constituents, texture, carbon concentration, elastic strain, and dislocation density: bulk average or three-dimensional (3D) distribution by scanning technique, although it requires scan time.

(2) SANS: volume fraction, shape and size of the second phase: bulk average.

(3) BE: phase volume fraction and carbon concentration of austenite, hopefully simultaneous 2D mapping of volume fraction, grain size, elastic strain and texture.

Hence, the combination of these methods would give us more fruitful information to understand microstructural evolution during processing for advanced steels.

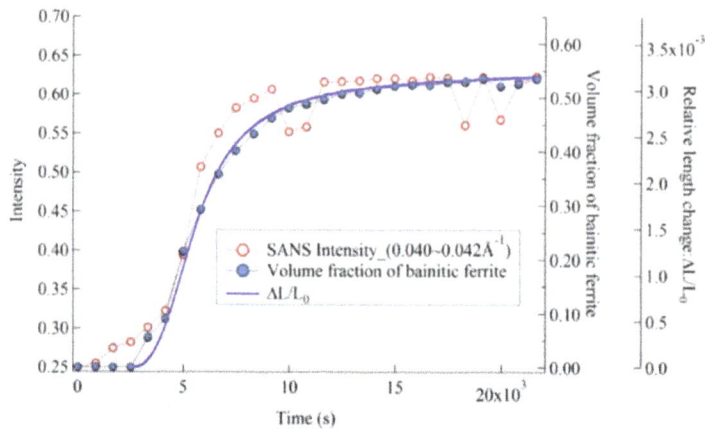

Figure 8. SANS intensity and volume fractions of ferrite determined by neutron diffraction and dilatometry as a function of holding time at 300 °C.

4. Conclusions

Bainite transformation was monitored by *in situ* measurements of conventional dilatometry as well as SANS and ND. The volume fraction of bainitic ferrite was estimated from the SANS intensity in the Porod region, showing good agreement with the results obtained by dilatometry and ND. A more advanced monitoring technique combining dilatometry, SANS, ND and BE was proposed.

Acknowledgments: We would like to thank T. Ishigaki, S. Koizumi, A. Hoshikawa and K. Iwase of Ibaraki University (BL20 (iMATERIA) instrument scientists) for their help to install a dilatometer. The SANS experiments at JRR-3 was performed through the proposal #2010-A-72 and a preliminary experiment at MLF/J-PARC through proposal #2012PM0003. This study was financially supported by a Grant-in-Aid for Scientific Research A (# 21246106). The authors sincerely thank H. Beladi of Deaken University for supplying the present steel and valuable discussion during his stay at NIMS.

Author Contributions: H.N. performed this research as her master degree thesis at Ibaraki University, Y.T. coordinated this research and wrote the manuscript with help from the other authors, Y. H. and W. G. helped with experiments and data analyses, and J.S. instructed the measurement method and analysis of SANS as an instrumental scientist.

Conflicts of Interest: The authors declare no conflict of interest.

References

1. Furukawa, T. Structure-property relationships of dual phase steels. *J. Jpn. Inst. Met.* **1980**, *19*, 439–446. [CrossRef]

2. Tomota, Y.; Tamura, I. Strength and deformation behavior of two-ductile- phase alloys. *J. Jpn. Inst. Met.* **1985**, *14*, 657–664. [CrossRef]

3. Tomota, Y.; Tamura, I. Mechanical Behavior of Steels Consisting of Two Ductile Phases. *Trans. ISIJ* **1982**, *22*, 665–677. [CrossRef]

4. Matsumura, O.; Sakuma, Y.; Takechi, H. Enhancement of elongation by retained austenite in intercritical annealed 0.4C–1.5Si–0.8Mn steel. *Trans. ISIJ* **1987**, *27*, 570–579. [CrossRef]

5. Tomota, Y.; Tokuda, H.; Adachi, Y.; Wakita, M.; Minakawa, N.; Moriai, A.; Morii, Y. Tensile Behavior of TRIP-Aided Multi-Phase Steels Studied by *in situ* Neutron Diffraction. *Acta Mater.* **2004**, *52*, 5737–5745. [CrossRef]

6. Caballero, F.G.; Bhadeshia, H.K.D.H.; Mawella, J.A.; Jones, D.G.; Brown, P. Very strong low temperature bainite. *Mater. Sci. Technol.* **2002**, *18*, 279–284. [CrossRef]

7. Garcia-Mateo, C.; Caballero, F.G.; Bhadeshia, H.K.D.H. Development of Hard Bainite. *ISIJ Int.* **2003**, *43*, 1238–1243. [CrossRef]

8.　Beladi, H.; Adachi, Y.; Timokhina, I.; Hodgson, P.D. Crystallographic analysis of nanobainitic steels. *Scr. Mater.* **2009**, *60*, 455–458. [CrossRef]

9.　Timokhina, I.; Beladi, H.; Xiong, X.Y.; Adachi, Y.; Hodgson, P.D. Nanoscale microstructural characterization of a nanobainitic steel. *Acta Mater.* **2011**, *59*, 5511–5522. [CrossRef]

10.　Carcia-Mateo, C.; Jimenz, J.A.; Yen, H.-W.; Miller, M.K.; Morales-Rivas, L.; Kuntz, M.; Ringer, S.P.; Yang, J.-R.; Caballero, F.G. Low temperature bainitic ferrite: Evidence of carbon super-saturation and tetragonality. *Acta Mater.* **2015**, *91*, 162–173. [CrossRef]

11.　Speer, J.G.; Streicher, A.M.; Matlock, D.K.; Rizzo, F.C.; Krauss, G. *Austenite Formation and Decomposition*; Damm, E.B., Merwinm, M., Eds.; TMS/ISS: Warrendale, PA, USA, 2003; pp. 502–522.

12.　Speer, J.; Matlock, D.; de Cooman, B.; Schroth, J. Carbon partitioning into austenite after martensite transformation. *Acta Mater.* **2003**, *51*, 2611–2622. [CrossRef]

13.　Edmonds, D.; He, K.; Rizzo, F.; de Cooman, B.; Matlock, D.; Speer, J. Quenching and partitioning martensite—A novel steel heat treatment. *Mater. Sci. Eng. A* **2006**, *438*, 25–34. [CrossRef]

14.　Santofimia, M.J.; Zhao, L.; Sietsma, J. Overview of mechanism involved during the quenching and partitioning process in steels. *Metall. Mater. Trans. A* **2011**, *42A*, 3620–3626. [CrossRef]

15.　Yuan, L.; Ponge, D.; Wittig, J.; Choi, P.; Jimenez, J.A.; Raabe, D. Nanoscale austenite reversion through partitioning, segregation and kinetic freezing: Example of a ductile 2 GPa Fe–Cr–C steel. *Acta Mater.* **2012**, *60*, 2790–2804. [CrossRef]

16.　Caballero, F.G.; Bhadeshia, H.K.D.H. Very Strong Bainite. *Curr. Opin. Solid State Mater. Sci.* **2004**, *8*, 251–257. [CrossRef]

17.　Koo, M.S.; Xu, P.; Suzuki, H.; Tomota, Y. Bainitic transformation behavior studied by simultaneous neutron diffraction and dilatometric measurement. *Scr. Mater.* **2009**, *61*, 797–800. [CrossRef]

18.　Gong, W.; Tomota, Y.; Koo, M.S.; Adachi, Y. Effect of ausforming on nanobainite steel. *Scri. Mater.* **2010**, *63*, 819–822. [CrossRef]

19.　Gong, W.; Tomota, Y.; Adachi, Y.; Paradowska, A.M.; Kelleher, J.F.; Zhang, S.Y. Effects of ausforming temperature on bainite transformation, microstructure and variant selection in nanobainite steel. *Acta Mater.* **2013**, *61*, 4142–4154. [CrossRef]

20.　Gong, W.; Tomota, Y.; Harjo, S.; Su, Y.H.; Aizawa, K. Effect of prior martensite on bainite transformation in nanobainite steel. *Acta Mater.* **2015**, *85*, 243–249. [CrossRef]

21.　Babu, S.S.; Specht, E.D.; David, S.A.; Karapetrova, E.; Zachack, P.; Peer, M.; Bhadeshia, H.K.D.H. *In situ* Observations of Lattice Parameter Fluctuations in Austenite and Transformation to Bainite. *Metall. Mater. Trans. A* **2005**, *36A*, 3281–3289. [CrossRef]

22.　Stone, H.J.; Peet, M.J.; Bhadeshia, H.K.D.H.; Withers, P.J.; Babu, S.S.; Specht, E.D. Synchrotron X-ray studies of austenite and bainitic ferrite. *Proc. Roy. Soc. A* **2008**, *464*, 1009–1027. [CrossRef]

23.　Gong, W. Transformation Kinetics and Crystallography of Nano-bainite Steel. Ph.D. Thesis, Ibaraki University, Hitachi, Japan, March 2012.

24.　Huang, J.; Vogel, S.C.; Poole, W.J.; Militzer, M.; Jacques, P. The study of low-temperature austenite decomposition in a Fe–C–Mn–Si steel using the neutron Bragg edge transmission technique. *Acta Mater.* **2007**, *55*, 2683–2691. [CrossRef]

25.　Muller, G.; Uhlemann, M.; Ulbricht, A.; Bohmert, J. Influence of hydrogen on the toughness of irradiated reactor pressure vessel steels. *J. Nucl. Mater.* **2006**, *359*, 114–121. [CrossRef]

26.　Ohnuma, M.; Suzuki, J.; Wei, F.G.; Tsuzaki, K. Direct observation of hydrogen trapped by NbC in steel using small-angle neutron scattering. *Scr. Mater.* **2008**, *58*, 142–145. [CrossRef]

27.　Buckley, C.E.; Birnbaum, H.K.; Bellmann, D.; Staron, P. Calculation of the radial distribution function of bubbles in the aluminum hydrogen system. *J. Alloy. Compd.* **1999**, *293–295*, 231–236. [CrossRef]

28.　Yasuhara, H.; Sato, K.; Toji, Y.; Ohnuma, M.; Suzuki, J.; Tomota, Y. Size Analysis of Nanometer Titanium Carbide in Steel by Using Small-Angle Neutron Scattering. *Tetsu-to-Hagane* **2010**, *96*, 545–549. [CrossRef]

29.　Su, Y.H.; Morooka, S.; Ohnuma, M.; Suzuki, J.; Tomota, Y. Quantitative Analyses on Cementite Spheroidization in Pearlite Structure by Small-Angle Neutron Scattering. *Metall. Mater. Trans. A* **2015**, *46A*, 1731–1740. [CrossRef]

30.　Oba, Y.; Koppoju, S.; Ohnuma, M.; Kinjo, Y.; Morooka, S.; Tomota, Y.; Suzuki, J.; Yamaguchi, D.; Koizumi, S.; Sato, M.; *et al.* Quantitative Analysis of Inclusions in Low Carbon Free Cutting Steel Using Small-Angle X-ray and Neutron Scattering. *ISIJ Int.* **2012**, *52*, 458–464. [CrossRef]

31. Schelten, J.; Schmatz, W. Multipe-Scattering Treatment for Small Angle Scattering Problem. *J. Appl. Cryst.* **1980**, *13*, 385–390. [CrossRef]

32. Järvinen, M. Texture Effect in X-ray Analysis of Retained Austenite in Steels. *Textures Microstruct.* **1996**, *26–27*, 93–101. [CrossRef]

33. Gnäuperl-Herold, T.; Creuziger, A. Diffraction study of the retained austenite content in TRIP steels. *Mater. Sc. Eng. A* **2011**, *528*, 3594–3600. [CrossRef]

34. Oishi, T.; Yonemura, M.; Morishima, T.; Oshikawa, A.; Torii, S.; Ishigaki, T.; Kamiyama, T. Application of matrix decomposition algorithms for singular matrices to Pawley method in Z-Rietveld. *J. Appl. Cryst.* **2012**, *45*, 299–308. [CrossRef]

9

Microstructure and High Temperature Deformation of Extruded Al-12Si-3Cu-Based Alloy

Seung-Baek Yu and Mok-Soon Kim *

Division of Materials Science and Engineering, Inha University, Incheon 22207, Korea; 22151518@inha.edu
* Correspondence: mskim@inha.ac.kr

Academic Editor: Hugo F. Lopez

Abstract: The high temperature deformation behavior of commercial Al-12Si-3Cu-2Ni-1Mg alloy (DM104™) which was fabricated by casting and subsequent hot extrusion was evaluated by compressive tests over the temperature range of 250–470 °C and strain rate range of 0.001–1/s. The extruded alloy had equiaxed grains, spherical Si particles and fine intermetallic phases, such as $\delta(Al_3NiCu)$ and $Q(Al_5Cu_2Mg_3Si_6)$. The true stress-true strain curves from the compressive tests exhibited steady-state flow after reaching the peak stress. A close relationship between the steady-state stress and a constitutive equation for high temperature deformation was observed. Fine equiaxed grains and a dislocation structure within the equiaxed grains were observed in the deformed specimens, suggesting the occurrence of dynamic recrystallization during high temperature deformation.

Keywords: piston alloy; extrusion; compressive test; plastic deformation; high temperature

---

1. Introduction

Al-Si-based alloys are used widely as a piston materials because of their good specific strength, excellent wear resistance and low coefficient of thermal expansion [1–4]. The addition of Cu and Mg to Al-Si alloys was reported to promote their mechanical properties through the precipitation of the Al_2Cu, Mg_2Si phases [5–8]. Furthermore, it was reported that Ni is a desirable element for improving the mechanical properties of Al-Si-based alloys at elevated temperatures by forming thermally stable Ni-containing intermetallics [9]. To meet the strength requirement of advanced aluminum material for high performance automotive engines, the Al-12Si-3Cu-2Ni-1Mg piston alloy (DM104™) was recently developed by DYP (Dong Yang Piston Co, Ltd., Gyeonggi-do, Korea). The DM104 alloy can be applicable as a forging stock for forged pistons and as a cast piston material. In the case of its use as a forging stock, a careful study of the plastic deformation behavior of the material up to the high temperature region is needed to determine the optimal processing conditions for forging. In this study, the plastic deformation behavior of the extruded Al-12Si-3Cu-2Ni-1Mg alloy was examined using compressive tests over a wide range of temperatures and strain rates. Plastic flow curves were obtained from compressive tests and used to evaluate a constitutive equation. Detailed microstructural observations were made for the specimens before and after deformation.

2. Experimental Section

The material used in this study was a commercial Al-12Si-3Cu-2Ni-1Mg (nominal composition, wt. %) alloy, designated as DM104™, which was fabricated by casting and subsequent hot extrusion. The extrusion billet had a cylindrical shape with a diameter of 90 mm.

Table 1 lists the chemical composition of the as-received alloy (hereafter, "the alloy" or "the billet" denotes DM104 extruded alloy). Cross-sections of the billet were ground using SiC paper up to a grit number of 2000, and polished using a diamond suspension and colloidal silica. The microstructure was observed using an Olympus PME3-313UN optical microscope (Olympus, Tokyo, Japan) after etching with modified Weak's reagent. The hardness was measured using a Akashi HM-124 Vickers hardness tester (Mitsudoyo, Tokyo, Japan) with a load of 0.1 kg and a dwell time of 10 s. The Vickers hardness was obtained by an average of at least 20 measurements. Cylindrical compression test specimens with a diameter of 5 mm and a height of 7.5 mm were fabricated from the billet. The samples were taken from the 3 cm radius location from the billet edge. Compressive tests were conducted over the temperature range, 250–470 °C and strain rate range 0.001–1/s. The compressive tests were performed up to the true strain of approximately 100% after holding for 15 min at the target temperature. After the compressive test, the sample was quenched in water and cut along the loading axis to observe the microstructure. Phase analysis was conducted by energy dispersive spectroscopy (EDS; Emax 132-10, Horiba, Kyoto, Japan) and X-ray diffraction (XRD; X'pert MPD PRO, Philips, Amsterdam, The Netherlands), and compared with the JMatPro simulation data.

Table 1. Chemical composition of the alloy studied.

Alloy Compositions (wt. %)								
Element	Si	Fe	Cu	Mg	Ni	Ti	V	Zr	Al
DM104	12	0.3	3.2	0.9	2.0	0.1	0.1	0.1	Bal.

3. Results and Discussion

3.1. Microstructure of the As-Received Alloy

Figure 1 shows the etched microstructure of a cross-section. As shown in the figure, the microstructure consisted of well-developed equiaxed grains (see arrow a), spherical Si particles (arrow b) and fine intermetallic phases (arrow c). The mean grain size of the equiaxed grains was measured to be approximately 5 μm. The Vickers hardness value of the matrix was measured to be 90 Hv.

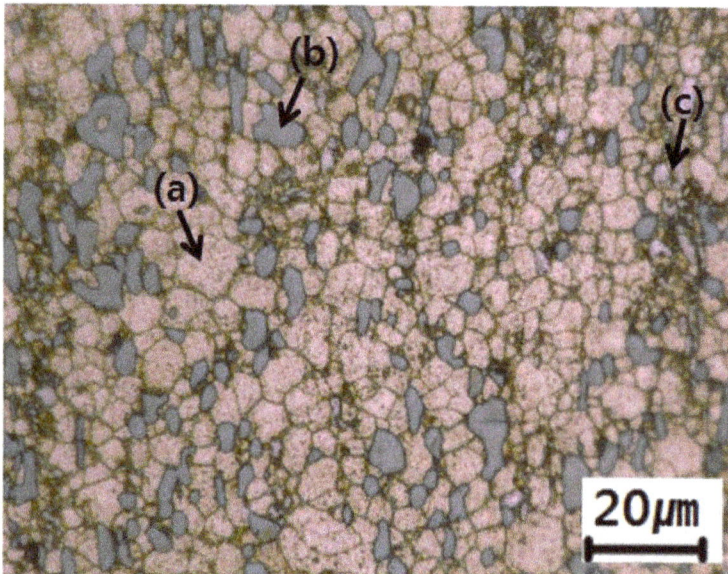

Figure 1. Optical micrograph of the cross-section for as-received alloy.

Figure 2 shows the results of the EBSD analysis data for the cross-section of the alloy. In the EBSD map (Figure 2b), low-angle boundaries (LABs) (misorientation < 15°) and high-angle boundaries (HABs) (misorientation > 15°) are shown with red and blue colors, respectively. Figure 2b also shows that the LABs which are known to show evidence of the sub-grain boundaries are mostly distributed within HABs. From Figure 2d, the fraction of HABs was measured to be approximately 60%. These results suggest that the alloy had experienced dynamic recrystallization during the extrusion process [10].

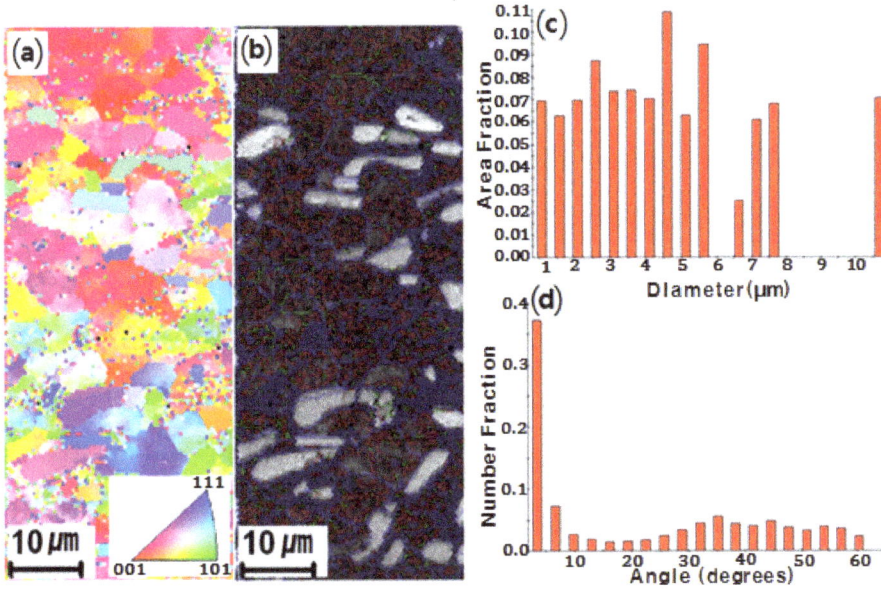

Figure 2. EBSD analysis data of the as-received alloy: (**a**) EBSD map (the insert in bottom right corner of (**a**) showing the representation of the color key used to identify the crystallographic orientations on a standard stereographic projection); (**b**) grain boundary misorientation map; (**c**) grain size distribution; (**d**) misorientation angle distribution.

3.2. Phase Analysis

Figure 3 presents the JMatPro simulation data for the DM104 alloy, indicating the existing phases and their weight fraction during the solidification process. In the case of the equilibrium state (Figure 3a), eutectic Si and intermetallic T phase (Al_9FeNi) solidify at 560 °C and 550 °C, respectively. As the temperature decreased to less than 532 °C, the weight fraction of the T phase decreased with the formation of the δ(Al_3Ni_2 or Al_3NiCu) and Q($Al_8Cu_2Mg_8Si_6$) phases. The weight fraction of the δ phase decreased with increasing formation of the T and γ(Al_2Cu_4Ni) phase under 375 °C. At the end of the solidification process, the Si phase comprised approximately 11% of the weight fraction and every intermetallic phase comprised less than 5% of the weight fraction. Although the solidification temperatures of the main phases appeared to be similar among the results in Figure 3a,b, the weight fraction of each phase in the non-equilibrium state (Figure 3b) maintained a specific value until the end of solidification. Figure 3b shows that the δ, T and Q phases have considerable weight fractions.

Figure 4 shows the XRD patterns of the as-received alloy, indicating the peaks of δ(Al_3Ni_2 or Al_3NiCu) and Q($Al_8Cu_2Mg_8Si_6$) phases in addition to the intensive peaks of Al and Si phases.

Figure 5a shows a SEM image of the alloy. The phases labeled in Figure 5a were identified by EDS, and their chemical compositions are given in Figure 5b. The chemical compositions of points ② and ③ appear to correspond to the stoichiometric composition of the Q($Al_8Cu_2Mg_8Si_6$) and δ(Al_3NiCu)

phases, respectively. The δ and Q phases are frequently detected by SEM (Figure 5a), and these results are in accordance with both the XRD pattern in Figure 4 and the JMatPro simulation data in Figure 3.

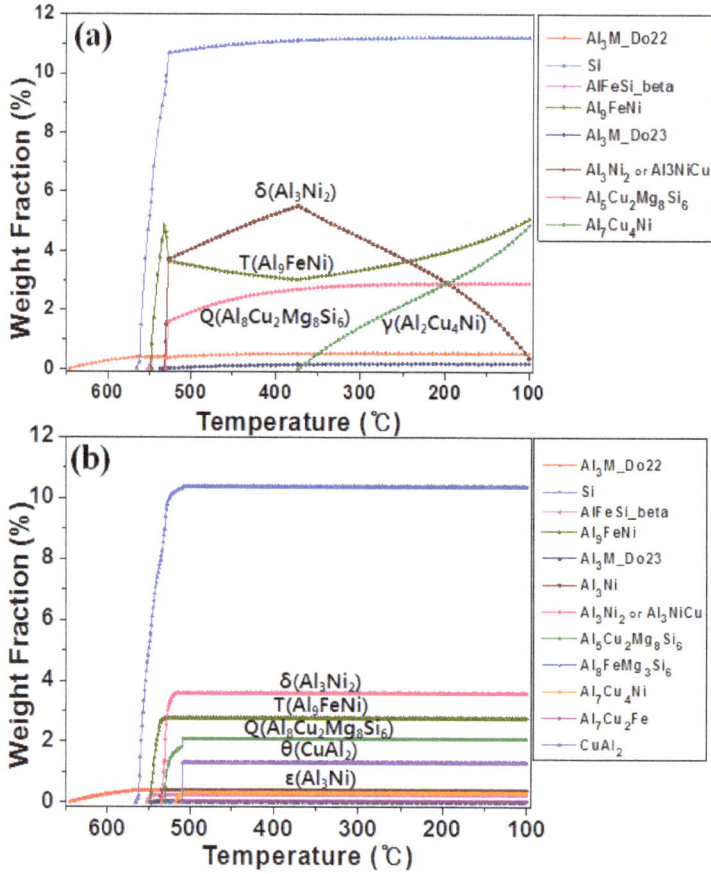

Figure 3. JMatPro simulation data of the as-received alloy for (**a**) equilibrium; (**b**) non-equilibrium state.

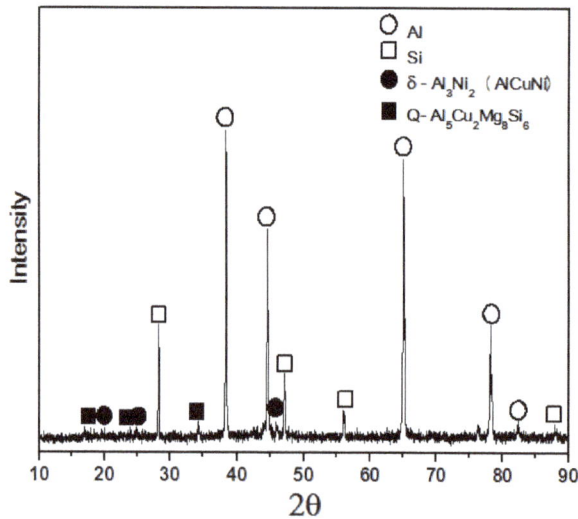

Figure 4. XRD result of the as-received alloy.

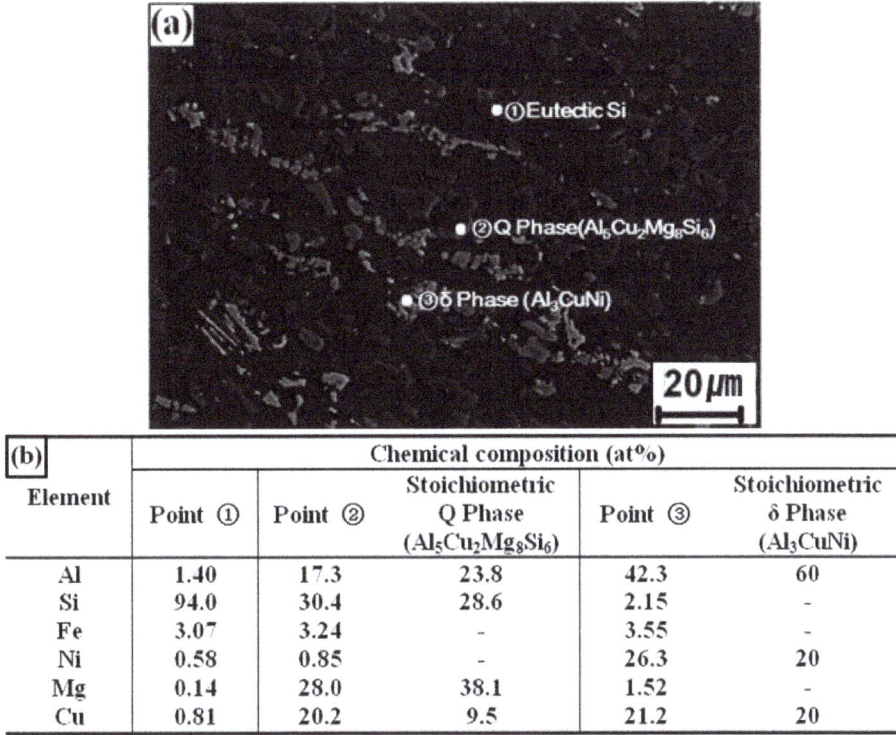

(b)	Chemical composition (at%)				
Element	Point ①	Point ②	Stoichiometric Q Phase (Al$_5$Cu$_2$Mg$_8$Si$_6$)	Point ③	Stoichiometric δ Phase (Al$_3$CuNi)
Al	1.40	17.3	23.8	42.3	60
Si	94.0	30.4	28.6	2.15	-
Fe	3.07	3.24	-	3.55	-
Ni	0.58	0.85	-	26.3	20
Mg	0.14	28.0	38.1	1.52	-
Cu	0.81	20.2	9.5	21.2	20

Figure 5. (a) SEM image; (b) EDS analysis data of the as-received alloy.

3.3. Compressive Deformation

Figure 6 presents the true stress-true strain curves of the specimens compressive-tested at temperatures between 250 and 470 °C and strain rates between 0.001 and 1/s. The flow stress increased sharply with increasing strain at the initial stages of deformation. The flow stress then decreased and finally reached a steady-state with further increases in strain. The steady-state behavior was initiated at a lower strain range of deformation at a higher test temperature.

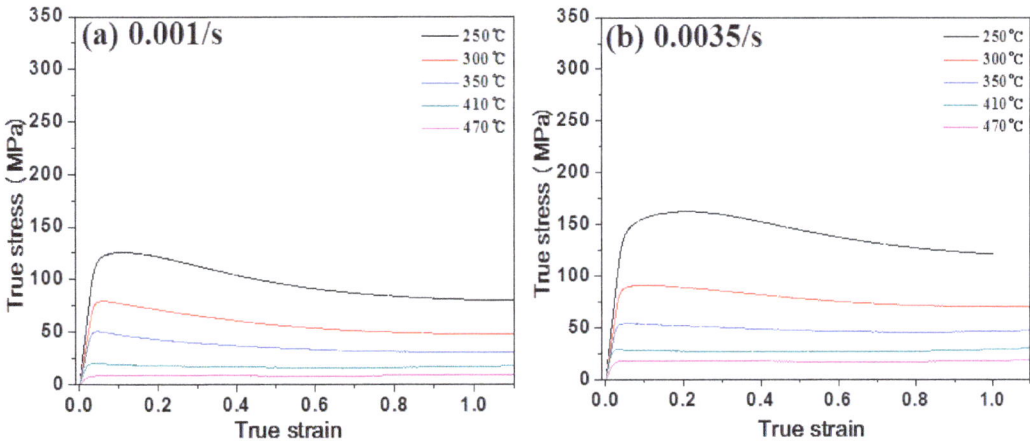

Figure 6. *Cont.*

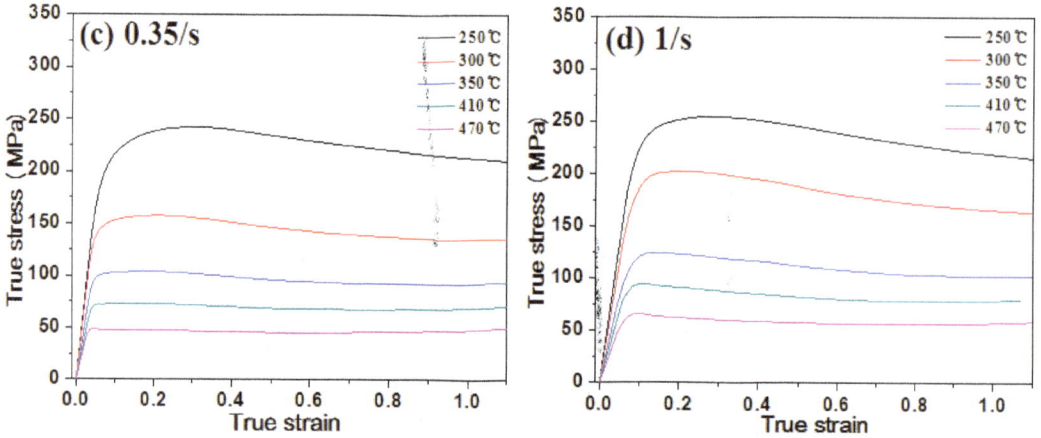

Figure 6. True stress-true strain curves of the specimens compressed at various temperatures under a strain rate of (a) 0.001/s; (b) 0.0035/s; (c) 0.35/s; (d) 1/s.

Figure 7 shows the temperature-dependence of the steady-state stress for different strain rates. At a given temperature, the steady-state stress increased with increasing strain rate, and at a given strain rate, the steady-state stress decreased with increasing temperature.

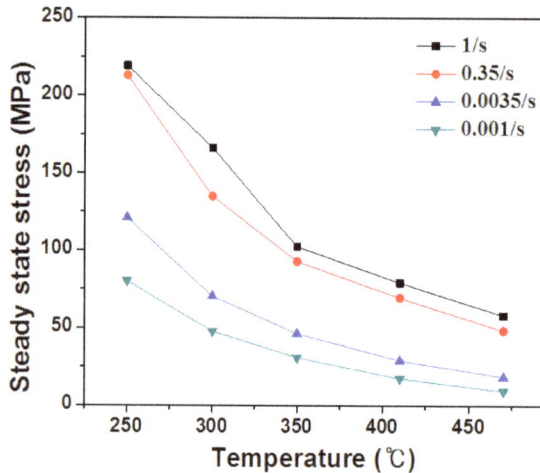

Figure 7. Variation of the steady-state stress with temperature at different strain rates.

3.4. Constitutive Equations

$$\dot{\varepsilon} = A_1 \sigma^n \exp(-\frac{Q}{RT}) : \text{Power law} \tag{1}$$

$$\dot{\varepsilon} = A_2 \exp(\alpha\sigma) \exp(-\frac{Q}{RT}) : \text{Exponential law} \tag{2}$$

$$\dot{\varepsilon} = A_3 [\sinh(\beta\sigma)]^{n'} \exp(-\frac{Q}{RT}) : \text{Hyperbolic sine law} \tag{3}$$

Constitutive equations, such as Equations (1)–(3), have been used to describe the relationship between the stress (σ), temperature (T) and strain rate ($\dot{\varepsilon}$) during the high temperature deformation of an Al-based alloy [11–13]. A_1, A_2, A_3, n, n', α, and β ($=\alpha/n$) are material constants. In the present study, constitutive equations were deduced using the steady-state stress values from Figure 6. The power law,

Equation (1), and exponential law, Equation (2), break down at the higher and lower stress regions, respectively. In contrast, the hyperbolic sine law, Equation (3), is suitable for stresses over a wide range.

Figure 8a,b show linear plots of $\ln(\dot{\varepsilon})$-$\ln(\sigma)$ and $\ln(\dot{\varepsilon})$-σ for the deformed alloys, respectively. The average values of n and α were calculated from the slope in Figure 8a,b, respectively. The value of β ($\beta = \alpha/n$) was applied to the hyperbolic sine law, Equation (3). Figure 8c,d present linear plots of $\ln(\dot{\varepsilon})$-$\ln[\sinh(\alpha\sigma)]$ and $\ln[\sinh(\alpha\sigma)]$-$1000/T$ for the deformed alloys, respectively. The mean values of n' and Q (activation energy) were calculated from the slope in Figure 8c,d, respectively. Table 2 presents the calculated constants from the above equations. As a result, the hyperbolic sine law of the alloy can be expressed as Equation (4).

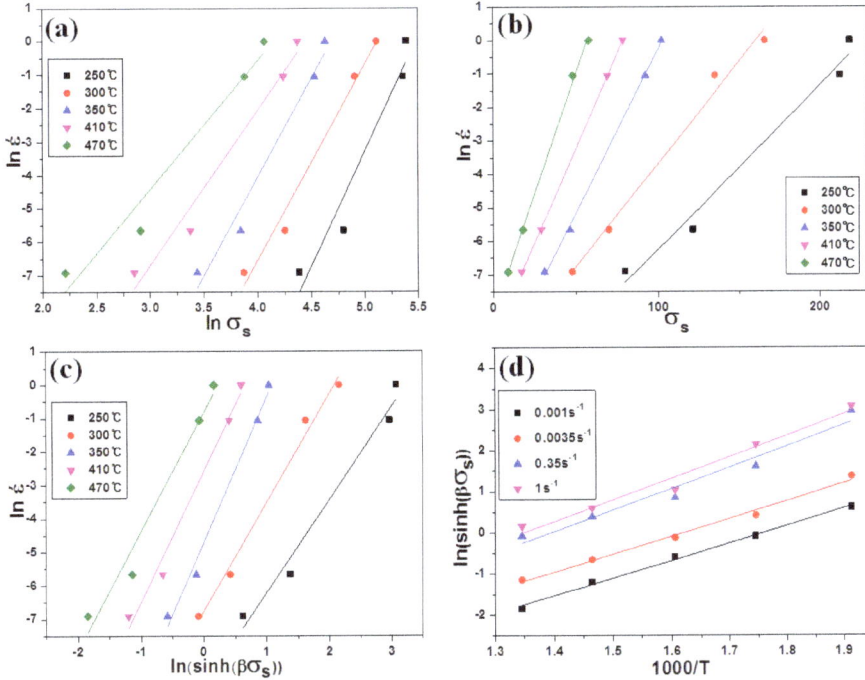

Figure 8. Relationship between steady-state stress, strain rate and temperature: (a) $\ln(\dot{\varepsilon})$ *versus* $\ln(\sigma)$; (b) $\ln(\dot{\varepsilon})$ *versus* σ; (c) $\ln(\dot{\varepsilon})$ *versus* $\ln[\sinh(\beta\sigma)]$; (d) $\ln[\sinh(\beta\sigma)]$ *versus* $1000/T$.

$$\text{Hyperbolic sine law}: \quad \dot{\varepsilon} = 5.3 \times 10^9 \{\sinh(0.017\sigma_s)\}^{3.57} \exp(-142000/RT) \qquad (4)$$

Table 2. Material constants for the alloy, which were evaluated from the constitutive equations.

Constant	n	α	β	A_3 (s^{-1})	n'	Q (kJ/mol)
Value	5.41	0.093	0.017	5.3×10^9	3.57	142

3.5. Zener-Hollomon Parameter

The deformation temperature and strain rate are incorporated into the Zener-Hollomon (Z) parameter, Equation (5).

$$Z = \dot{\varepsilon}\exp(\frac{Q}{RT}) : \text{Zener-Hollomon } (Z) \text{ parameter} \qquad (5)$$

$$Z = 5.3 \times 10^9 \{\sinh(0.017\sigma_s)\}^{3.57} \qquad (6)$$

Equation (6), which gives the relationship between $\sinh(\beta\sigma)$ and the Z parameter, can be obtained by combining Equation (4) with Equation (5).

The values of $\sinh(\beta\sigma)$ calculated using Equation (6) are presented as a linear line in Figure 9. In this figure, the experimentally obtained $\sinh(\beta\sigma)$ values are also included. As shown in Figure 9, the comparison between the experimental and calculated data revealed high similarity between the two datasets. Therefore, the constitutive equation, Equation (6), is reasonable for explaining the high temperature deformation behavior of the present alloy.

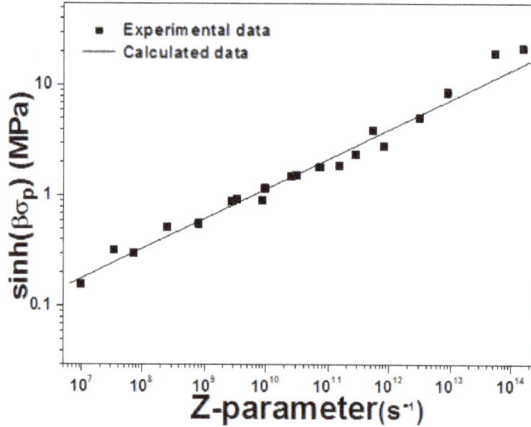

Figure 9. Comparison between the experimental and calculated data.

3.6. Microstructure of the Deformed Specimens

Figure 10 presents optical micrographs of the alloy deformed at a strain rate of 1/s and temperatures from 250 °C to 470 °C. Elongated grains were observed in the specimens deformed compressively at 250 °C (Figure 10a) and 300 °C (Figure 10b). In contrast, at 350 °C and above, equiaxed grains were visible near the eutectic Si particles and intermetallics, and the mean grain size of the equiaxed grains increased with increasing temperature. The large precipitates in Figure 10c–e were primary Si particles which may be generated by non-equilibrium solidification (hypereutectic solidification). Brittle primary Si particles were normally broken during the extrusion process, although some of them could remain after the extrusion.

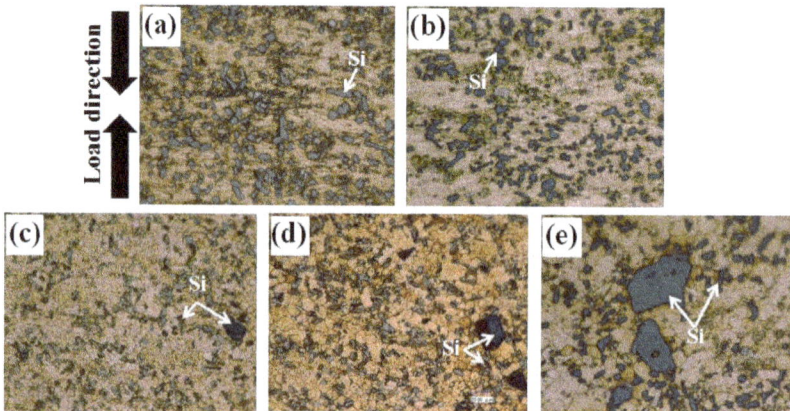

Figure 10. Optical micrographs of the deformed specimens at a strain rate of 1 /s at (a) 250 °C; (b) 300 °C; (c) 350 °C; (d) 410 °C; (e) 470 °C.

Table 3 lists the mean grain size after deformation under different test conditions. The mean grain sizes were in the range between 2.9 and 7.2 μm.

Table 3. Mean grain size (μm) after deformation under different test conditions.

Strain Rate (/s)	Temperature (°C)		
	350	410	470
1	2.9	3.2	4.1
0.35	3.3	3.4	4.5
0.0035	4.8	5.9	6.7
0.001	5.1	5.7	7.2

Figure 11a shows a TEM bright field image of the specimen deformed to 20% true-strain at 470 °C under a strain rate of 1/s. The 20% strain corresponds to a steady-state deformation region after reaching the peak stress (see Figure 6d). As shown in Figure 11a, fine dispersoids were distributed as indicated by points ② and ③. Figure 11b shows the EDS analysis data of points ② and ③, which appear to correspond to the stoichiometric composition of the δ(Al₃NiCu) phase. Figure 11a shows the dislocations within the equiaxed grains. The existence of both equiaxed grains and a dislocation structure within the equiaxed grains indicates that dynamic recrystallization occurred during the course of high temperature deformation of the present alloy.

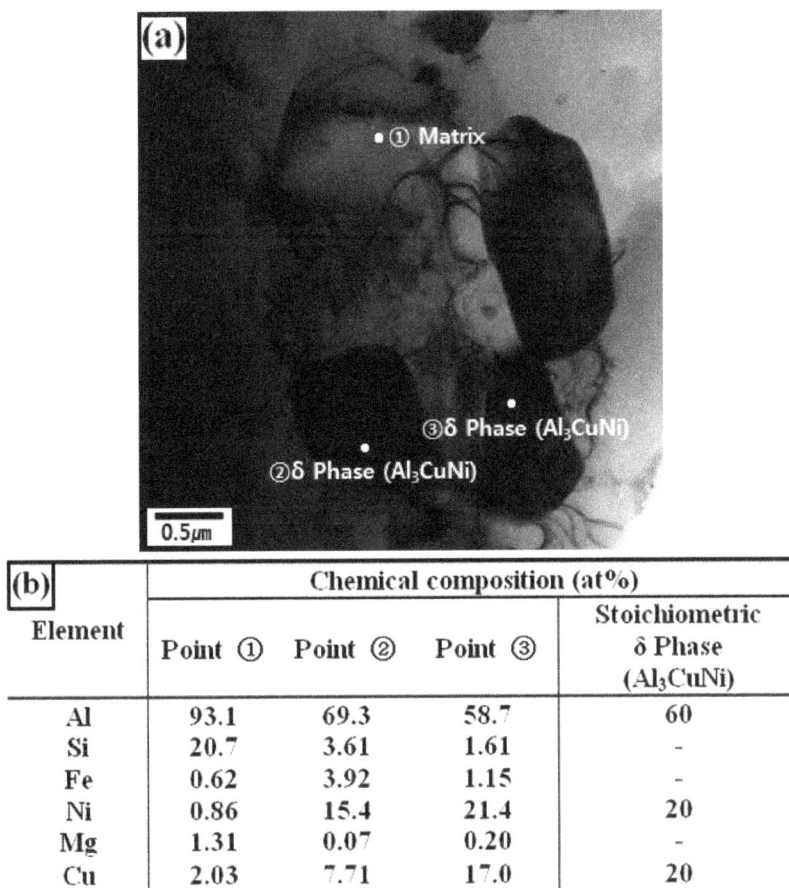

(b)	Chemical composition (at%)			
Element	Point ①	Point ②	Point ③	Stoichiometric δ Phase (Al₃CuNi)
Al	93.1	69.3	58.7	60
Si	20.7	3.61	1.61	-
Fe	0.62	3.92	1.15	-
Ni	0.86	15.4	21.4	20
Mg	1.31	0.07	0.20	-
Cu	2.03	7.71	17.0	20

Figure 11. (a) TEM image and (b) EDS data of the specimen deformed to 20% of true-strain at a strain rate of 1/s at 470 °C.

4. Conclusions

Compressive tests of the commercial Al-12Si-3Cu-2Ni-1Mg alloy (DM104™) fabricated by casting and subsequent extrusion were conducted over the temperature range, 250–470 °C, and strain rate range of 0.001–1/s. The following conclusions were obtained:

1. The as-received alloy had equiaxed grains with a mean grain size of approximately 5 μm, spherical Si particles and fine intermetallic phases including $\delta(Al_3NiCu)$ and $Q(Al_5Cu_2Mg_3Si_6)$.
2. EBSD analysis of the as-received alloy showed that the fraction of high-angle boundaries (HABs) was approximately 60%, and the low-angle boundaries (LABs) are distributed mostly as sub-grain boundaries within the HABs.
3. The true stress-true strain curves obtained from the compressive tests exhibited steady-state flow after reaching the peak stress.
4. The steady-state stress decreased with increasing temperature and decreasing strain rate. A close relationship was observed between the steady-state stress and a constitutive equation for high temperature deformation.
5. Fine equiaxed grains and a dislocation structure within the equiaxed grains were observed in the deformed specimens, indicating the occurrence of dynamic recrystallization during high temperature deformation.

Acknowledgments: The authors gratefully acknowledge the financial supports from The Small & Medium Business Administration (Korea) through Grant No. S2317902, and Inha University Research Grant. The authors are also grateful to Dong Yang Piston Co., Ltd. for provision of alloy studied in this work.

Author Contributions: Study design: Mok-Soon Kim; Experimental work: Seung-Baek Yu; Results analysis & Manuscript preparation: All authors; Manuscript proof and Submission: All authors.

Conflicts of Interest: The authors declare no conflict of interest.

References

1.   Manasijevic, S.; Radisa, R.; Markovic, S.; Acimovic-Pavlovic, Z.; Raic, K. Thermal analysis and microscopic characterization of the piston alloy AlSi13Cu4Ni2Mg. *Intermetallics* **2011**, *19*, 486–492. [CrossRef]
2.   Chen, C.-L.; Thomson, R.C. The combined use of EBSD and EDX analyses for the identification of complex intermetallic phases in multicomponent Al-Si piston alloys. *J. Alloys Compd.* **2010**, *490*, 293–300. [CrossRef]
3.   Kim, J.; Jang, G.S.; Kim, M.S.; Lee, J.K. Microstructure and compressive deformation of hypereutectic Al-Si-Fe based P/M alloys fabricated by spark plasma sintering. *Trans. Nonferrous Met. Soc. China* **2014**, *24*, 2346–2351. [CrossRef]
4.   Park, S.-C.; Kim, M.-S.; Kim, K.-T.; Shin, S.-Y.; Lee, J.-K.; Ryu, K.-H. Compressive Deformation Behavior of Al-10Si-5Fe-1Zr Powder alloys Consolidated by Spark Plasma Sintering Precess. *Korean J. Met. Mater.* **2011**, *49*, 853–859.
5.   Sjölander, E.; Seigeddine, S. The heat treatment of Al-Si-Cu-Mg casting alloys. *J. Mater. Processs. Technol.* **2010**, *210*, 1249–1259. [CrossRef]
6.   Aguilera-luna, I.; Castro-Roman, M.J.; Escobedo-bocardo, J.C.; Garcia-Pastor, F.A.; Herrera-Trejo, M. Effect of cooling rate and Mg content on the Al-Si eutectic for Al-Si-Cu-Mg alloys. *Mater. Charact.* **2014**, *95*, 211–218. [CrossRef]
7.   Zamani, M.; Seifeddine, S.; Jarfors, A.E.W. High temperature tensile deformation behavior and failure mechanisms of an Al-Si-Cu-Mg cast alloy—The microstrcutural scale effect. *Mater. Des.* **2015**, *86*, 361–370.
8.   Cho, H.S.; Kim, M.S. High Temperature Deformation Behavior of Al-16Si-5Fe based alloys Produced from Rapidly Solidified Powders. *J. Korean Inst. Met. Mater.* **1999**, *37*, 1191–1197.
9.   Li, Y.; Yang, Y.; Wu, Y.; Wang, L.; Liu, X. Quantitative comparison of three Ni-containing phases to the elevated-temperature properties of Al-Si piston alloys. *Mater. Sci. Eng. A* **2010**, *527*, 7132–7137. [CrossRef]
10. Wu, Y.; Liao, H.C.; Yang, J.; Zhou, K. Effect of Si content on Dynamic Recrystallization of Al-Si-Mg alloys During Hot Extrusion. *J. Mater. Sci. Technol.* **2014**, *30*, 1271–1277. [CrossRef]

11. Hu, H.E.; Wang, X.; Deng, L. High temperature deformation behavior and optimal hot processing parameters of Al-Si eutectic alloy. *Mater. Sci. Eng.* **2013**, *576*, 45–51. [CrossRef]

12. Zhang, H.; Li, L.; Yuan, D.; Peng, D. Hot deformation behavior of the new Al-Mg-Si-Cu aluminum alloy during compression at elevated temperatures. *Mater. Charact.* **2007**, *58*, 168–173. [CrossRef]

13. Liao, H.; Wu, Y.; Zhou, K.; Yang, J. Hot deformation behavior and processing map of Al-Si-Mg alloys containing different amount of silicon based on Gleebe-3500 hot compression simulation. *Mater. Des.* **2015**, *65*, 1091–1099. [CrossRef]

10

The Eh-pH Diagram and Its Advances

Hsin-Hsiung Huang

Academic Editors: Suresh Bhargava, Mark Pownceby and Rahul Ram

Metallurgical and Materials Engineering, Montana Tech, Butte, MT 59701, USA; hhuang@mtech.edu

Abstract: Since Pourbaix presented Eh *versus* pH diagrams in his "Atlas of Electrochemical Equilibria in Aqueous Solution", diagrams have become extremely popular and are now used in almost every scientific area related to aqueous chemistry. Due to advances in personal computers, such diagrams can now show effects not only of Eh and pH, but also of variables, including ligand(s), temperature and pressure. Examples from various fields are illustrated in this paper. Examples include geochemical formation, corrosion and passivation, precipitation and adsorption for water treatment and leaching and metal recovery for hydrometallurgy. Two basic methods were developed to construct an Eh-pH diagram concerning the ligand component(s). The first method calculates and draws a line between two adjacent species based on their given activities. The second method performs equilibrium calculations over an array of points (500 × 800 or higher are preferred), each representing one Eh and one pH value for the whole system, then combines areas of each dominant species for the diagram. These two methods may produce different diagrams. The fundamental theories, illustrated results, comparison and required conditions behind these two methods are presented and discussed in this paper. The Gibbs phase rule equation for an Eh-pH diagram was derived and verified from actual plots. Besides indicating the stability area of water, an Eh-pH diagram normally shows only half of an overall reaction. However, merging two or more related diagrams together reveals more clearly the possibility of the reactions involved. For instance, leaching of Au with cyanide followed by cementing Au with Zn (Merrill-Crowe process) can be illustrated by combining Au-CN and Zn-CN diagrams together. A second example of the galvanic conversion of chalcopyrite can be explained by merging S, Fe–S and Cu–Fe–S diagrams. The calculation of an Eh-pH diagram can be extended easily into another dimension, such as the concentration of a given ligand, temperature or showing the solubility of stable solids. A personal computer is capable of drawing the diagram by utilizing a 3D program, such as ParaView, or VisIt, or MATLAB. Two 3D wireframe volume plots of a Uranium-carbonate system from Garrels and Christ were used to verify the Eh-pH calculation and the presentation from ParaView. Although a two-dimensional drawing is still much clearer to read, a 3D graph can allow one to visualize an entire system by executing rotation, clipping, slicing and making a movie.

Keywords: Pourbaix diagram; Eh-pH diagram; Eh-pH applications; ligand component; equilibrium line; mass balance point; Gibbs phase rule; 3D Eh-pH diagrams; ParaView; VisIt; MATLAB

1. Introduction

All Eh-pH diagrams are constructed under the assumption that the system is in equilibrium with water or rather with water's three essential components, $H(+1)$, $O(-2)$ and $e(-1)$; the oxidation states are presented using Arabic numbers with a + or a − sign. The diagrams are divided into areas, each of which represents a locally-predominant species. Eh represents the oxidation-reduction potential based on the standard hydrogen potential (SHE), while pH represents the activity of the hydrogen ion (H^+, also known as a proton). An Eh-pH diagram can describe not only the effects of potential and pH, but also of complexes, temperature and pressures. By convention, Eh-pH diagrams always show the thermodynamically-stable area of water by two dashed diagonal lines.

Two typical Eh-pH diagrams, both based on thermodynamic data from the NBS database [1], are presented. Figure 1 shows an Eh-pH diagram for one component (excluding three essential H(+1), O(−2) and e(−1) components) of metal, in this case manganese, Mn, while Figure 2 is that of another component of mineral acid phosphorus, P. Both diagrams show that oxidized species reside in high Eh areas, while reduced species are in low Eh areas. The metal diagram starts, at the left edge, from metal ions (Mn^{2+}) at low pH, which progressively react with OH^- as pH increases to produce metal hydroxides ($Mn(OH)_2$) or oxides. The diagram for the mineral acid starts, again from the left, with acid (H_3PO_4) and progressively deprotonates due to reactions with OH^- to finally produce phosphate ion (PO_4^{3-}) at high pH. Figure 1 also illustrates the tendencies to transition between species.

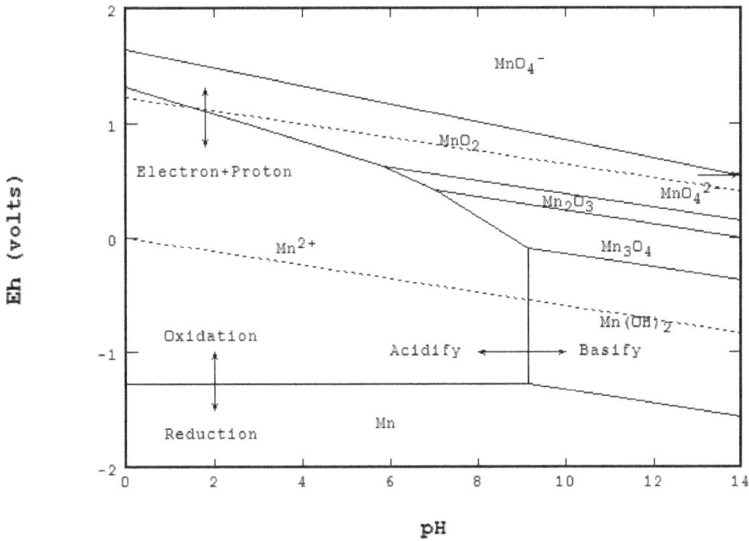

Figure 1. Eh-pH diagram of a Mn–water system. Dissolved manganese concentration, [Mn] = 0.001 M.

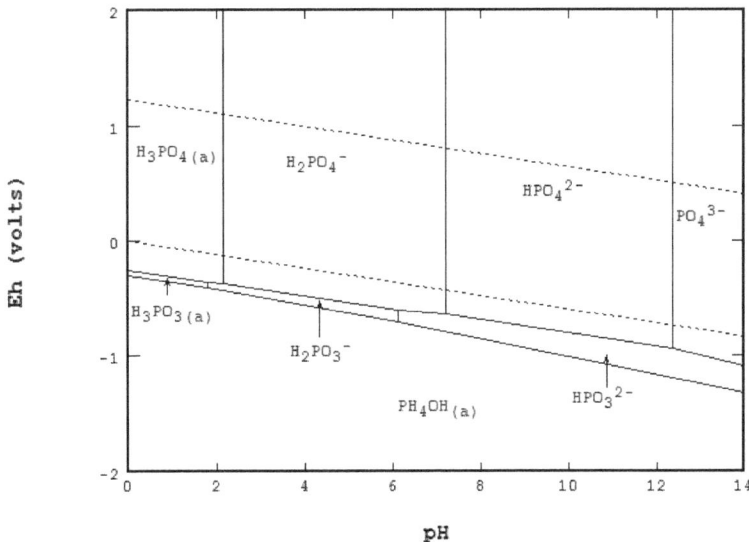

Figure 2. Eh pH diagram of a P-water system. Dissolved phosphorus species, [P] = 0.001 M.

Scope of the Paper

This paper illustrates some ways to improve a basic Eh-pH diagram for better visualization of species and stability regions. The demonstrated methods are all calculated and constructed with an ordinary PC, without a high-end graphics card, using Windows 7 or a higher version. All diagrams can be obtained in a short time. The fundamentals underlying the calculations are briefly described and/or available in the literature and listed as references. Discussions include:

1. Examples of applications: geochemical formation, corrosion and passivation, leaching and metal recovery, water treatment precipitation and adsorption.
2. Development of equilibrium line and mass balance point methods to handle ligand component(s): the theory, illustration and result comparison are presented; both methods satisfy the Gibbs phase rule derived for the Eh-pH diagram.
3. Examples by merging two or more diagrams for better illustration of the overall reactions involved in a process.
4. Demonstrations using a third party program to produce 3D diagrams with the addition of a third axis. The axis can represent the solubility of stable solids, ligand concentration or temperature. Two 3D wireframe volume plots of the Uranium-carbonate system based on a classic Garrels and Christ [2] work were used to verify the Eh-pH calculation and the presentation from ParaView.

This paper is not intended to discuss the following topics in detail:

1. Comparison among existent computer programs listed from the literature that directly or indirectly construct an Eh-pH diagram.
2. Effects from temperature, pressure, ionic strength and surface complexation for aqueous chemistry.
3. The algorithm and flow sheet to construct the diagram used by the author: they are available and referenced elsewhere; no source codes of the programs are presented.
4. Comparison or comments on third party 3D programs used by the author.

Note: The diagrams shown in this paper are solely for illustration. Unless specified, all were constructed at a temperature of 25 °C and zero ionic strength. The molarity is used for a dissolved species as [species], and Σcomponent is used to represent the sum of all mass from one component. Various thermodynamic databases were used as was convenient. Except as noted for 3D plots, all diagrams were constructed by STABCAL [3] running on the Windows operating system using win8.1 64 bit, Pentium i7, 4.3 GHz with 16 GB RAM hardware, and 1680×1050 resolution monitor.

2. Crucial Developments of the Eh-pH Diagram

Chapter 2 of the Pourbaix Atlas [4] presented the method of calculation and the procedure of the construction of an Eh-pH diagram. The process was relatively simple since only one component was considered.

Garrels and Christ [2] dedicated a full chapter to the Eh-pH diagrams. Several diagrams related to geochemical systems were not only presented, but also explained. They laid out a procedure to construct the diagrams when ligand(s) were involved, such as illustrated in the Fe–S and Cu–Fe–S systems. They also presented two 3D wireframe volume diagrams for the Eh-pH-CO_2 system, which will be discussed later in this paper.

A crucial development in constructing an Eh-pH diagram was in deciding how to handle a system when a ligand component was involved. Two completely different approaches were evolved.

2.1. Development of the Equilibrium Line Method

The equilibrium line method was originally used by Pourbaix for simple metal-hydroxide systems. Each line equation is derived from an electrochemical and/or acid-base reaction between species.

Garrels and Christ used Fe–S as an example to show that the same procedure presented by Pourbaix could be applied to a multicomponent system. Basically, it involved two separate steps: domain areas of ligand S were first constructed, then all Fe species (including Fe–S complexes) were distributed in each isolated area of the ligand species. Huang and Cuentas [5] presented a computer algorithm to construct this type of diagram using an early personal computer.

2.2. Development of the Mass Balance Point Method

Forssberg *et al.* [6] constructed several Eh-pH diagrams related to chalcopyrite, CuFeS$_2$, by performing equilibrium calculations for the whole system at once at each given Eh and pH. By doing so, the Cu:Fe:S ratio could be strictly maintained to 1:1:2 at all points. They used the SOLGASWATER program developed by Eriksson [7] to perform the calculation. This point-by-point mass balance method identifies the predominant species at each given point of Eh and pH. Points of the same species were combined into an area for the final diagram. The SOLGASWATER program used free energy minimization, which is commonly used for equilibrium calculation. Woods *et al.* [8] also presented diagrams for the Cu–S system using SOLGASWATER.

The mass balance method can also be computed considering the law of mass action (Huang *et al.* [9]). This approach simultaneously solves all equations, equilibria and mass balances, at each given point of Eh and pH. As with the free energy minimization method, the final diagram has to be plotted by grouping calculated results together. presented later, was reconstructed using the law of mass action for Cu–S and matched with from Woods *et al.* [8].

Besides matching the mass input, these diagrams reveal the presence of multiple solids as restricted only by the Gibbs phase rule. The key to the success of using the point-by-point method, however, is the resolution of the grids used in the calculation. Except for 3D diagrams, all mass balance diagrams in this paper were constructed using grids of at least 400 × 800.

3. Applications for the Diagrams

Eh-pH diagrams are widely used in many areas where an aqueous system is affected by oxidation-reduction and/or acid-base reactions, ligand complexation, temperature or pressure. The following three examples are presented to illustrate these effects.

3.1. Geochemical Formation

Copper porphyry ore deposits occur throughout the world and are very important sources of copper, silver and gold. These deposits initially consist of disseminated sulfide minerals in a rock matrix, but near-surface weathering oxidizes the sulfides and leaches dissolved metals from the residual mass. These leached metals in solution percolate downward and are often reprecipitated in an enrichment zone overlying unreacted sulfide protore. The near-surface weathered, oxidized portion of the deposit corresponds to the oxidizing region of an Eh-pH diagram, while the non-oxidizing reduced enrichment zone corresponds to the reducing diagram region. Figure 3 is a geologic sketch of an idealized porphyry deposit *versus* the depth from the surface, while Figure 4 is a copper Eh-pH diagram in which iron, sulfur and carbonate, besides copper, are considered in the calculations. The minerals predicted in the diagram, solely from thermodynamic considerations, correspond extremely well with minerals observed in these deposits and with the relationships between these minerals. In the oxidized and weathered zone, the original copper and iron sulfides are not stable, while copper carbonates (antlerite, malachite, azurite) and oxides (tenorite, cuprite) form instead. In the enrichment zone, the copper-only sulfides covellite (CuS) and chalcocite (Cu$_2$S) are dominant, with native copper seen to occur in both oxidized and enriched zones.

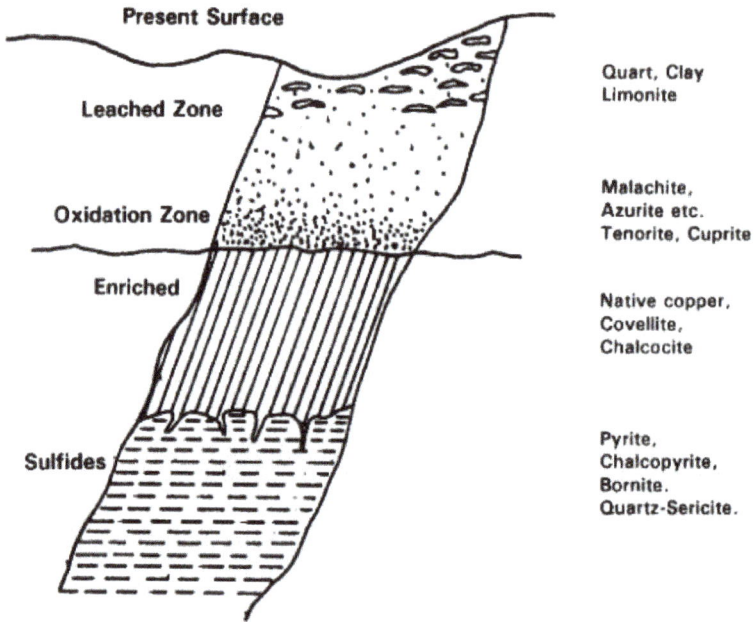

Figure 3. Illustrated copper ore deposit for comparison to the Eh-pH diagram to the right (Dudas *et al.* [10]).

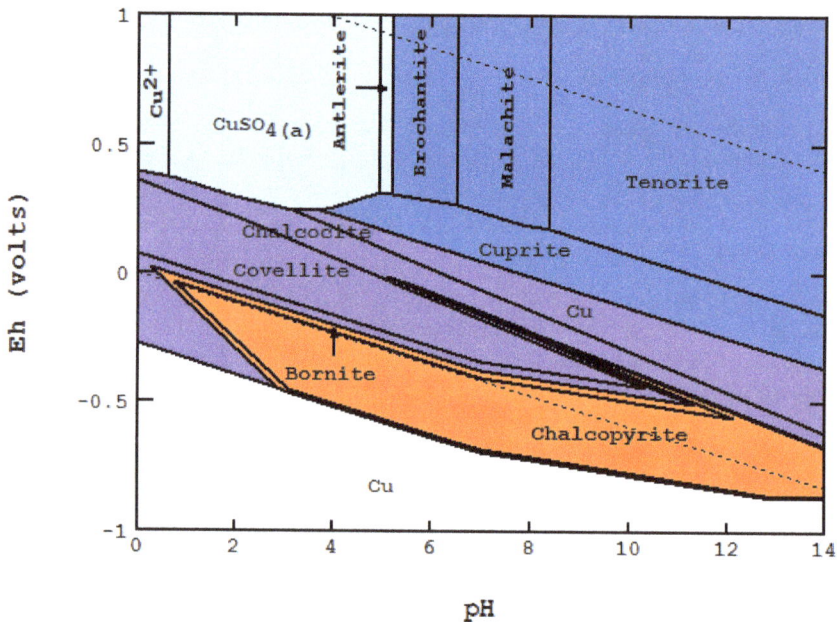

Figure 4. Eh-pH diagram Cu–CO_2–Fe–S in water. pCO_2 = 0.1 atm, [S] = 0.01 M, [Fe] = [Cu] = 0.001 M. Species were taken from the LLnL database [11].

Another geochemical example is the Eh-pH diagram modeling metamorphic conditions. In order to show the effect of high pressure, a database such as SUPCRT (Johnson *et al.* [12]) is required. See the reference from Kontny *et al.* [13] for a Fe–S diagram at 300 °C and 1500 bars pressure or Huang [14] for more calculations and examples using SUPCRT-related databases.

3.2. Corrosion and Passivation

Metallic corrosions are widespread problems of great importance in virtually all physical structures. Corrosion chiefly occurs when metal electrochemical dissolution is favored. One way to protect the metal from corrosion is to form a passivated layer, which may simply be a metal oxide. Some metal oxides, such as PbO, exhibit relatively high solubility and provide little corrosion protection.

The distribution-pH diagram (Figure 5) shows the concentrations of dissolved Pb species, as well as the solubility of PbO, *versus* pH. Formation of metal-carbonate, as shown in the Eh-pH diagram of Figure 6, offers a wider passivation region. Both diagrams were constructed using the LLnL [11] database. Pourbaix in his lectures [15] presented a similar case for using CO_2 to passivate Zn metal.

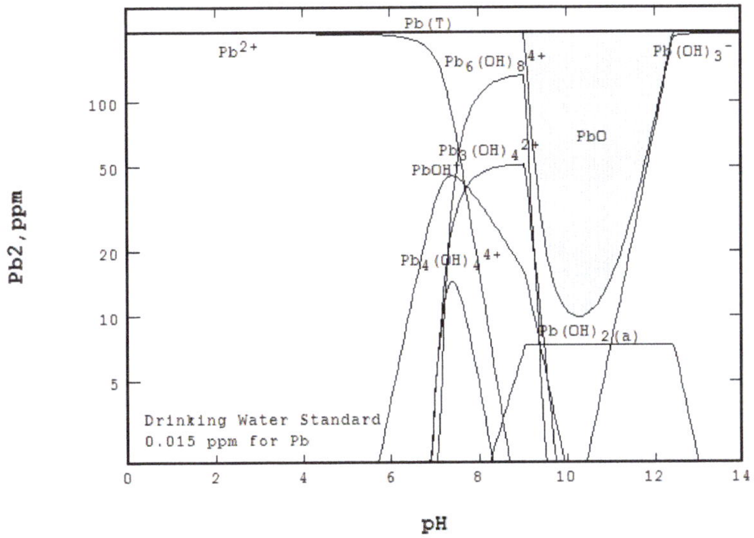

Figure 5. Solubility of PbO (shaded) *versus* pH. PbO does not provide good corrosion protection, even at elevated pHs.

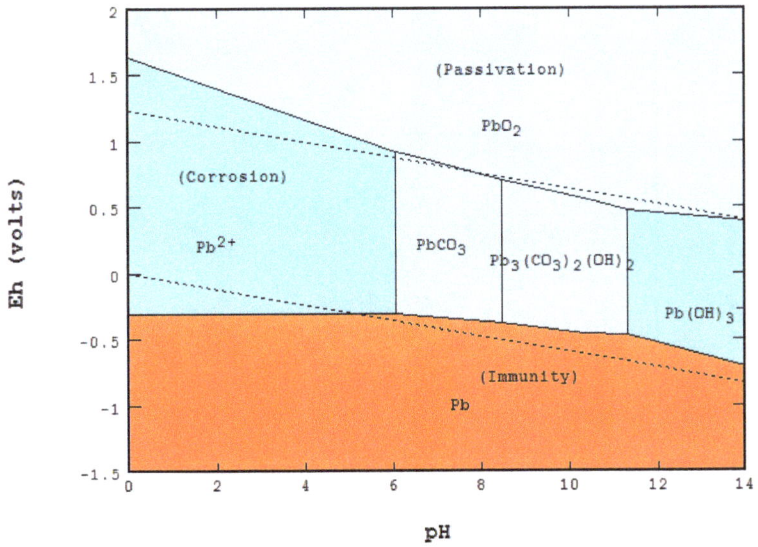

Figure 6. Eh-pH of the $PbCO_3$–water system. [Pb] = 1×10^{-6} and [CO_3] = 0.001 M. Pb carbonate phases do provide corrosion resistance.

3.3. Water Treatment and Adsorption

Water discharge standards almost always include concentration limits for the acid, base and heavy metals. When feasible, precipitation of a solid, followed by a liquid-solid separation is usually the preferred means of achieving these limits, but often, stringent standards are difficult, if not impossible, to comply with by this means. Adsorption onto metal oxides/hydroxides sometimes provides an alternative means of removing these metals from the discharge solution. The adsorption of arsenic (As) by ferrihydrite is demonstrated in Figure 7 using data from Nishimura *et al.* [16]. For this particular experiment, the initial conditions were $\Sigma As = 37.5$ mg/L with a Fe/As mole ratio of 10. The source of ferric iron was dissolved $Fe_2(SO_4)_3:5H_2O$.

The species considered and their thermodynamic values were also taken from the LLnL database [11]. The equilibrium calculation included adsorption using a surface complexation model. Potentially adsorbed species onto ferrihydrite are three arsenates, one arsenite, two sulfates, hydrogen ion and hydroxide. Their equilibrium constants, $\log K_{ads}^{int}$, were obtained from Dzombak and Morel [17]. In order to better fit the experimental data, some modifying changes were made:

1. Type 2 site density for ferrihydrite was changed from 0.2 to 0.3 mole As/mole Fe due to co-precipitation,
2. The $\log K_1^{int}$ for adsorbed species $\equiv FeH_2AsO_4$ was changed from 29.31 to 31.67,
3. The adsorbed species $\equiv FeAsO_4^{2-}$ and its $\log K_3^{int} = 21.404$ were added and
4. Solid scorodite ($FeAsO_4:2H_2O$) and its $\Delta G^0_{25C} = -297.5$ kcal/mole were included with the LLnL dbase.

Figure 7 is the resulting distribution-pH diagram, of the same type as Figure 5, for arsenate As(V). The adsorption model nicely matches the experimental data, demonstrating effectively what the arsenic removal should be. The adsorption of arsenite As(III), while not shown, also matches the experimental data. Figure 8 is presented to illustrate the Eh-pH diagram for the As–Fe–S–water system constructed using the mass-balanced (600 × 800 grids) method. The areas in light blue show solids and adsorbed species to a dissolved concentration less than 0.1 ppm.

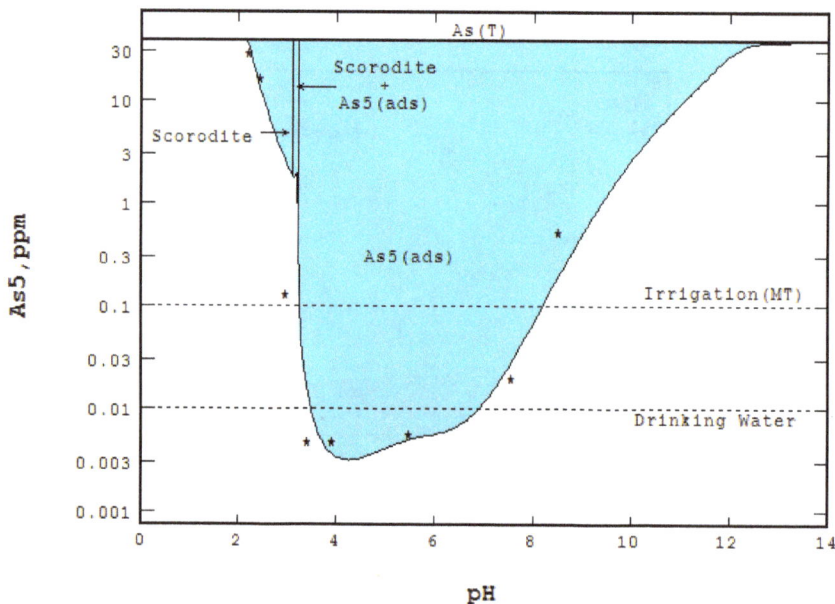

Figure 7. Distribution of As(V) *vs.* pH diagram when Fe/As = 10. Asterisks are experimentally-observed values. Drinking water standard from EPA (2001).

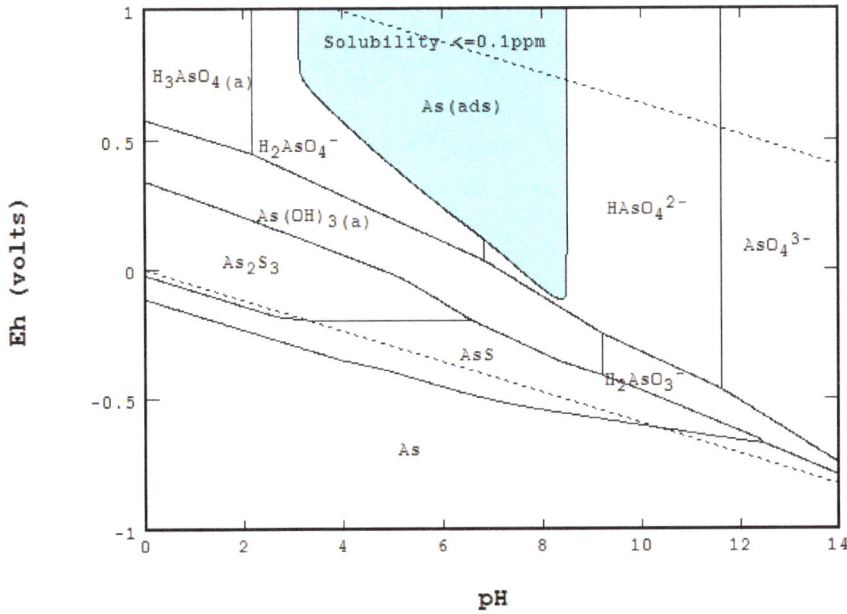

Figure 8. Eh-pH of As–Fe–S water where the mole ratio of Fe/As = 10. The colored area indicates less than 0.1 ppm concentration of As by adsorption.

3.4. Hydrometallurgical Leaching and Metal Recovery

Three applications for hydrometallurgy are presented in more detail later in the section titled "Enhancing the Eh-pH Diagrams by Merging Two or More Diagrams". These are:

1. Cyanidation of Au and cementation with Zn Metal,
2. Cementation of copper with elemental Fe, and
3. Galvanic conversion of chalcopyrite with Cu metal with two construction methods for Eh-pH diagrams to handle ligand components.

4. Descriptions and Comparison between These Two Crucial Methods

4.1. Equilibrium Equations for Eh-pH Diagrams

The chemical equation between Species A and B in the water system, with or without electron involvement, can be expressed as:

$$aA + cC \leftrightarrow bB + dD + hH^+ + wH_2O\,(+ne^-) \tag{1}$$

Species C and D are ligand and complexes produced with ligand. The stoichiometric coefficient of a species is taken as positive if it is on the right-hand side of the equation, and *vice versa*. Species H^+, H_2O and e^- may not always be on the right-had side of the equation. Because so many equations and species are involved while performing equilibrium calculations for an Eh-pH diagram, it is easier to use the free energy of formation of each involved species, ΔG_i^0, then to calculate the free energy of reaction as,

$$\Delta G_{rex} = \sum (v_i \times \Delta G_i^0) \tag{2}$$

where v_i represents the stoichiometric coefficient of species i.

Depending on whether or not the reaction involves an electron and/or hydrogen ion, the equations are:

The Nernst equation for oxidation-reduction reaction with or without acid-base:

$$\text{Eh} = \text{Eh}^0 + \frac{\ln(10)RT}{(n \times F)} \times \left[\log \left(\frac{\{B\}^b \{D\}^d}{\{A\}^a \{C\}^c} \right) - h\text{pH} \right] \qquad (3)$$

where $\text{Eh}^0 = \dfrac{\Delta G\text{rex}}{(n \times F)}$, where R is the universal gas constant, $8.314472(15)$ J/(K· mol); T is in kelvins; F is the Faraday constant $96,485.3399(24)$ J/(V· equivalent); and $\{A\}$ and the others species are defined as the activities of Species A. The activities of solid and liquid are normally assumed to be one; gas is taken as the atmosphere (atm). The activity of an aqueous solution is the multiplication of the concentration in mol/L, symbolized as [A], with its activity coefficient. The coefficient can be computed from one of the appropriate models. Without having the acid-base, the "hpH" term in the equation will be dropped out.

The equilibrium equation for acid-base reaction without redox reaction:

$$\text{pH} = \frac{1}{h} \times \left[\log \left(\frac{\{B\}^b \{D\}^d}{\{A\}^a \{C\}^c} \right) + \frac{\Delta G\text{rex}}{\ln(10) \times RT} \right] \qquad (4)$$

The equation for reaction involves neither an electron nor a hydrogen ion:

$$\log Q - \log K = \log \left(\frac{\{B\}^b \{D\}^d}{\{A\}^a \{C\}^c} \right) + \frac{\Delta G\text{rex}}{\ln(10) \times RT} \qquad (5)$$

Species A will be favored if $\log Q - \log K$ is positive, and *vice versa*.

As mentioned earlier, two different approaches may be used to construct an Eh-pH diagram. One is to calculate equilibrium equations between pairs of species and to construct the diagram by plotting the resulting equilibrium lines. The other is to perform equilibrium calculations from all involved species at each point in a grid, then selecting the predominant species at each point. Regardless of which method is used, these equilibrium equations have to be satisfied.

4.2. Line Method Using Equilibrium Concentration [5]

The diagram is constructed by computing the equilibrium between two adjacent species from their activities. The concentration or activity of aqueous species has to be given. Figure 9 shows the Eh-pH diagram for Cu at three different concentrations. The case where a ligand component is also involved is demonstrated in Figure 10, for the Cu–S–water system using [S] = 0.001 and [Cu] = 0.001 mol/L. The areas of predominance for the various ligand S species (labeled in light blue) were first constructed. The distribution of Cu species, including Cu–S complexes, in each S domain (such as the area of H_2S shaded with light blue) was then constructed. The final diagram of Cu species was determined by combining all of the areas from S ligands. It should be noted that the total concentration of ΣS may change depending on whether or not Cu is complexed with S. When Cu species are not complexed with S, ΣS would be 0.001 mol/L, as described. However, when Cu species are complexed with S, as in the formation of CuS, ΣS will be the molar sum of the S concentration plus the CuS concentration, which will be equal to 0.002 mol/L. In such a case, the mass of total S may not be constant, as originally assigned.

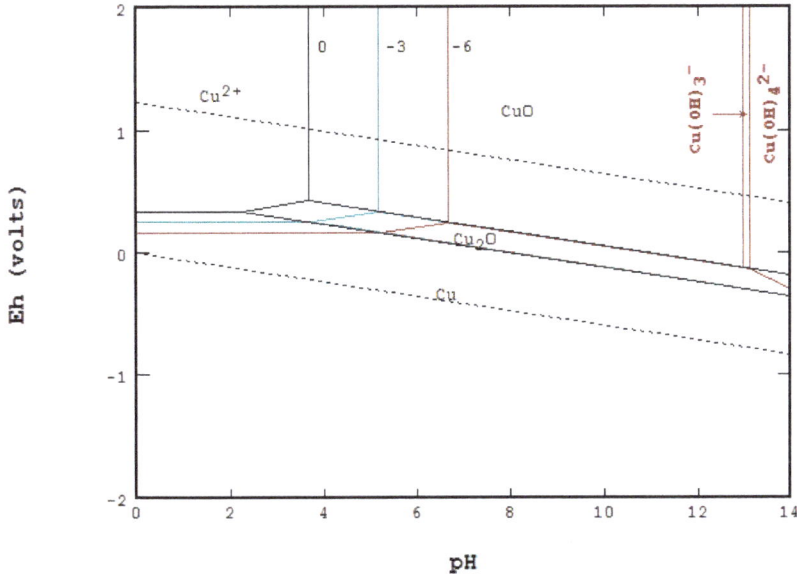

Figure 9. Eh-pH of Cu-water constructed by the line method where three concentrations in log scale are plotted. Data taken from NBS [1].

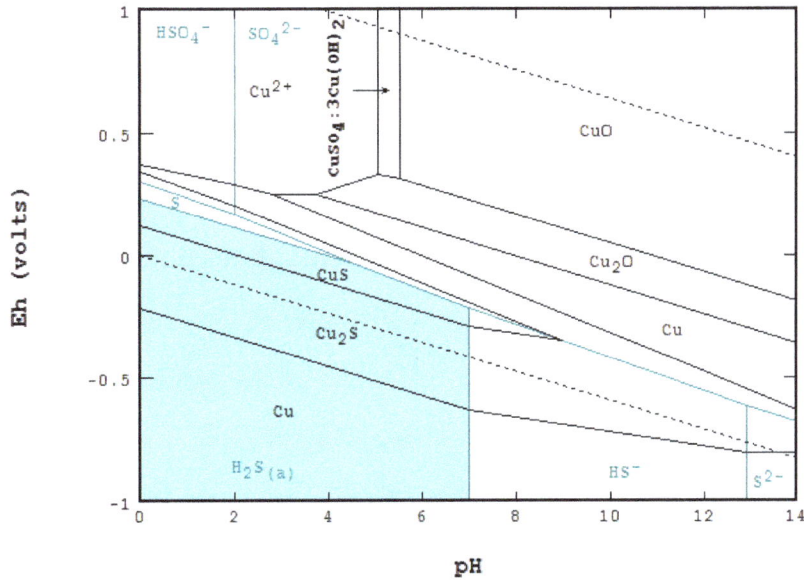

Figure 10. Eh-pH of Cu(main)–S (ligand). The line method plots the ligand first, shown in blue color. The distribution of Cu species for each domain of S is then determined.

4.3. Point-by-Point Method Using Mass Balance [9,18]

In this method, mass balances are considered and calculated with all of the equilibria from all of the components at once from every point of the grid. Unlike the line method where the concentration or activity of aqueous species is specified, this method requires knowing the total mass of each component. The calculation requires not only satisfying all equilibrium equations, but also matching all of the mass balances. The results are sorted out in order to plot the diagram for each specific component. This type of diagram is particularly important for Eh-pH diagrams, which include solids with composition ratios that are close to mineral formation ratios (such as 3:1:4 mole ratios for enargite Cu_3AsS_4)

Mass inputs, including masses of ligands, are crucial for determining critical areas of these diagrams. Figures 11 and 12 illustrate this for the Cu–S–water system (data from NBS [1]). Figure 11 shows the case where S is stoichiometrically slightly less than copper, *i.e.*, $\Sigma Cu = 0.001$ and $\Sigma S = 0.0009$ M, while Figure 12 shows the case where S is slightly in excess. The higher mass of S leads to a larger area of predominance for CuS.

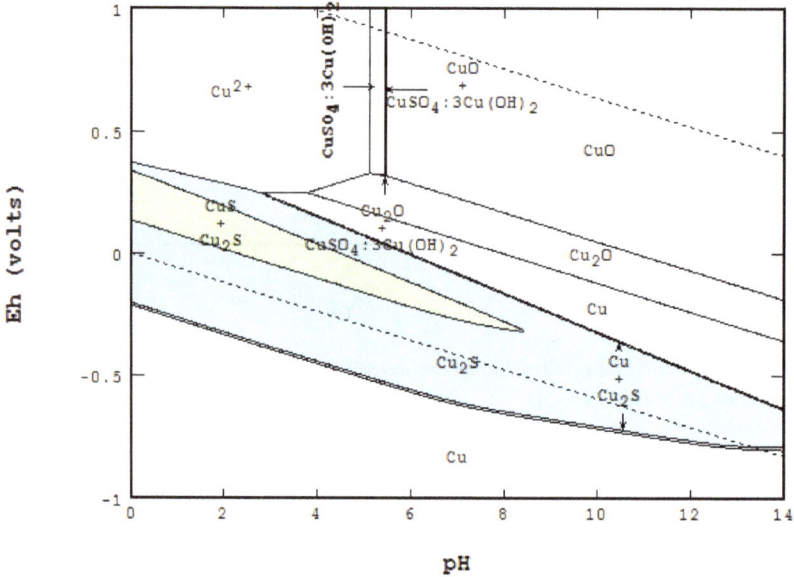

Figure 11. Mass-balanced Eh-pH diagram for the Cu–S–water system with copper slightly in excess. $\Sigma Cu = 0.001$ M and $\Sigma S = 0.0009$ M.

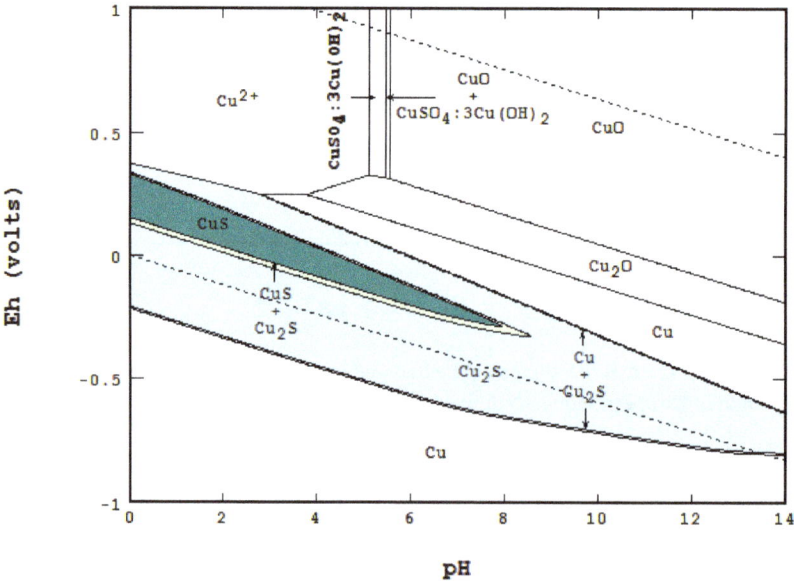

Figure 12. Mass-balanced Eh-pH diagram for the Cu–S–water system with copper slightly in deficit. $\Sigma Cu = 0.001$ M and $\Sigma S = 0.0011$ M.

4.4. Differences and Comparison between the Methods

Different results of the two methods can be seen by comparing diagrams constructed within the Cu–S system, with sulfate species not shown due to unfavorable kinetics (Woods *et al.* [8]). Figure 13 was constructed by the equilibrium line method where [Cu] = 0.118 and [S] = 0.059, and Figure 14 was constructed by the mass-balanced point method where ΣCu = 0.118 and ΣS = 0.059. Free energy data were taken from Woods *et al.* [8]. Crucial differences can be seen in the general area of Cu–S solids. As can be seen from Figure 13, even though the concentration ratio of Cu/S is specified as two to one, CuS, not Cu_2S, is the predominant species.

The mass-balanced point calculation involved points on a 400×800 grid to a precision of 1×10^{-10}, but took less than three minutes for a PC from creating a worksheet for input to plotting the final diagram.

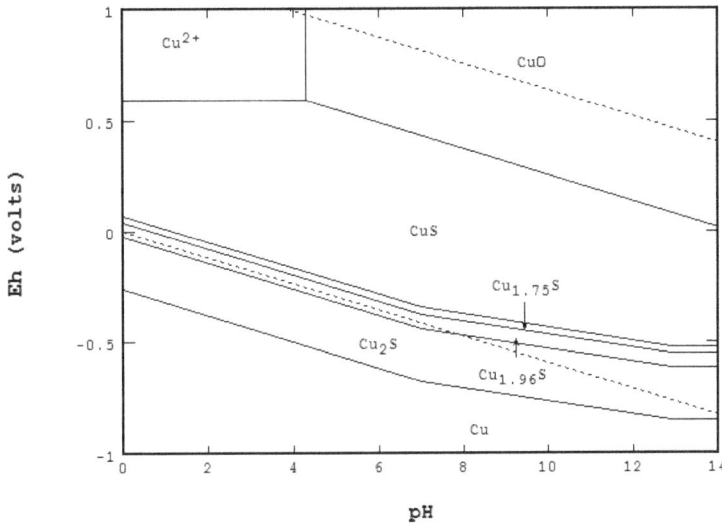

Figure 13. Eh-pH diagram for the Cu–S–water system constructed by the equilibrium line method.

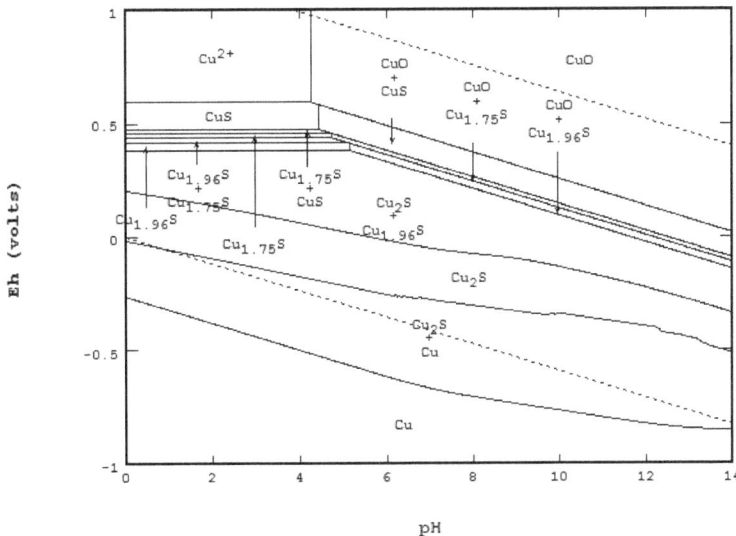

Figure 14. Eh-pH diagram for the Cu–S–water system constructed by the mass-balance point-by-point method. This diagram matches Figure 5 of Wood *et al.* [8].

The equilibrium line method was favored in the past due to its relative ease of construction. When diagrams were constructed using manual calculation (as by Pourbaix), the equilibrium line method was the only practical approach. As greater computational power became available, the mass-balanced point-by-point method came into favor. The following list includes some areas where the mass balance method should be considered over the line method.

1. When the exact composition of the system is needed: Examples include leaching and flotation studies. See Huang and Young [18] for more examples. The Eh-pH diagram of enargite (Cu_3FeS_4) (Figure 15) was constructed using data collected by Gow [19].

2. When a system is required to specify total concentration, not equilibrium concentration nor activity.

3. When the adsorption by solids, such as ferrihydrite, is considered (refer to Figure 8).

4. When multiple phases of a solid need to be shown: Figure 16 was constructed by showing the coexistence of schwertmannite with various forms of jarosites in Berkeley pit water. Water samples were taken and analyzed from 1987 to 2012 by the Montana Bureau of Mines and Geology [20], and the thermodynamic data for the solids species were regression estimated by Srivastave [21].

5. A diagram will most likely be mass balanced if a speciation program, such as PHREEQC (USGS) [22], was used to construct it. Results from the program were collected manually or electronically, then combined into an Eh-pH diagram.

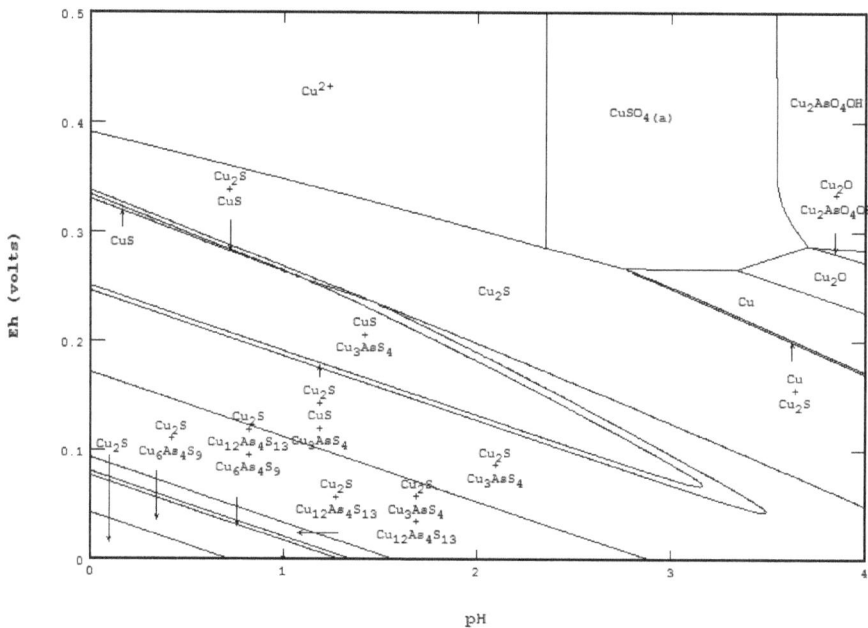

Figure 15. Mass balanced Eh-pH diagram for the enargite Cu_3AsS_4 system. The mass ratio is 0.75:0.25:1 for Cu, As and S. The diagram shows only the copper species in acid solution.

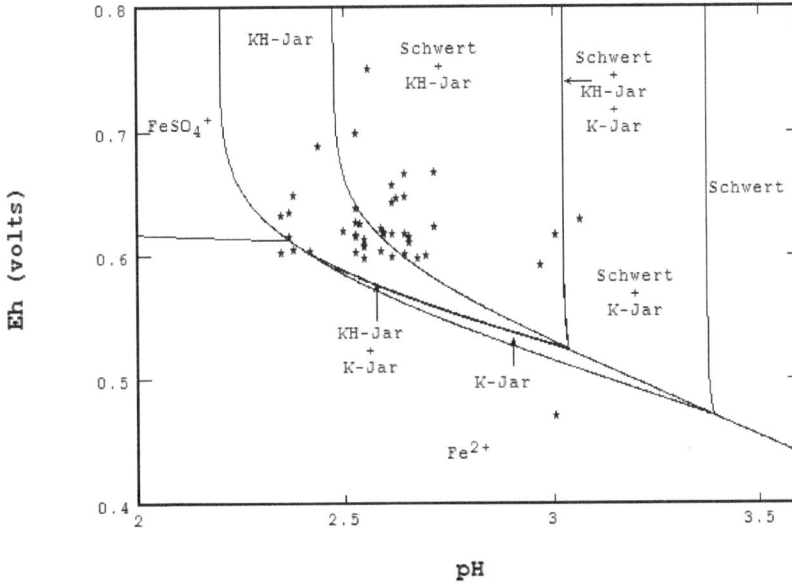

Figure 16. Mass balanced Eh-pH diagram for the Fe–K–S system at 7 °C. This diagram shows the coexistence among schwertmannite, K-jarosite and KH-jarosite. * represents the data analyzed from the sampled water.

Both methods, however, can produce identical diagrams under the following conditions:

1. A one-component system, such as Figure 1 for Mn and Figure 2 for P,
2. The concentration of ligand component(s) is much greater than the main component, such as metal corrosion by sea water, and
3. Gas is the only ligand, such as the Fe–CO_2(g) system.

4.5. Gibbs Phase Rule Applied to an Eh-pH Diagram

An Eh-pH diagram constructed using either method must follow the Gibbs phase rule. The original phase rule equation, $P + F = C + 2$, was developed for considerations of temperature and pressure. It can be refined for use in constructing an Eh-pH diagram by implementing some concepts and restrictions.

Mass-balanced method: This method calculates equilibrium from all components at once. The variables are as follows:

1. P is the total number of phases = 1 (liquid water) + 1 (gas if considered) + N (maximum number of solids/liquids),
2. F is the degree of freedom on the diagram, which is two for an open area, one on a boundary line and zero on a triple point,
3. C is the total number of components = 3 + EC (extra components). Three components are essential for Eh-pH calculation in an aqueous system. These are H(+1), O(−2) and e(−1). The extra components include the main component to be plotted, as well as all ligands.
4. The term of +2 is for temperature and pressure variables. Since both are considered to be constant, +2 will be dropped off. If any system involved a gaseous species, +1 should be used, but it will be canceled out with one extra gaseous phase to the equation.

The phase rule equation for an Eh-pH diagram, best expressed as the maximum number of solids plus liquids excluding the liquid phase of water, thus becomes:

$$N_{\text{maxsolid}} = C - F - 1 \qquad (6)$$

Example 1, Cu and S two-component system: The incorporation of the rule is illustrated in Figure 17, in which all solids containing Cu, as well as S components are presented. Since the method computes equilibria from all components involved at once, $C = 3 + 2$ and $N_{\text{maxsolid}} = 5 - F - 1$ or $4 - F$.

1a. In an open area of the diagram where $F = 2$, N_{maxsolid} will be equal to two. The co-existence of two solids can be seen in many places on the diagram,

1b. On a boundary line where $F = 1$, N_{maxsolid} becomes three. For instance, while each of the light blue areas contains two solids, the line between them represents the presence of three: CuO, Cu_2S and $Cu_{1.96}S$,

1c. On a triple point where three lines meet, $F = 0$, $N_{\text{maxsolid}} = 4$. At the point labeled A, for instance, even though four areas meet, only three solids are coexistent at the point: CuO, $Cu_{1.75}S$ and $Cu_{1.96}S$.

Example 2, Pb-S-KEX (potassium ethyl xanthate) three-component system: Pb-S-KEX was also used to illustrate the phase rule. Figure 18 was constructed using data taken from Pritzker and Yoon [23]. The plot illustrates a small, but intricate area, Eh from -0.5 to -0.3 and pH from 10 to 13, with a resolution of 600×800. A small pink area shows three stable solids: PbS, Pb and PbX_2. This number agrees with the phase rule equation for the Eh-pH diagram of $N_{\text{maxsolid}} = 5 - F$, where F is equal to two, being inside an open area. There are four solids (PbS, Pb, PbX_2 and $Pb(OH)_2$) along the line between this pink area and the area right above it. The N_{maxsolid} for all three corners of this area was no greater than five, as described by the rule.

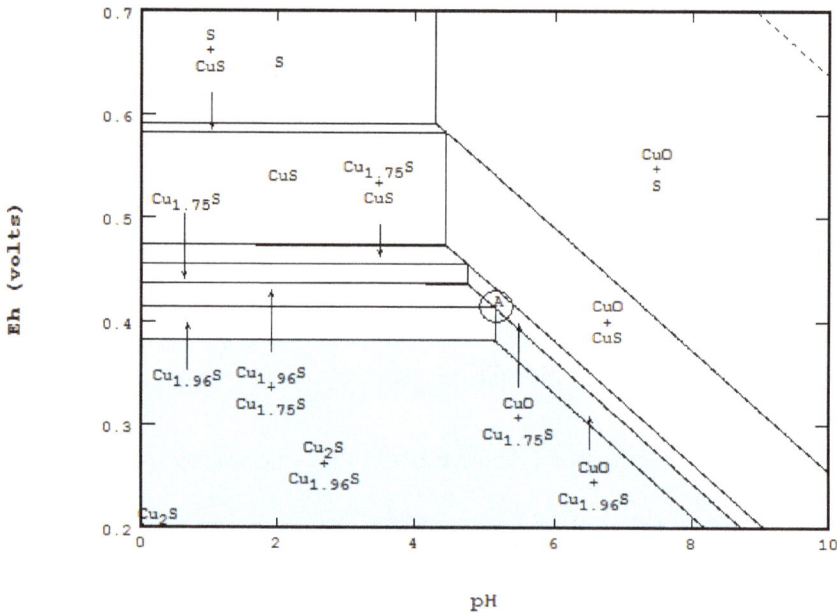

Figure 17. Mass-balanced Eh-pH diagram of the Cu–S–water system to illustrate the phase rule. This is a zoomed-in detail from Figure 14; stable solids include elemental S from the S component. $\Sigma Cu = 0.118$ and $\Sigma S = 0.059$ M.

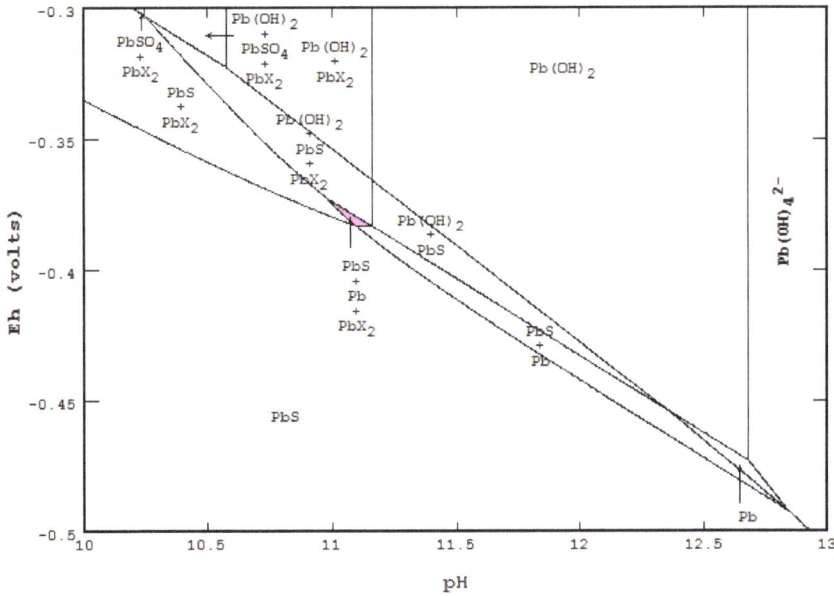

Figure 18. Mass-balanced Eh–pH diagram of the Pb–S–potassium ethyl xanthate (KEX) water system. $\Sigma Pb = \Sigma S = 0.45$ and $\Sigma X = 0.0001$ M. This zoomed-in detail has a resolution of 600×800 to verify the Gibbs phase rule.

Equilibrium line method: Since this method works on one component at a time, $N_{maxsolid} = 4 - F - 1 = 3 - F$. Referring to Figure 13, in any open area of the diagram, $N_{maxsolid} = 1$, which means only one solid is allowed. On any boundary line, $N_{maxsolid} = 2$, as shown by the line between CuO and CuS. If a triple point is formed, $N_{maxsolid}$ will be equal to three.

5. Enhancing the Eh-pH Plot by Merging Two or More Diagrams

Most Eh-pH diagrams indicate the stability of water by two dashed lines: this is a typical example of merging two diagrams together. Other examples include showing several solubilities of solid species (see Figure 9), showing dissolved species in areas dominated by solids and showing ligands in addition to the main component (see Figure 10).

An Eh-pH diagram, including the examples listed above, often shows only half of a reaction. To illustrate the whole process, merging another relevant diagram may be necessary. The combined diagram can be re-plotted or overlaid by a graphics program, such as MS PowerPoint. It is strongly suggested to use different colors for each merged diagram.

5.1. Cyanidation of Gold and Cementation with Zn Metal

The combination of a Au–CN diagram with a Zn–CN diagram is shown as Figure 19. The up-arrow indicates leaching of Au using CN as the complexing ligand and O_2 as the oxidant. The down-arrow indicates the cementation of Au replaced by Zn metal.

5.2. Cementation of Copper with Metallic Iron

Figure 20 shows the combination of a Cu–water diagram and a Fe-water diagram and illustrates cementation of copper by iron metal. Additionally, the diagram shows other reactions of interest, such as those that represent wasteful consumption of iron metal by reactions involving Fe^{3+}, $O_2(g)$ and $H^+(a)$. Although not part of the diagram calculations, the rest potential between the Cu^{2+}/Cu–Fe^{2+}/Fe electrodes is also shown.

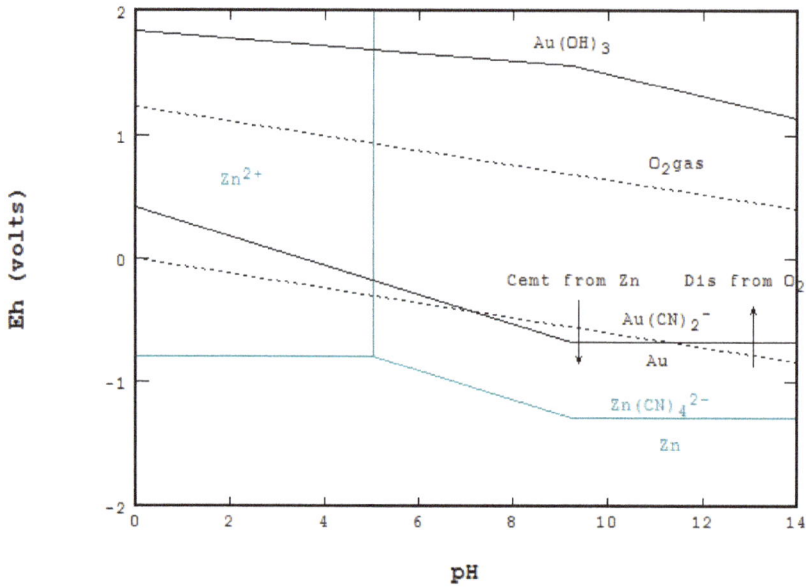

Figure 19. Combined Au–CN and Zn–CN Eh-pH diagrams would show leaching and cementation for the gold cyanidation process.

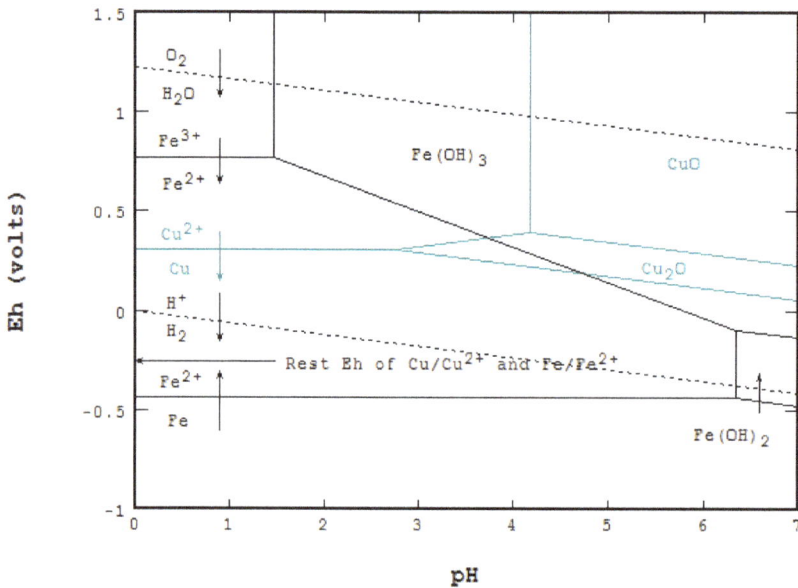

Figure 20. Combined Cu and Fe Eh-pH diagrams would show that reactions occur during copper cementation using metallic iron.

5.3. Galvanic Conversion of Chalcopyrite with Copper Metal [24]

When in contact with Cu metal, chalcopyrite reacts cathodically in an effect known as galvanic conversion:

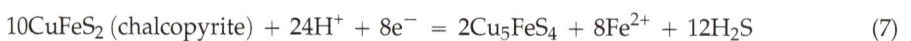

$$10CuFeS_2 \text{ (chalcopyrite)} + 24H^+ + 8e^- = 2Cu_5FeS_4 + 8Fe^{2+} + 12H_2S \qquad (7)$$

$$2Cu_5FeS_4 \text{ (bornite)} + 6H^+ + 2e^- = 5Cu_2S + 2Fe^{2+} + 3H_2S \qquad (8)$$

An anodic reaction takes place on the metallic copper as:

$$2Cu + H_2S = Cu_2S + 2H^+ + 2e^- \tag{9}$$

A schematic diagram for all these reactions is shown as Figure 21. In order to present all of the species involved, three Eh-pH diagrams are superimposed and shown as Figure 22.

1. The three diagrams used are: S species in cyan, Fe and Fe–S in red and Cu–Fe–S in black.
2. Areas of predominance are shown as: chalcopyrite in yellow, bornite in gray, chalcocite in light blue and metallic copper in orange.
3. The down-arrow indicates where galvanic conversion occurs down from chalcopyrite to bornite and, finally, to Cu_2S. The up-arrow indicates where the anodic reaction occurs up from metallic copper to Cu_2S.
4. The diagram indicates that both cathodic and anodic reactions lead to the formation of Cu_2S, and other final species match what Hiskey and Wadsworth [24] described.

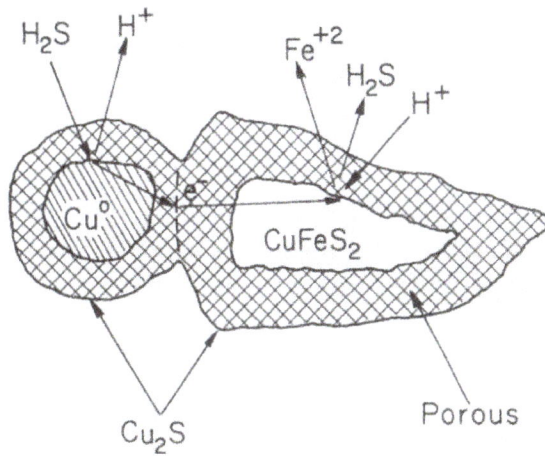

Figure 21. Schematic diagram of reactions occurring upon galvanic conversion of chalcopyrite with metallic copper [24].

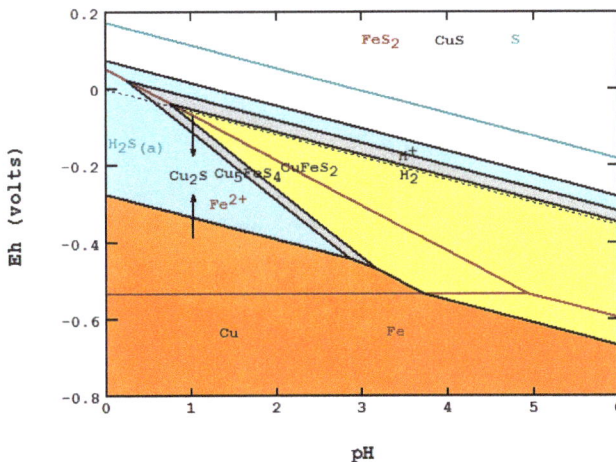

Figure 22. Combination of three Eh-pH diagrams showing the galvanic conversion reactions of chalcopyrite with metallic copper.

6. Third Dimension to an Eh-pH Diagram

Even more information may be shown by adding a third dimension to a base Eh-pH diagram. The third dimension can be the simple solubility of solid or an independent variable, such as temperature or ligand concentration. First, the data needed for a 3D Eh-pH diagram must be calculated. Thereafter, 3D programs for PC, such as ParaView [25], VisIt [26] or MATLAB [27], combine all of the data into a single diagram. These programs can also provide other functions, such as animated rotation, clipping and slicing. This section presents some 3D examples by considering extension of the Eh-pH diagram into a third dimension. Data creation and setup input files for a 3D program are briefly presented. Three areas are illustrated:

1.   Eh-pH along with the solubility of stable solids; two example diagrams are illustrated: passivation of lead (Figure 6) and adsorption of As(III) and As(V) onto ferrihydrite (Figure 8).
2.   Eh-pH with an extra axis for ligand CO_2: two wireframe volume diagrams of Eh-pH-CO_2 taken from Garrels and Christ [2] are used for verifying the results; these two are:

   (a)   Figure 7.32b: in order to match the given ΣCO_2 for the third axis, the mass balance method has to be used; the output of 3D and discussion for this case are presented in more detail.
   (b)   Figure 7.32a: since the third axis is given as the pressure of $CO_2(g)$, the equilibrium line method can be applied; the time required for the Eh-pH calculation was much less.

3.   Presentation of a system in which two or more solid phases, such as CuS and Cu_2S, can coexist.

6.1. Eh-pH with Solubility

Including the solubility of solids in an Eh-pH diagram can give a much clearer view of what can happen to the solid. The following two diagrams constructed by MATLAB extend the 2D Eh-pH diagrams presented earlier. In order for MATLAB to plot 3D solubility diagrams, the Eh-pH program needs to create two files: the (name).m file contains instructions to be executed by MATLAB, and the data file contains solubility from each Eh and pH from the grid. A MATLAB plot can show the matching contour (iso-solubility) lines below the 3D feature.

Figure 23 is an extension of Figure 6, showing the solubility of Pb(II) as the third dimension. The red areas indicate conditions where corrosion can be expected to occur. The gulch area in yellow indicates conditions where Pb(II) is passivated by CO_2.

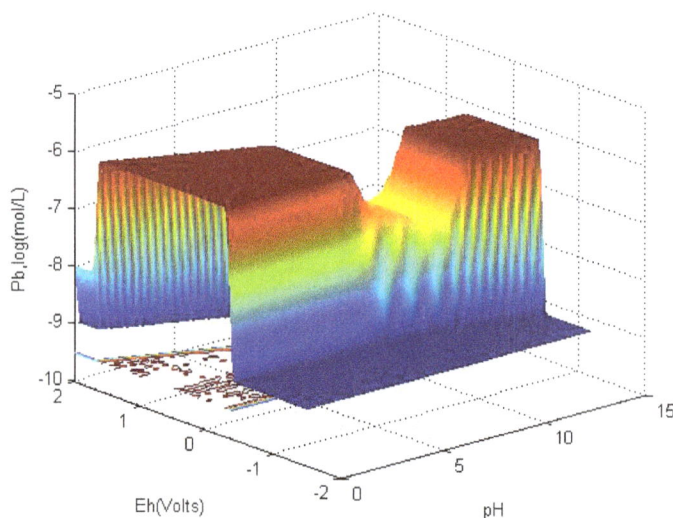

Figure 23. The Eh-pH plus solubility diagram for Pb–CO_3–water. The yellow gulch area in the middle is where Pb(II) is likely passivated by CO_3.

Figure 24 illustrates the same system as Figure 8, which showed where As(V) and (III) can be adsorbed upon the formation of ferrihydrite. The solubility diagram shows where the lowest concentrations of As(V) and (III) can be achieved. The deep blue area at the low Eh of the diagram is where arsenic metal becomes stable and immunized from corrosion.

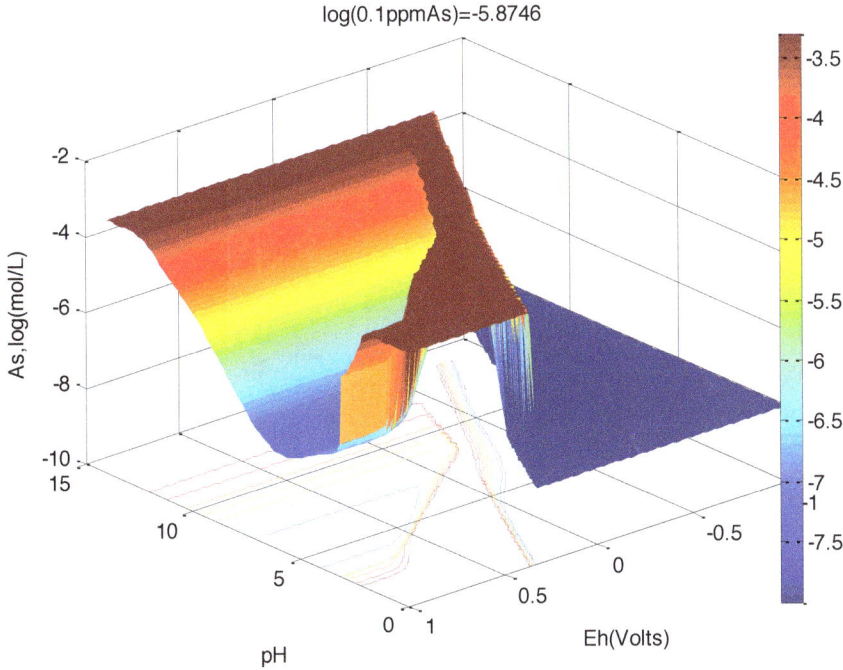

Figure 24. The Eh-pH plus solubility diagram for As–Fe–water. The valley area on the left is where the greatest adsorption of arsenic can occur.

6.2. 3D Eh-pH, Uranium with ΣCO_2: Using the Mass-Balance Point Method

Example system: Figure 7.23b from Garrels and Christ [2] is one of the earliest three-dimensional diagrams for the U–CO_2–water system. It is an Eh-pH diagram with the concentration of CO_2 used for the third dimension. See the duplicated plot from Figure 25. In it, each predominant species is enclosed by the faces of adjacent species. A semi-transparency presentation can be a more advanced graphical method, but it had not been developed at the time. A comparison to the Garrels and Christ plot was generated, considering the same species and their ΔG^0s (Free energy of formation). Other considerations required are:

1. One species in one volume: Since total carbonate is given, assuming total CO_2 means all carbonates, including dissolved, solids and complexes with U, the mass-balanced method for Eh-pH diagram is used. However, to match the Garrels and Christ plot, only one single solid in each volume was selected, with no regions of mixed solids allowed.

2. One missing species: One species on the Garrels and Christ diagram, indicated by a red letter A in Figure 25, seems to have been mislabeled as $UO_2(CO_3)^{4-}$. Judging from its high pH and carbonate location, and being sandwiched between U(IV) and U(VI), the species $UO_2(CO_3)_3^{5-}$ [28] seems to be a good fit. Figure 26 is the regenerated 2D Eh-pH diagram using $\log\Sigma CO_2 = -1$ M.

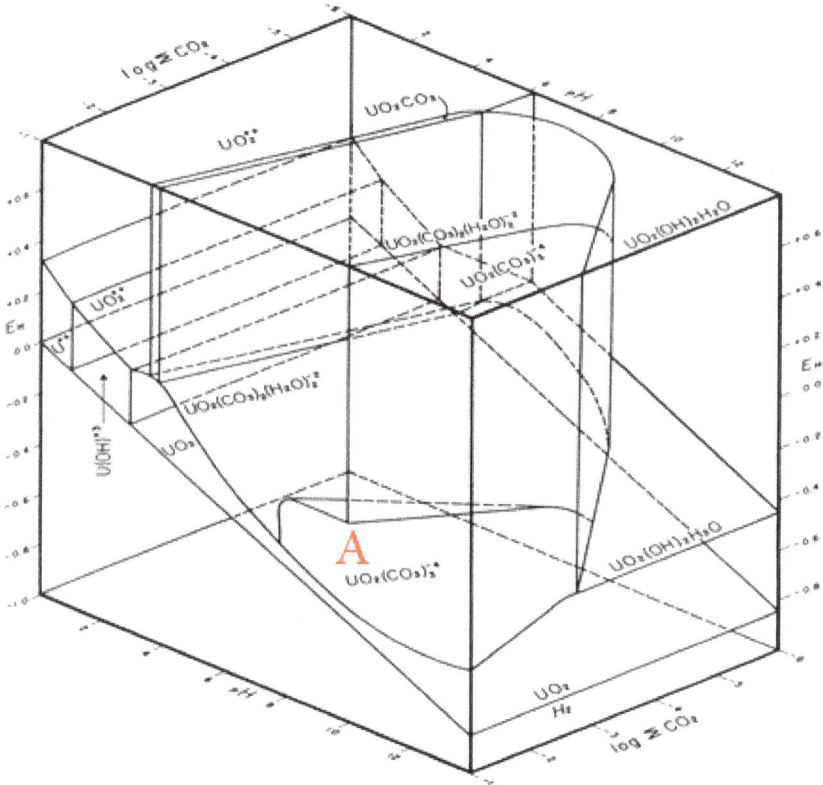

Figure 25. 3D diagram after Garrels and Christ. Label A is where the questionable species is located.

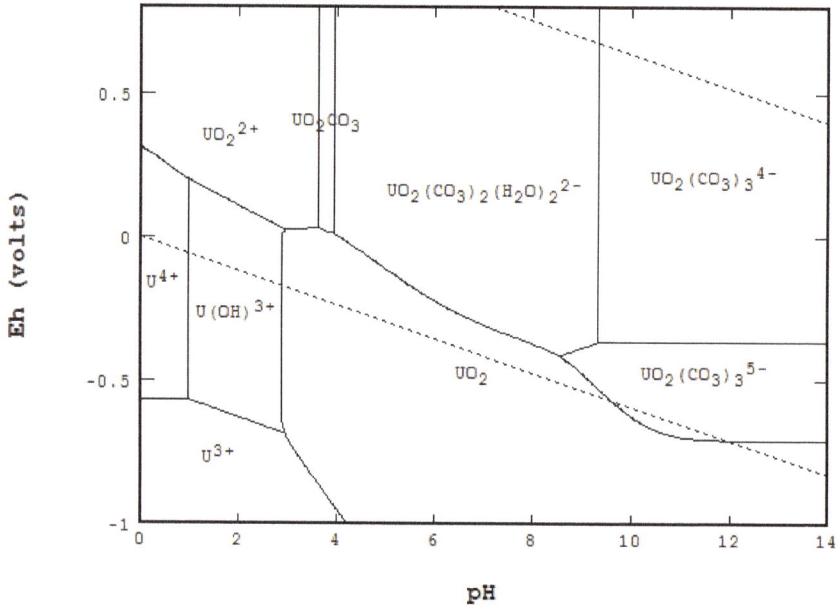

Figure 26. The Eh-pH diagram of U with $\log\Sigma CO_2 = -1$. The inclusion of $UO_2(CO_3)_3{}^{5-}$ clarifies a region that Garrels and Christ might have misrepresented.

Example diagrams and conditions: Figures 27–32 are three-dimensional diagrams created using ParaView (Version 4.3.1) based on the data generated by the STABCAL program. Although the complete ParaView diagram allows such functions as continuous rotation, clipping and slicing, these static diagrams demonstrate the range of features that can be achieved. A color map (bar), shown in Figure 27, indicates that the names of the species should be added at least once to one of the figures. Computational conditional limits include:

- Ranges: Eh = −1 to 0.8, pH = 0 to 14 and $\log\Sigma CO_2$ = −6 to −1,
- Grid point: 250 × 250 × 250,
- Computer: 64 bit, 3.40 GHz, 16 G of RAM,
- Program algorithm: mass balance point method using mass action law,
- Accuracy (sum of squared residual) <1 × 10^{-8} and
- Time to complete the calculation: 1:36:25 (h:mm:ss) from i7 PC or 2:00:51 from i5; contrary to using the line method, shown later, which took less than 30 s.

In order for ParaView to plot a 3D diagram, the Eh-pH program needs to create a (name).vtk file [29] that specifies the type of grid, the values of all X, Y and Z coordinates followed by all of the point data that indicate which species are to be plotted for each point on all three axes.

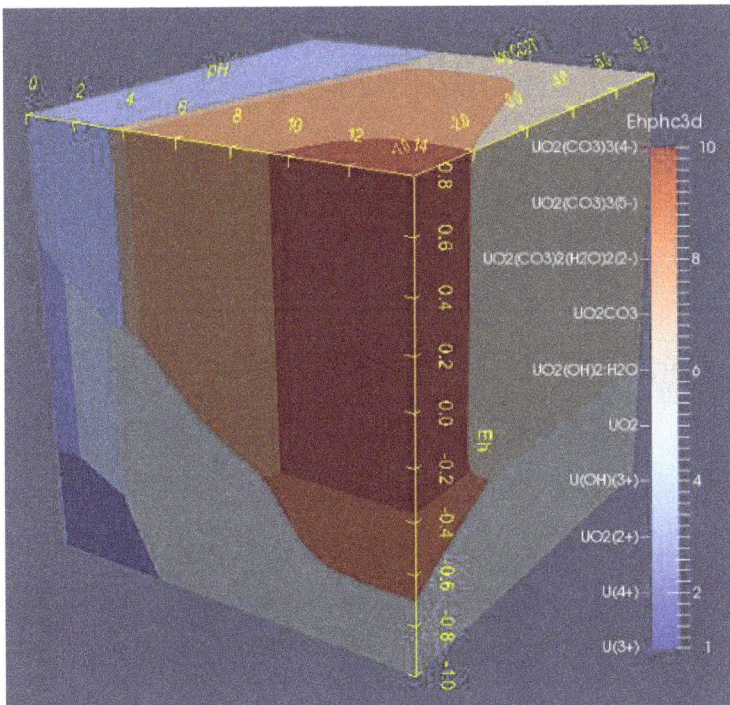

Figure 27. Eh-pH-$\log\Sigma CO_2$ of uranium where the species are shown by colors. A color bar is inserted to show the corresponding species.

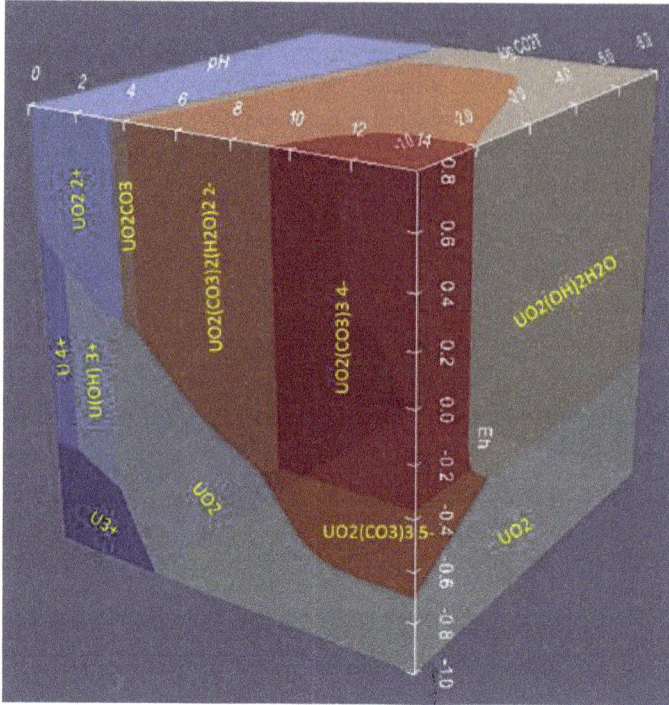

Figure 28. The transparent view of the 3D plot. Each colored area is labeled with the name of the species.

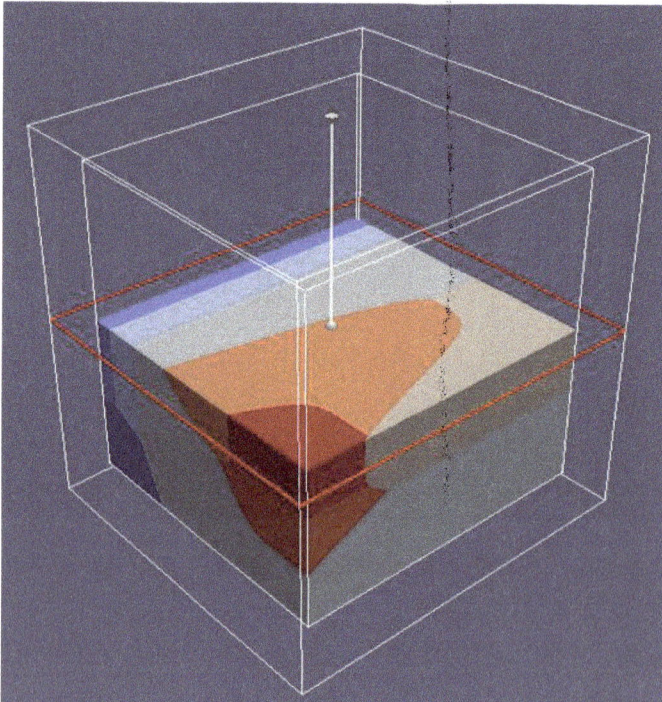

Figure 29. The clip plot shows only regions below the constraint of Eh -0.1 V.

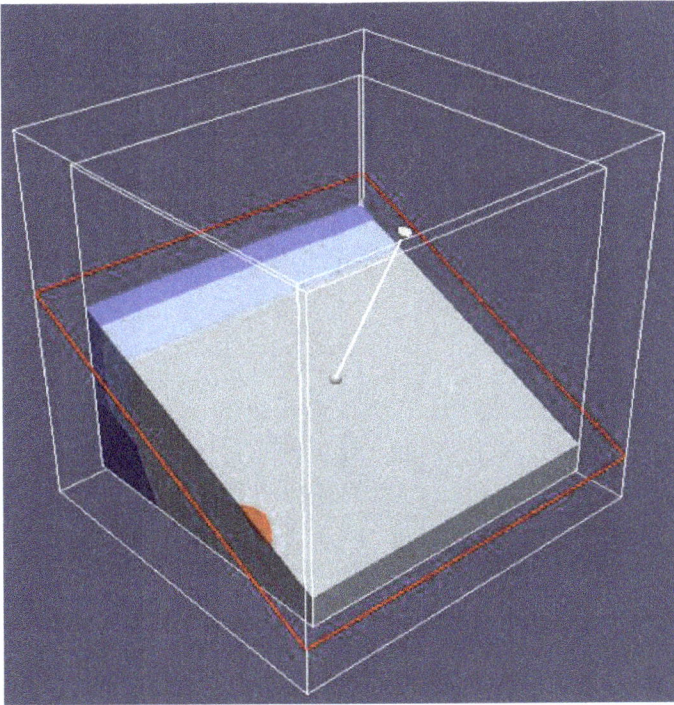

Figure 30. The clip plot shows species below the plane of theoretic water stability between H_2O and $H_2(g)$. Compare to the lower dashed water line in Figure 26 or the lower plane where no species were shown in Figure 25.

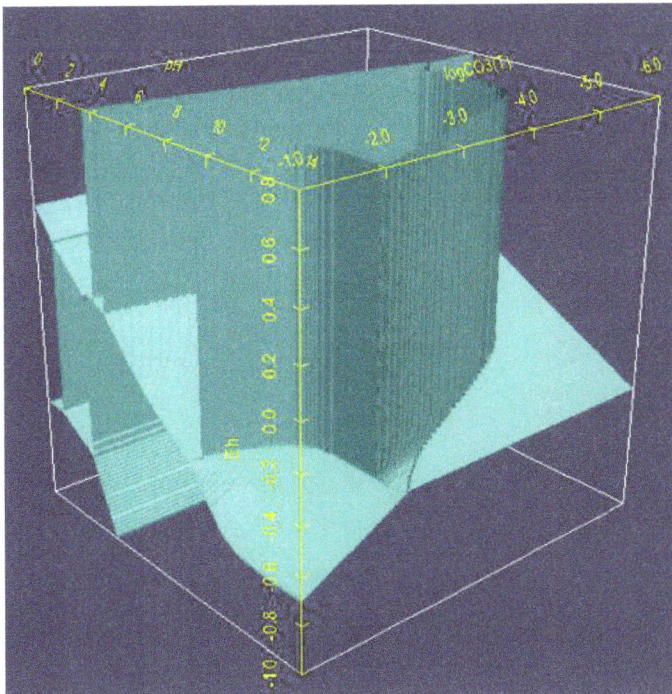

Figure 31. 3D plot that shows the boundaries (contour) between species.

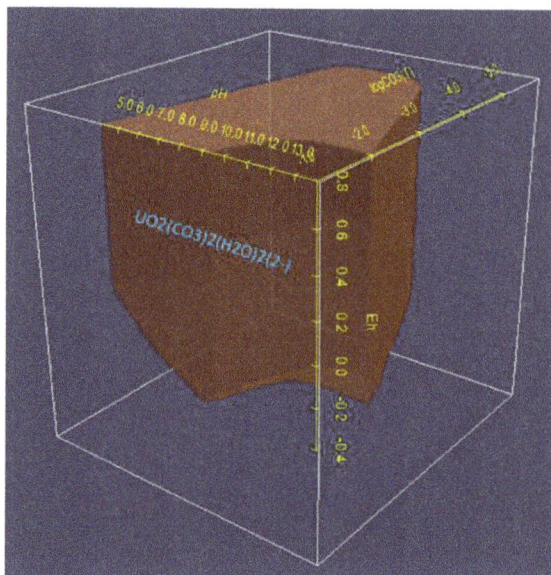

Figure 32. Semi-transparent plot that shows the stability region of $UO_2(CO_3)_2(H_2O)_2^{2-}$ species.

A three-dimensional plot has the great advantage of being able to show the effects of multiple variables at once. Tools, such as rotate, slice and clip, can easily identify areas of particular interest. For a complicated system, however, a three-dimensional image may not be as clear as a two-dimensional drawing due to the memory and screen resolution imposed by an ordinary PC. It is therefore wise to choose or combine the use of 2D and 3D to have the best presentation.

A simple way to turn a set of line-drawn diagrams into a three-dimensional object is to arrange six two-dimensional surface plots (two from each of Eh-pH, Eh-ligand and pH-ligand) into a cube. It may be necessary to reverse the x- or y-axis direction for this purpose. Simple plastic cubes, intended for use with photographs, are readily available. See Figure 33, where the section on the left (three plots combined) is the top-front view and the section on the right is the bottom-rear view. The two sections are jointed along the Eh axis, at pH = 14 and log ΣCO_3 = −1 M. Rotating the right sections to the left will make a complete cube. To show the continuity between plots, species UO_2 was purposely colored in light blue.

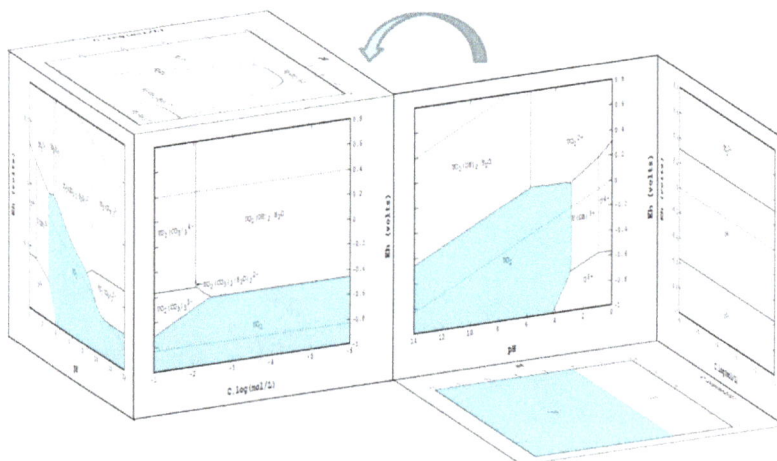

Figure 33. Combination of six surface plots of the U Eh-pH-logΣCO_2 system to form a cube.

6.3. 3D Eh-pH, Uranium with Pressure CO₂(g): Using the Equilibrium Line Method

This example used Figure 7.32a from Garrels and Christ for U with the CO_2 system. Since the extra axis is the pressure of $CO_2(g)$, the equilibrium line method was used. The Eh-pH program took less than 30 s to create necessary data for making the (name).vtk file for ParaView. Without having to repeat the same features presented earlier, only the semi-transparent surfaces and the boundaries between species are shown on Figures 34 and 35 respectively. These two plots agree with the original diagram presented by Garrels and Christ.

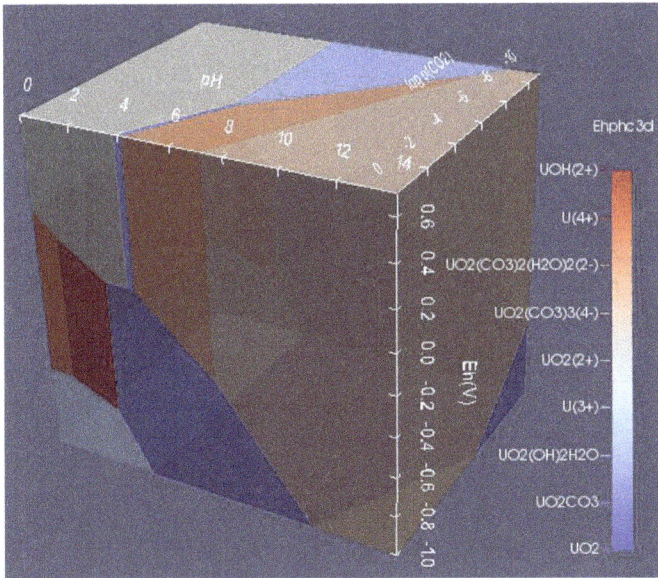

Figure 34. Eh-pH-logpCO₂(g) of uranium where the species are shown by colors. This diagram matches Figure 7.32a of Garrels and Christ.

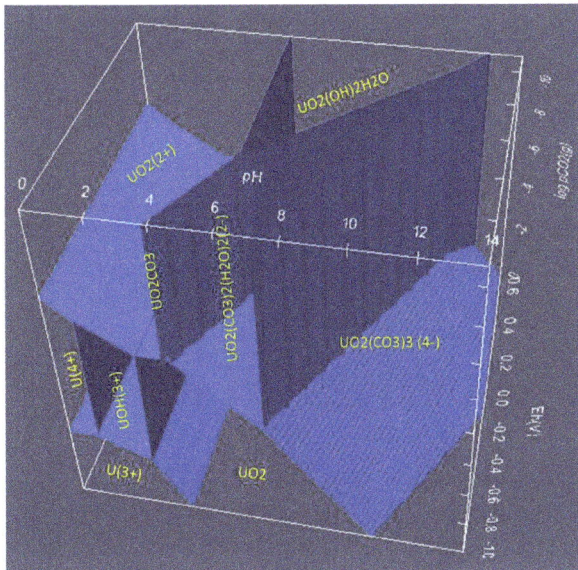

Figure 35. 3D plot that shows the boundaries (contour) between species. This diagram used data from the line method where logpCO₂(g) was given.

6.4. How to Show Two or More Solids in One Area

ParaView is capable of showing areas occupied by more than one solid phase. As with single solid regions, multi-phase areas are also colored and shown on the color map. Using the same conditions as for Figures 11 and 12, Figures 36 and 37 are plotted by ParaView indicating four two-solid phases. The areas occupied by CuS plus Cu_2S are indicated in red and pointed to by an arrow.

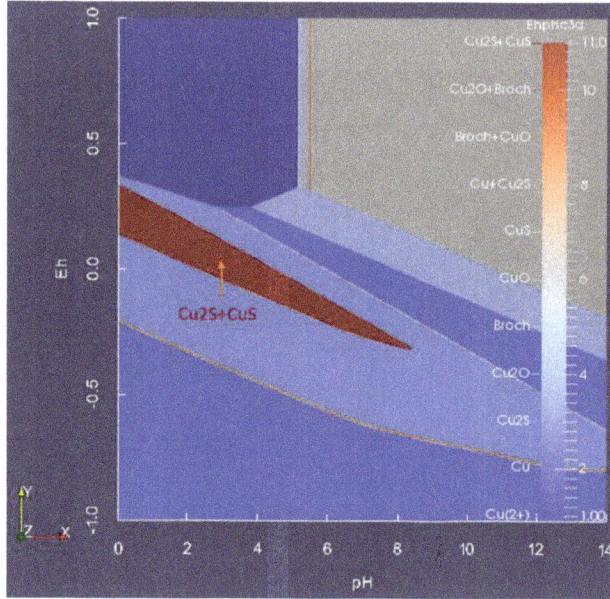

Figure 36. ParaView of Cu–S–water, where $\Sigma Cu = 0.001$ and $\Sigma S = 0.0009$ M. See the 2D plot of Figure 11 for a comparison.

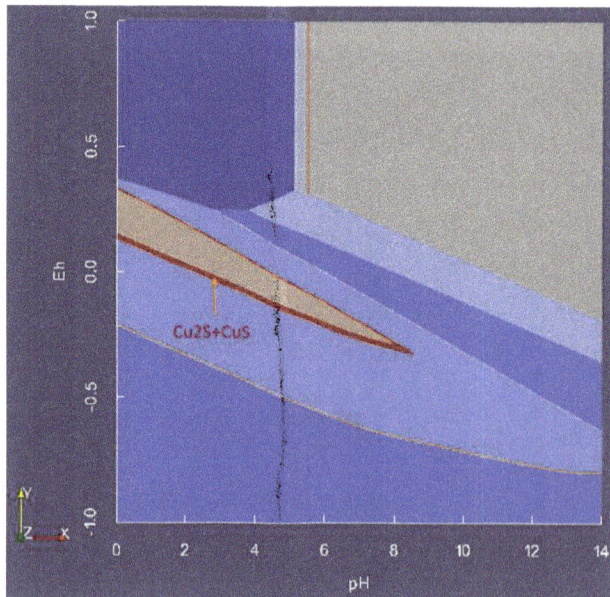

Figure 37. ParaView of Cu–S–water, where $\Sigma Cu = 0.001$ and $\Sigma S = 0.0011$ M. See the 2D plot of Figure 12 for a comparison.

7. Summary

An Eh-pH diagram, commonly known as a Pourbaix diagram, is an effective way of presenting the effects from oxidation-reduction potential, acid and base, complexing ligands, temperature and pressure for an aqueous system. It can be used in many scientific fields, including hydro- and electro-metallurgy, geo and solution chemistry and corrosion science. Diagrams describing natural copper deposits, lead corrosion prevention and arsenic adsorption by ferrihydrite and leaching and metal recovery were illustrated. The fundamental principles behind the diagram were briefly described.

To handle ligand components for these diagrams, two separate methods developed over time, the line of equilibrium concentration and the point of complete mass balances, were described and illustrated. Both satisfy the Gibbs phase rule in their own way of calculation. The comparison and applications from these two methods are mentioned and illustrated.

Many advances of the Eh-pH diagrams are presented. These are:

Merging diagrams: Most of the Eh-pH diagrams describe only half of the reactions. Merging several same-sized Eh-pH diagrams together can better illustrate the overall system. Examples include cyanidation of Au plus cementation with Zn metal, cementation of Cu with metallic Fe and galvanic conversion of chalcopyrite with metallic Cu.

Creating 3D diagrams: Expanding the Eh-pH program to create data from an extra axis is relatively simple. However, a more professional third party 3D program is the best choice for drawing the final diagram. The author has tested ParaView, VisIt and MATLAB for 3D Eh-pH diagrams. Most of these programs can perform animations, such as rotation, clipping and slicing. The diagram can be semi-transparent or show the boundary for better illustration. The following examples are demonstrated:

1. An Eh-pH diagram that shows solubility. Examples include: lead corrosion prevention with CO_2 and concentrations of arsenic adsorption by ferrihydrite; MATLAB was used for more colorful pictures.

2. An Eh-pH with an independent variable; diagrams created by ParaView were illustrated for its functionality, and example are:

    The mass balance point method for the Uranium system where the extra axis is $\log(\Sigma CO_3)$;
    The equilibrium line method for the Uranium system where the third axis is $\log p(CO_2(g))$;
    The Cu–S system for showing the coexistence of two or more solids in one volume.

When a 3D picture becomes too complex to label all of the species involved, two-variable diagrams (Eh-pH, Eh-ligand, Eh-temperature or pH-ligand) can be presented side-by-side for clarity.

Acknowledgments: The author would like to thank Dave Tahija, Hecla Greens Creek Mining, for editing this manuscript and Chen-Luh Lin, Metallurgical Engineering University of Utah, for introducing the use of ParaView. Thanks to colleagues in the Department of Metallurgical and Materials Engineering at Montana Tech for their support, in particular Courtney A. Young, Chairman of the Department, for introducing the mass balance method and encouragement for continuing development of the STABCAL program, and to Steve McGrath for his valuable discussion on the 3D Uranium-carbonate diagrams from Garrels and Christ.

Conflicts of Interest: The author declares no conflict of interest.

References

1. Wagman, D.D.; Evans, W.H.; Parker, V.B.; Schumm, R.H.; Halow, I.; Bailey, S.M.; Churney, K.L.; Nuttall, R.L. The NBS tables of chemical thermodynamic properties. *J. Phys. Chem. Ref. Data* **1982**. [CrossRef]
2. Garrels, R.M.; Christ, C.L. Eh-pH. In *Solutions, Minerals, and Equilibria*; Freeman, Cooper & Company: New York, NY, USA, 1975; Chapter 7; pp. 172–277.
3. *STABCAL*, version 2015; Stability Calculation for Aqueous and Nonaqueous System; Montana Tech: Butte, MT, USA, 2015.
4. Pourbaix, M. *Atlas of Electrochemical Equilibria in Aqueous Solution*, 1st ed.; Pergamon Press: Bristol, UK, 1966.

5. Huang, H.H.; Cuentas, L. Construction of Eh-pH and Other Stability Diagrams of Uranium in a Multicomponent system with a Microcomputer—I. Domains of Predominance Diagram. *Can. Metall. Q.* **1989**, *28*, 225–234. [CrossRef]

6. Forssberg, E.; Antti, B.-M.; Palsson, B. Computer-assisted Calculations of thermodynamic equilibria in the Chalcopyrite-ethyl xanthate system. In *Reagents in the Minerals Industry*; Jones, M.J., Oblatt, R., Eds.; The Institution of Mining and Metallurgy: London, UK, 1984; pp. 251–264.

7. Eriksson, G.A. An algorithm for the computation of aqueous multicomponent, multiphase equilibria. *Anal. Chim. Acta* **1979**, *112*, 375–383. [CrossRef]

8. Woods, R.; Yoon, R.H.; Young, C.A. Eh-pH diagrams for stable and metastable phases in copper-sulfur-water system. *Int. J. Miner. Process.* **1987**, *20*, 109–120. [CrossRef]

9. Huang, H.H.; Twidwell, L.G.; Young, C.A. Speciation for aqueous system—An equilibrium calculation approach. In *Computational Analysis in Hydrometallurgy—35th Annual Hydrometallurgy Meeting*; Dixon, D.G., Dry, M.J., Eds.; CIM: Calgary, AB, Canada, 2005; pp. 295–310.

10. Dudas, L.; Maass, H.; Bhappu, R. Role of mineralogy in heap and *in situ* leaching of copper ores. In *Solution Mining Symposium*; AIME: New York, NY, USA, 1974; pp. 193–201.

11. *LLnL database*, version V8.R6.230; Lawrence Livermore National Laboratory: Livermore, CA, USA, 2010.

12. Johnson, J.; Oelkers, E.H.; Helgeson, H.C. SUPCRT92: A software package for calculating the Standard molal thermodynamic properties of mineral, gases, aqueous species, and reactions from 1 to 5000 bar and 0 to 1000 °C. *Comput. Geosci.* **1992**, *18*, 899–947. The database for the program has be updated as slop98.dat from Geopig group. Available online: https://www.asu.edu/sites/default/files/slop98.dat (accessed on 4 July 2006). [CrossRef]

13. Kontny, A.; Friedrich, A.; Behr, H.J.; de Wall, H.; Horn, E.E. Formation of ore minerals in metamorphic rocks of the German continental deep drilling site (KTB). *J. Geophys. Res.* **1997**, *102*, 18323–18336. [CrossRef]

14. Huang, H.H. The Application of Revised HKF Model for Thermodynamically Describing Elevated Pressure and Temperature processes such as Treatment of Gold Bearing Materials in Autoclaves. In *Hydrometallurgy 2008—Proceedings of the Sixth International Sixth International Symposium*; Young, C.A., Taylor, P.R., Anderson, C.G., Choi, Y., Eds.; Society for Mining, Metallurgy, and Exploration (SME): Littleton, CO, USA, 2008; pp. 1066–1077.

15. Pourbaix, M. *Lectures on Electrochemical Corrosion*; Plenum Press: New York, NY, USA, 1973; pp. 143–144.

16. Wang, Q.; Nishimura, T.; Umetsu, Y. Oxidative Precipitation for Arsenic Removal in Effluent Treatment. In *SME Minor Elements 2000 Processing and Environmental Aspects of As, Sb, Se, Te and Bi*; Young, C.A., Ed.; SME: Littleton, CO, USA, 2000; pp. 39–52.

17. Dzombak, D.A.; Morel, F.M.M. *Surface Complexation Modeling: Hydrous Ferric Oxide*; John Wiley & Sons: New York, NY, USA, 1990.

18. Huang, H.H.; Young, C.A. Mass-Balanced Calculations of Eh-pH diagrams using STABCAL. In *Mineral and Metal Processing IV*; Woods, R., Ed.; Electrochemical Society: Pennington, NJ, USA, 1996; pp. 227–233.

19. Gow, R.N.V. Spectroelectrochemistry and Modelling of Enargite (Cu_3AsS_4) Reactivity under Atmospheric Conditions. Ph.D. Thesis, The University of Montana, Butte, MT, USA, June 2015.

20. Duaime, T.E.; Tucci, N.J. Butte Mine Flooding Operable Unit: Water-Level Monitoring and Water-Quality Sampling 2011. Available online: http://www.pitwatch.org/download/mbmgannual/BMF-2011.pdf (accessed on 25 Marth 2014).

21. Srivastave, R. Estimation and Thermodynamic Modeling of Solid Iron Species in the Berkeley Pit water. M.Sc. Thesis, The University of Montana, Butte, MT, USA, June 2015.

22. Parkhust, D.L.; Appelo, C.A.A. PHREEQC Computer Program for Speciation, Batch-Reaction, One-Dimensional Transport, and Inverse Geochemical Calculations version 3.3.3. Available online: http://wwwbrr.cr.usgs.gov/projects/GWC_coupled/phreeqc/ (accessed on 13 March 2014).

23. Pritzker, M.D.; Yoon, R.H. Thermodynamic Calculations on Sulfide Flotation System: I. Galena-Ethyl Xanthate System in the Absence of Metastable Species. *Int. J. Miner. Process.* **1984**, *12*, 95–125. [CrossRef]

24. Hiskey, J.B.; Wadsworth, M.E. Galvanic Conversion of Chalcopyrite. *Metall. Trans. B* **1975**, *6*, 183–190. [CrossRef]

25. ParaView version 4.3.1 64-bit. Available online: http://www.paraview.org/download/ (accessed on 11 February 2015).

26. VisIt 2015 Version 2.4.2. Available online: https://wci.llnl.gov/simulation/computer-codes/visit/ (accessed on 23 April 2012).
27. *MATLAB*, Version R2013a; Mathworks Computer program: Natick, MA, USA, 2013.
28. Grossmann, K.; Arnold, T.; Lkeda-Ohno, A.; Steudtner, R.; Geipel, G.; Bernhard, G. Fluorescence properties of a uranyl(V)-carbonate species $[U(V)O_2(CO_3)_3]^{5-}$ at low temperature. *Spectrochim. Acta A* **2009**, *72*, 449–453. [CrossRef] [PubMed]
29. VTK Format. The VTK User's Guide, Version 4.2, Kitware. 2010. Available online: http://www.vtk.org/wp-content/uploads/2015/04/file-formats.pdf (accessed on 19 February 2015).

11

Study on Austenitization Kinetics of SA508 Gr.3 Steel Based on Isoconversional Method

Xiaomeng Luo, Lizhan Han and Jianfeng Gu *

Academic Editor: Hugo Lopez

Institute of Materials Modification and Modeling, School of Material Science and Engineering, Shanghai Jiaotong University, 800 Dongchuan Road, Shanghai 200240, China; luoxiaomeng1983@163.com (X.L.); victory_han@sjtu.edu.cn (L.H.)
* Correspondence: gujf@sjtu.edu.cn

Abstract: The austenitization kinetics of SA508 Gr.3 steel during heating was studied using the isoconversional method combined with continuous thermal dilatometric tests for the first time. The model-free austenitization kinetics was built and the effective activation energy as a function of transformed austenite fraction was determined without transformation models. Then, the corresponding regression validation was carried out. The time-temperature-austenitization (TTA) diagram of SA508 Gr.3 steel, which is very difficult to be obtained using experiment measures, was constructed. Finally, the austenitization kinetics in a more realistic case, *i.e.*, under non-constant heating rates, was predicted, which is found to agree well with the experimental results.

Keywords: isoconversional method; austenitization kinetics; SA508 Gr.3 steel; dilatometry; time-temperature-austenitization (TTA) diagram

1. Introduction

Austenitization is an important solid-state phase transformation involved in almost all of the hot-working processes of steels, such as heat treatment, welding, surface cladding, forging, *etc.* In recent years, several quantitative models have been proposed to describe the austenitization kinetics during heating, which were all based on the Johnson-Mehl-Avrami-Kolmogorov (JMAK) equation [1]:

$$\alpha = 1 - \exp[-(k(T)t)^n] \tag{1}$$

where α represents the transformed austenite fraction, $k(T)$ and n are the parameters depending on temperature T. With the aid of the nucleation and growth functions reported by Roosz *et al.* [2], different models describing the kinetics of austenite formation from different initial microstructures (pearlite, ferrite, bainite, martensite, or their mixtures) during heating were developed. Garcíia de Andrés *et al.* [3] derived a mathematical model of the pearlite-austenite transformation during continuous heating in a eutectoid steel with a fully pearlitic initial microstructure. Caballero *et al.* [4] presented a model to describe the ferrite-austenite transformation in Armco iron and three low-carbon/low-manganese steels starting with a fully ferritic initial microstructure. This model can be used to calculate the volume fractions of austenite and ferrite during transformation as a function of temperature. Caballero *et al.* [5] used theoretical knowledge regarding the isothermal formation of austenite from pure and mixed initial microstructures to develop a model for the non-isothermal austenite formation in a wide range of steels with an initial microstructure consisting of ferrite and/or pearlite. In addition to the initial microstructures, more details of some other factors were also taken into account in the study of austenitization kinetics. Gaude-Fugarolas and Bhadeshia [6] suggested a model that considered the effect of various factors, such as the steel composition, grain size, pearlite interlamellar spacing, and heating rate. Surm et al. [7] investigated the effect of heating rate on austenitization kinetics and modeled the austenitization with non-constant heating rate in hypereutectoid steels.

The dilatometric testing has long been employed in transformation kinetics investigation, which collects the information of diameter or length change of a sample during isothermal or continuous thermal process. In isothermal dilatometric testing, the transformation is usually expected to be fulfilled completely so that the kinetics information can be collected, but the limitation is also obvious as it is very difficult to keep the whole transformation process in an isothemal condition. For example, the austenitization process of some steels can be finished in very short period, i.e., tens of second, or even several second, making transformation starts during the heating-up stage and only part of the transformation happens under isothermal condition [8,9]. From this point of view, the continuous thermal dilatometry technique is an alternative method. In addition, it is worth mentioning that some new methods have been used for austenite formation study recent years. For example, Savran et al. [10] studied the nucleation and growth of austenite from ferrite/pearlite structures using a three-dimensional high-energy X-ray diffraction (3-D XRD) microscope. Additionally, Esin et al. [11] investigated the austenitization during both slow and fast heating for different microstructures of a selected low-alloy steel by simultaneously using in situ high-energy synchrotron X-ray diffraction and dilatometry.

The isoconversional method, namely "model-free" method, which is derived from the isoconversional principle [12], can evaluate the dependence of the effective activation energy on the degree of conversion without any assumption on the conversion mechanism. Since not making use of any approximations about transformation models, this method would potentially be more accurate and might avoid the risk due to wrong model assumptions. An important advantage of the isoconversional method is that it is possible to correctly and easily describe the kinetics of complex conversions in model-free way, i.e., model-free kinetics.

A large number of isoconversional methods has been constructed, such as Flynn-Wall-Ozawa (FWO) Method [13,14], Kissinger-Akahira-Sunose (KAS) Method [15,16], Vyazovkin's Advanced Isoconversional (VA) Method [17], etc. The most commonly used isoconversional method is the differential isoconversional method of Friedman, namely Friedman method [18,19].

This method has widely been used in the field of thermal analysis, but in very few studies of the solid-state phase transformation kinetics of metals, for example, aluminum alloys [20], hypo-eutectoid Cu-Al alloys [21], and Cu-Al-Mn alloys [22]. In this work, we tried to use, for the first time, the isoconversional method to investigate the austenitization kinetics of a hypo-eutectoid steel, based on the data collected from the continuous thermal dilatometric experiments.

2. Theoretical Background and Experimental Procedures

2.1. General Equation for Kinetics

Measurement and parameterization of the phase transformation rate is one of the focused subjects in modern kinetic theory. For solid-state phase transformations, e.g., austenitization, the transformation rate is expressed by the equation below [23,24]:

$$\frac{d\alpha}{dt} = k(T)f(\alpha) = A\exp\left(\frac{-E}{RT}\right)f(\alpha) \tag{2}$$

where $k(T)$ is the rate constant which represents the temperature dependence of the transformation rate and can be described by Arrhenius Equation $k(T) = A\exp(-E/RT)$, $f(\alpha)$ indicates the transformation model, A is the pre-exponential factor, E denotes the effective activation energy, and R is the universal gas constant (8.314 J/mol·K).

For the linear non-isothermal condition, Equation (2) is frequently used in the form of [10]:

$$\beta \frac{d\alpha}{dt} = A\exp\left(\frac{-E}{RT}\right) f(\alpha) \tag{3}$$

where β is the heating rate.

2.2. JMAK Equation

Equation (2) is a general equation of kinetic theory and can be used to describe different conversion processes with different explicit expressions of $f(\alpha)$ [25]. The JMAK equation can be translated into Equation (2). Taking logarithmic derivative of Equation (1) leads to:

$$\frac{d\alpha}{dt} = nk(T)^n t^{n-1}(1-\alpha) \tag{4}$$

$$t^{n-1} = k(T)^{(1-n)}\ln\left(\frac{1}{1-\alpha}\right)^{\frac{n-1}{n}} \tag{5}$$

Substituting Equation (5) into Equation (4), and making simple rearrangements, yields the JMAK rate equation (Equation (6)), namely the n-dimensional Avrami equation [26]:

$$\frac{d\alpha}{dt} = k(T) \cdot n \cdot (1-\alpha)[-\ln(1-\alpha)]^{\left(1-\frac{1}{n}\right)} \tag{6}$$

Compared with Equation (2), the transformation model $f(\alpha)$ for solid-state phase transformation, based on the JMAK equation, can be written as [27]:

$$f(\alpha) = n(1-\alpha)[-\ln(1-\alpha)]^{\left(1-\frac{1}{n}\right)} \tag{7}$$

2.3. Isoconversional Method—Model-Free Method

Generally speaking, for some specified transformation processes, the explicit expression of $f(\alpha)$ based on some physical model is needed to be given. However, based on the isoconversional principle, the "model free" isoconversional methods can be constructed, in which the explicit form of $f(\alpha)$ or, in other words, the detail of the chemical-physical change in the transformation process, needs not to be known.

The isoconversional principle [12] states that the transformation rate at constant extent of conversion α (*i.e.*, the isoconversional rate) is only a function of temperature, which can be illustrated by taking logarithmic derivative of the transformation rate (Equation (2)) at α = const:

$$\left[\frac{d\ln\left(\frac{d\alpha}{dt}\right)}{d\frac{1}{T}}\right]_\alpha = \left[\frac{d\ln(k(T))}{d\frac{1}{T}}\right]_\alpha + \left[\frac{d\ln(f(\alpha))}{d\frac{1}{T}}\right]_\alpha \tag{8}$$

where the subscript α indicates the isoconversional values, *i.e.*, the values related to a given extent of conversion. Because α = const, $f(\alpha)$ is also constant, and the last term in the right hand side of Equation (8) is zero, thus,

$$\left[\frac{d\ln\left(\frac{d\alpha}{dt}\right)}{d\frac{1}{T}}\right]_\alpha = -\frac{E_\alpha(\alpha)}{R} \tag{9}$$

It follows from Equation (9) that the temperature dependence of the isoconversional rate can be used to evaluate isoconversional values of the effective activation energy, $E_\alpha(\alpha)$, without assuming or determining any particular form of the function $f(\alpha)$. For this reason, the isoconversional method is commonly called the "model-free" method. Nevertheless, that does not mean one can take this term literally. Vyazovkin *et al.* [12] indicated that, although these methods do not need to identify the transformation model, they do assume that the conversion dependence of the transformation rate obeys some $f(\alpha)$ model.

Friedman proposed to apply logarithm of the transformation rate $d\alpha/dt$ as a function of the reciprocal temperature $1/T$ at any extent of conversion α. As mentioned above, the transformation rate can be described in the form of Equation (2). Replacing $Af(\alpha)$ with the modified pre-exponential factor $A'(\alpha)$ in Equation (2) leads to:

$$\frac{d\alpha}{dt} = A'(\alpha)\exp\left(\frac{-E_\alpha(\alpha)}{RT(t)}\right) \tag{10}$$

Taking logarithmic of Equation (10) yields:

$$\ln\left(\frac{d\alpha}{dt}\right)_\alpha = \ln\left(A'(\alpha)\right) - \frac{E_\alpha(\alpha)}{RT(t)} \tag{11}$$

where the subscript α is a given extent of conversion, $E_\alpha(\alpha)$ is the effective activation energy, and $T(t)$ is temperature. It follows from Equation (11) that the dependence of the $\ln(d\alpha/dt)$ on the reciprocal temperature $1/T$ shows a straight line with the slope $m = -(E/R)$, and the intercept equals to $\ln(A'(\alpha)) = \ln(A(\alpha)f(\alpha))$ at any fixed α.

2.4. Material and Dilatometric Testing

The hypo-eutectoid steel, SA508 Gr.3, was investigated with the chemical composition listed Table 1. Figure 1 shows its initial microstructure consisted of pearlite (about 35.8 vol. %) and ferrite determined using quantitative metallography.

Table 1. Chemical composition of SA508 Gr.3 steel.

Element	C	Mn	Si	Ni	Cr	Mo	V	Al	N	Fe
Composition (wt. %)	0.20	1.47	0.17	0.89	0.13	0.51	0.001	0.039	0.014	Bal.

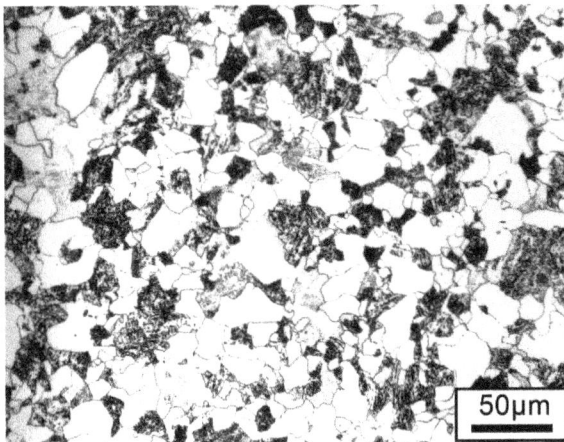

Figure 1. Initial microstructure of SA508 Gr.3 steel.

Both the dilatometric tests and validation experiments were carried out on a Gleeble-3500 thermal simulation testing machine, using cylindrical samples with diameter of 6 mm and length of 82 mm. In the dilatometric tests, the variation of diameter is measured. K-Type thermocouples (0.25 mm diameter) are used for the temperature measurements, which are spot-welded on the surface of the middle section of the sample (Figure 2). The sample is heated by resistance heating using low frequency current in the Gleeble system. Due to small skin effect, there is an isothermal section vertical to the length, *i.e.*, the temperature distribution in the central cross section of the sample is homogeneous. The thermocouple measures and controls the temperature of this section, instead of whole specimen, by a closed-loop control system. The continuous thermal dilatometric curves were obtained from 293 K to 1173 K, with 15 different heating rates (β) in a range of 0.008~500 K/s. When the related results are presented in the following sections, only limited representative data are marked in detail, instead of all of them being included, in order to avoid crowding in the figures, *etc.*

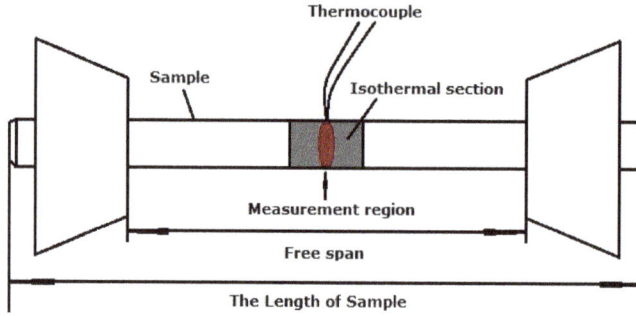

Figure 2. Sketch of dilatation measurement using Gleeble-3500 thermal simulation testing machine.

In order to demonstrate that the isoconversional method can be used to accurately predict the austenitization during the practical heating processes, a heating process with non-constant heating rates was investigated in this study, and the specific heating process is shown in Figure 3.

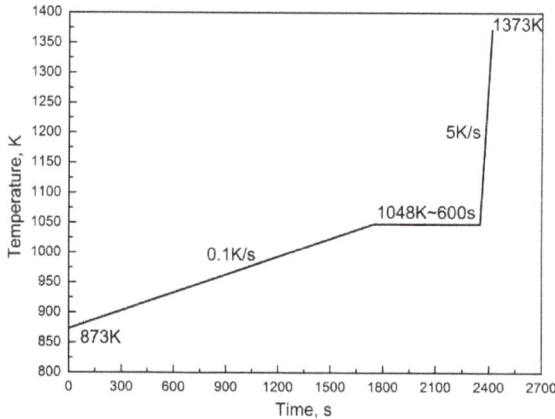

Figure 3. A specific continuous heating process with non-constant heating rates designed for SA508 Gr.3 steel.

3. Results and Discussion

3.1. Data Processing

Figure 4a shows the recorded dilatometric curves with different heating rates (β). The volume fraction (α) of austenite formed during continuous heating was calculated by applying the lever rule, as shown in Figure 4b; Figure 4c is the differential from of the $\alpha–T$ dependence. It is found that most of the $d\alpha/dt$ vs. T curves for different β exhibit two peaks, indicating two maximum phase transformation rates corresponding to the two phase transformation stages, the dissolution of pearlite and ferrite transforming to austenite, respectively, occurred during austenitization. The temperatures of the maximum phase transformation rate increase with β, suggesting that a quicker heating rate results in a delay of the austenitization start temperature.

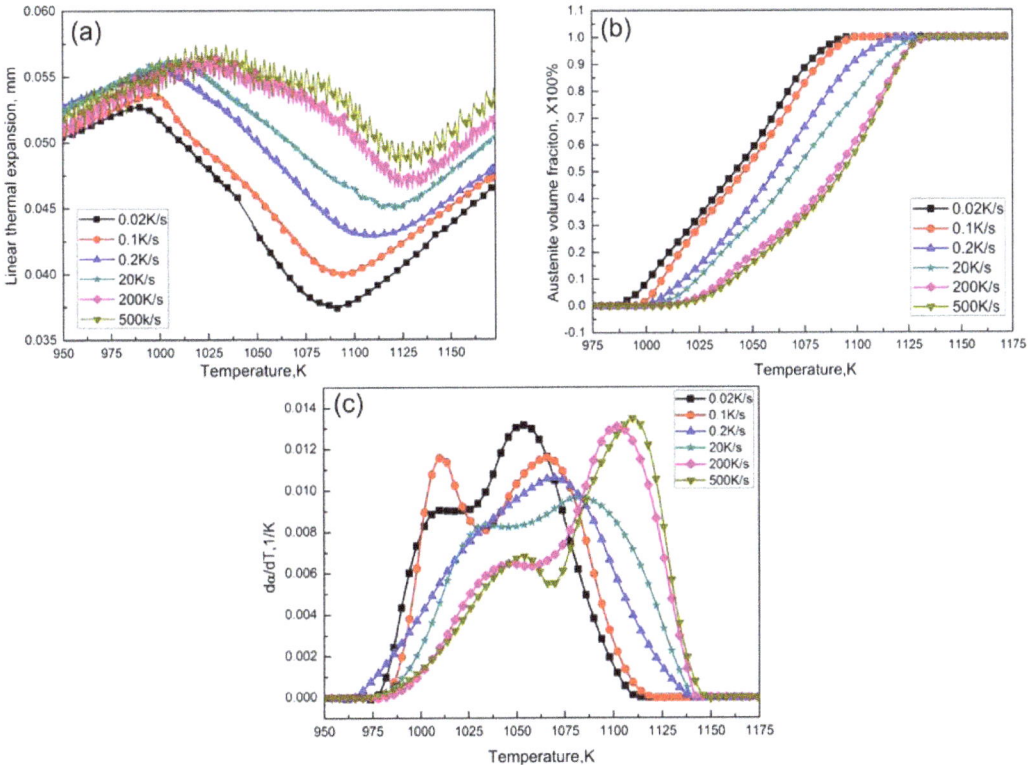

Figure 4. (a) The measured dilatometric curves; (b) the calculated transformation fraction α; (c) and the $d\alpha/dt$ vs. T curves of SA508 Gr.3 steel at different heating rates, as marked.

Since the β is constant in individual, continuous heating experiments, the transformation rate can be expressed as:

$$\frac{d\alpha}{dt} = \frac{d\alpha}{dT} \times \beta \tag{12}$$

Taking logarithmic of Equation (12) and substituting it into Equation (11), the relationship between $\ln(d\alpha/dt)$ and $1/RT(t)$ is established. The method of determination of $E_\alpha(\alpha)$ and $\ln(A'(\alpha))$ is described as follows: Considering $\alpha = 0.5$ as an example, the temperatures T reaching $\alpha = 0.5$ for individually specified β are found in Figure 4b, and the corresponding $\ln(d\alpha/dt)$ and $1/RT(t)$ can be read in Figure 4c. Taking into account all 15 sets of continuous heating experiments, the values of $\ln(d\alpha/dt)$ and $1/RT(t)$ for a fixed α are collected. Based on these data, an Arrhenius plot describing the dependence of $\ln(d\alpha/dt)$ on $1/RT(t)$ can be established using linear fitting (Figure 5). Hence, according to the isoconversional method (Equation (11)), the values of $E_\alpha(\alpha)$ and $\ln(A'(\alpha))$ for $\alpha = 0.5$ can be determined from the slope and intercept of the fitted line, respectively. In the present study, 25 various α dispersed in a wide range of 0 to 1.0 were considered, and, for a simplified presentation, the Arrhenius plots at selected heating rates $\alpha = 0.1, 0.35, 0.5,$ and 0.7 are shown in Figure 5.

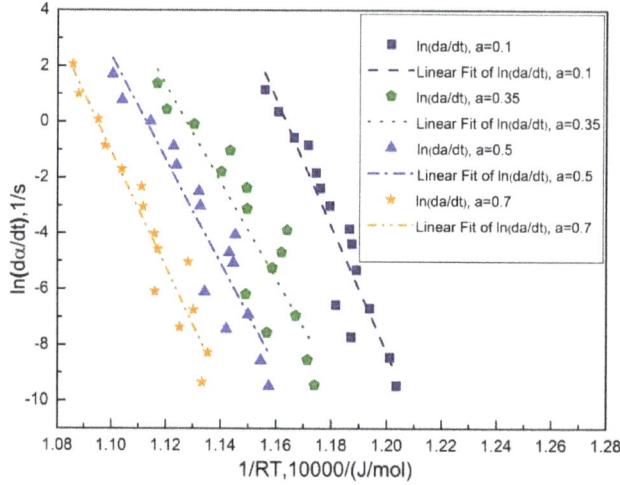

Figure 5. The Arrhenius plots at different heating rates when α = 0.1, 0.35, 0.5, and 0.7.

3.2. Determination of Model-Free Austenitization Kinetics

The effective activation energy $E_\alpha(\alpha)$ and modified pre-exponential $\ln(A'(\alpha))$ are the two input parameters necessary for predicting phase transformation kinetics. Based on the isoconversional method, and that of data processing, the dependence of $E_\alpha(\alpha)$ and $\ln(A'(\alpha))$ on α were determined without explicitly assuming a particular form of the transformation model $f(\alpha)$. That is to say, the model-free austenitization kinetics can be determined using this method.

As illustrated in Figure 6, $E_\alpha(\alpha)$ is not a constant value, but changes with the extent of the austenitization process. The curve of $E_\alpha(\alpha)$ *vs.* α is characterized with a roughly "V" shape. $E_\alpha(\alpha)$ declines with the increasing extent of transformation at the early stage of austenitization ($0 \leqslant \alpha \leqslant 0.35$), and then increases until the completion of austenitization. The minimum $E_\alpha(\alpha)$ is at $\alpha \approx 0.35$, which is in accord with the initial microstructure composition of the SA508 Gr.3 steel, namely, 35.8 vol. % pearlite and 64.2 vol. % ferrite. The reason for such a V-shaped change trend of $E_\alpha(\alpha)$ is due to the two distinct stages, dissolution of pearlite and transformation of ferrite to austenite, occurring successively in the process of austenitization. As shown in Figure 6, the calculated $E_\alpha(\alpha)$ varies in a range of 1760~2670 kJ/mol during the progress of the austenitization process, similar to, but rather higher, than those reported in the literature (e.g., 2367 kJ/mol [28], 840 kJ/mol and 470 kJ/mol [29]). This can be explained by the differences between the two different forms of the JMAK equation.

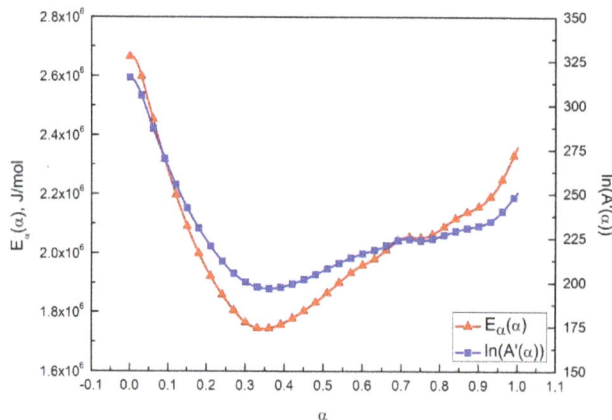

Figure 6. The model-free austenitization kinetics determined by the isoconversional method.

As previously mentioned, the kinetics of austenitization in general is mostly described by the JMAK equation. However, there are two commonly used different forms, and the form (I) (Equation (1)) is used in our study. Another form (form (II)) is the original one proposed by Avrami, and is more popular, which can be expressed as follows:

$$\alpha = 1 - \exp[-k(T)t^n] \tag{13}$$

According to Kohout [1], if the JMAK equation is used in combination with the Arrhenius Equation $k(T) = A\exp(-E/RT)$, the types of equations are not dependent on the choice between JMAK equation forms (I) and (II). The difference consists only in the values of parameters A and E. The application of forms (I) and (II) leads to different values of activation energies: The value $E_{(II)}$ corresponding to form (II) is equal to the product $nE_{(I)}$, where the value $E_{(I)}$ corresponds to form (I). Therefore, the value of effective activation energy of austenitization for SA508 Gr.3 steel, obtained by form (I) in our study, is not comparable to that of the real one using form (II). The real one should be calculated based on form (II). In this case, the real effective activation energy of austenitization for SA508 Gr.3 steel can be obtained from the activation energy using form (I). When "high effective activation energy" (1760~2670 kJ/mol) multiplies n, where the value of n for the whole process of austenitization extracted from our experimental data is about 0.13~0.18 (Figure 7), we can get the real effective activation energy within the range of 229 to 480 kJ/mol. These data are more close to the "commonly referred" effective activation energy value of transformation in steels, and should be reasonably acceptable. The reason for using form (I) in our paper is due to the convenience of data processing in order to acquire the effective activation energy. The effective activation energy, acting as an important parameter in austenization simulation based on the isoconversional method, is extracted from the experiment data.

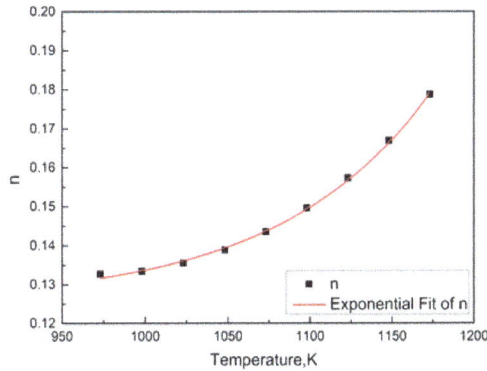

Figure 7. The parameters n in the JMAK equation for the whole austenitization process of SA508 Gr.3 steel.

The modified pre-exponential $\ln(A'(\alpha))$ shows a changing trend, similar to that of the effective activation energy, as seen in Figure 6. When the transformation faction is below ~0.35, the $\ln(A'(\alpha))$ decreases with an increasing extent of transformation, and then increases gradually until the completion of austenization completion.

3.3. Regression Validation

The transformation progress can be calculated using the integral form of the isoconversional method, the following equation (Equation (14)), which is:

$$\int_0^\alpha \frac{d\alpha}{f(\alpha)} = \frac{A(\alpha)}{\beta} \int_{T_{\alpha 0}}^T \exp\left[-\frac{E_\alpha(\alpha)}{RT}\right] dT \tag{14}$$

For the infinitesimal range of transformation progress $\Delta\alpha$, Equation (14) can be rewritten as:

$$\int_{\alpha \to \Delta\alpha}^{\alpha} \frac{d\alpha}{f(\alpha)} = \frac{A(\alpha)}{\beta} \int_{T_{\alpha \to \Delta\alpha}}^{T} \exp\left[-\frac{E_\alpha(\alpha)}{RT}\right] dT \qquad (15)$$

If $\Delta\alpha$ is very small, the effective activation energy $E_\alpha(\alpha)$ can be assumed to be constant. Based on the theorem of the average value, for $\Delta\alpha \to 0$, Equation (15) can be transformed into:

$$\alpha = \sum_{i=0}^{m} A'(\alpha_i)\exp\left[-\frac{E_\alpha(\alpha_i)}{RT_i}\right]\Delta t \qquad (16)$$

Using this equation, the transformation progress of austenitization in a specific thermal process can be calculated.

To validate regression the model-free austenitization kinetics ($E_\alpha(\alpha)$ and $A'(\alpha)$) derived from the experimental data, the dependence of transformation fraction α on temperature T was calculated for the thermal programs, identical to the continuous thermal dilatometric experiments. As shown in Figure 8, a good agreement is found to exist between them and, hence, indicates that the isoconversional method allows an accurate determination of the model-free austenitization kinetics.

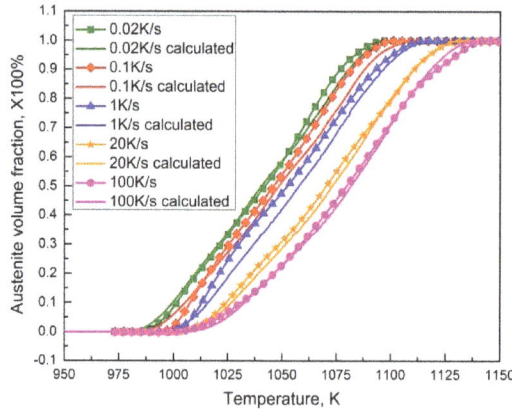

Figure 8. Comparison between experimental and calculated results of transformation fractions α during continuous heating with various constant heating rates.

3.4. Calculation of TTA Diagram

Generally, it is difficult to obtain the TTA diagram of steels by experiments. However, it can be conveniently calculated, using the isoconversional method. Taking integration of Equation (10), it yields [30]:

$$t_\alpha = \int_0^t dt = \int_{\alpha_0}^{\alpha} \frac{d\alpha}{A'(\alpha)\exp\left[-\dfrac{E_\alpha(\alpha)}{RT(t)}\right]} \qquad (17)$$

where t_α is the time to reach a given transformation fraction α. Based on Equation (17), the kinetic predictions for any temperature is possible.

For SA508 Gr.3 steel, when temperature is between the A_1 and A_3, two phases, ferrite and austenite, coexist, and, thus, the maximum transformation fraction of austenitization at a specified temperature has a limit. Considering the influence of alloying elements contained in the present SA508 Gr.3 steel on the Fe–C Phase Diagram, the thermodynamic software JMatPro® was used to recalculate the Ae$_3$ line, as seen in Figure 9. Then, based on the calculated Ae$_3$, using the lever rule, the maximum equilibrium transformation fraction γ_{max} of the steel was also calculated at a specific temperature, shown as the blue dotted line in Figure 9.

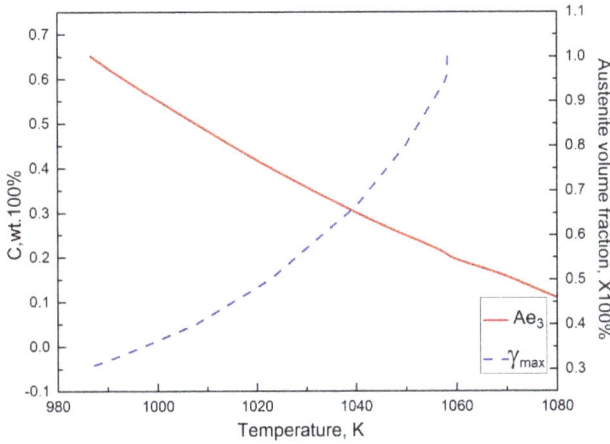

Figure 9. The Ae$_3$ and maximum austenite transformation fraction γ_{max} of SA508 Gr.3 steel as the function of temperature T.

Finally, taking the influence of the maximum transformation fraction into account, the TTA diagram of SA508 Gr.3 steel (Figure 10) was constructed using the isoconversional method (Equation (17)) and the model-free austenitization kinetics.

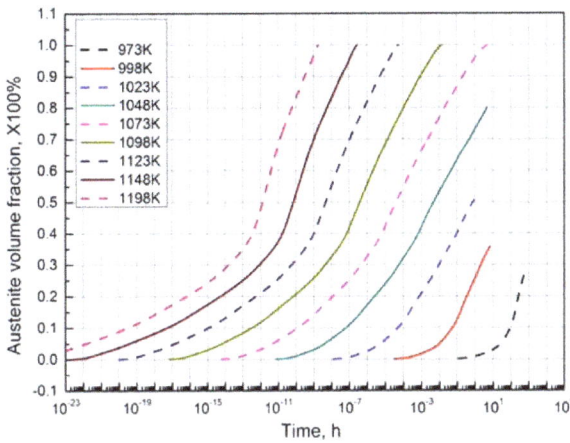

Figure 10. TTA diagram of SA508 Gr.3 steel.

3.5. Determination of the Parameters k(T) and n in JMAK Equation Based on TTA Diagram

The JMAK equation (Equation (1)) gives rise to the transformation fraction α as a function of time t during austenitization. Taking its logarithmic, the equation can be expressed as:

$$\ln\left(\frac{1}{1-\alpha}\right) = n[\ln(k(T)) + \ln t] \tag{18}$$

If using the JMAK equation, the parameters $k(T)$ and n should be given previously. Based on the TTA diagram predicted by the isoconversional method (Equation (17)), the parameters of $k(T)$ and n can be fitted according to Equation (18) and the results are shown in Figure 11, in which the subscripts P and F denote the stages of pearlite dissolution and ferrite to austenite transformation, respectively. It is seen that both the $\ln(k(T))$ for two stages increase with increasing temperature, but compared with $\ln(k(T))_F$, $\ln(k(T))_P$ increases faster at lower temperatures, but slower at higher temperatures. For parameter n, n_P decreases with increasing temperature, while n_F exhibits in an opposite way.

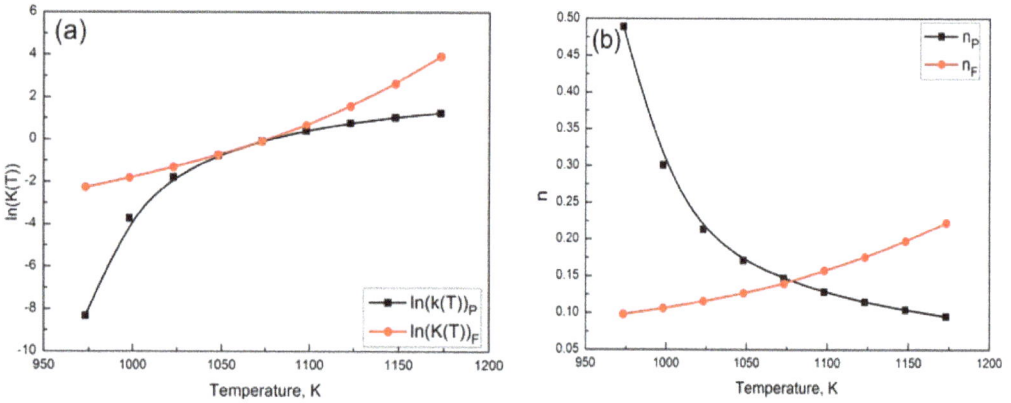

Figure 11. The parameters (a) $\ln(k(T))$ and (b) n in JMAK equation as a function of T.

3.6. Prediction of Austenitization Process during Heating with Non-Constant Heating Rates

In the heating processes considered above, only the ideal thermal conditions (isothermal or constant rate heating) were investigated. However, in real heating processes, such conditions are difficult to be practically performed. It must be very much anticipated to see whether the isoconversional method (Equations (14) and (17)) can also be used to predict the austenitization during the real heating processes accurately or not. Therefore, a specific continuous heating process with non-constant heating rates designed for SA508 Gr.3 steel has been investigated in this work, while the dilatometric curve obtained from the corresponding validation experiment is plotted in Figure 12. Using the isoconversional method, the dependencies of transformation fraction α on temperature T and time t during the heating process, corresponding to that in Figure 3, were predicted. The results are shown in Figure 13, together with those of the experimental measurements, respectively.

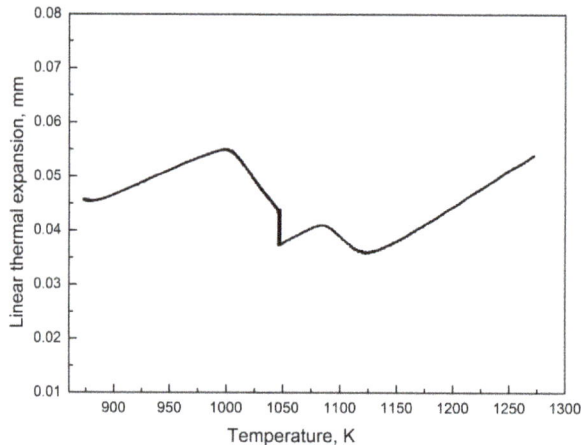

Figure 12. The dilatometric curve for SA508 Gr.3 steel during continuous heating with non-constant heating rates, as shown in Figure 3.

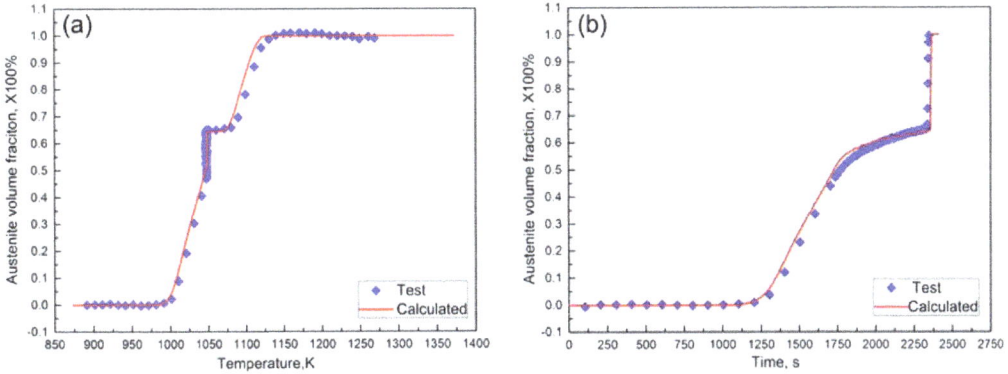

Figure 13. Comparison between experimental and predicted curves of transformation fraction α vs. (a) temperature T and (b) time t, respectively, of SA508 Gr.3 steel during continuous heating with non-constant heating rates, as shown in Figure 3.

The comparison clearly shows that the predicted and experimental results are in satisfactory agreement and, thus, the isoconversional method allows a reliable and feasible prediction for the austenitization kinetics, even under real heating processes.

4. Conclusions

In this paper, the isoconversional method, incorporating continuous thermal dilatometric tests, has been employed in the study of the austenitization kinetics for SA508 Gr.3 steel during heating. Based on the experimental investigation and numerical calculations, the following conclusions can be drawn:

(1) The model-free austenitization kinetics is built by the isoconversional method, and the effective activation energy $E_\alpha(\alpha)$ is also determined. The curve of $E_\alpha(\alpha)$ vs. α is characterized with a roughly "V" shape. The value of $E_\alpha(\alpha)$ is in the range of 1760 to 2670 kJ/mol, and reach a minimum at $\alpha \approx 0.35$, consistent with what expected from the initial dual phase (ferrite and pearlite) microstructure of the investigated steel.

(2) Further, the TTA diagram of SA508 Gr.3 steel has been obtained using the model-free austenitization kinetics and the isoconversional method, providing an effective way for the accurate determination of a TTA diagram, which is usually difficult to obtain from experiments.

(3) Under non-constant heating rate conditions, a satisfactory agreement between the curve of the phase transformation fraction using numerical prediction with the established model-free austenitization kinetics and that of the calculation from validation dilatometric curve, demonstrates that the isoconversional method can be used to characterize the austenitization kinetics during the heating process.

Acknowledgments: The authors wish to thank the financial support from the National Basic Research Program of China (973 Program, Grant No. 2011CB012904) and CNC machine tools and basic manufacturing equipment technology comments (Grant No. 2012ZX04012011).

Author Contributions: Xiaomeng Luo, Lizhan Han and Jianfeng Gu conceived and designed the experiments and the simulations; Xiaomeng Luo performed the experiments and simulations; Xiaomeng Luo and Lizhan Han analyzed the data; Xiaomeng Luo wrote the paper.

Conflicts of Interest: The authors declare no conflict of interest.

References

1. Kohout, J. An alternative to the JMAK equation for a better description of phase transformation kinetics. *J. Mater. Sci.* **2008**, *43*, 1334–1339. [CrossRef]

2. Roosz, A.; Garcia, Z.; Fuchs, E.G. Isothermal formation of austenite in eutectoid plain carbon steel. *Acta Mater.* **1983**, *31*, 509–517. [CrossRef]

3. De Andrés, C.G.; Caballero, F.G.; Capdevila, C.; Bhadeshia, H.K.D.H. Modelling of kinetics and dilatometric behavior of non-isothermal pearlite-to-austenite transformation in an eutectoid steel. *Scr. Mater.* **1998**, *39*, 791–796. [CrossRef]

4. Caballero, F.G.; Capdevila, C.; de Andrés, C.G. Kinetics and dilatometric behaviour of non-isothermal ferrite-austenite transformation. *Mater. Sci. Technol.* **2001**, *17*, 1114–1118. [CrossRef]

5. Caballero, F.G.; Capdevila, C.; de Andrés, C.G. Modelling of kinetics of austenite formation in steels with different initial microstructures. *ISIJ Int.* **2001**, *41*, 1093–1102. [CrossRef]

6. Gaude-Fugarolas, D.; Bhadeshia, H. A model for austenitisation of hypoeutectoid steels. *J. Mater. Sci.* **2003**, *38*, 1195–1201. [CrossRef]

7. Surm, H.; Kessler, O.; Hoffmann, F.; Zoch, H.W. Modelling of austenitising with non-constant heating rate in hypereutectoid steels. *Int. J. Microstruct. Mater. Prop.* **2008**, *3*, 35–48. [CrossRef]

8. Tszeng, T.C.; Shi, G. A global optimization technique to identify overall transformation kinetics using dilatometry data—Applications to austenitization of steels. *Mater. Sci. Eng. A* **2004**, *380*, 123–136. [CrossRef]

9. Tszeng, T.C.; Shi, G.; Purohit, S. Cementite and carbide dissolution in steels during austenitization at high heating rates. In Proceedings of the Modeling Control and Optimization in Ferrous and Nonferrous Industry Symposium, Chicago, IL, USA, 9–12 November 2003; pp. 379–379.

10. Savran, V.I.; Offerman, S.E.; Sietsma, J. Austenite Nucleation and Growth Observed on the Level of Individual Grains by Three-Dimensional X-Ray Diffraction Microscopy. *Metall. Mater. Trans. A* **2010**, *41*, 583–591. [CrossRef]

11. Esin, V.A.; Denand, B.; Bihan, L.Q.; Dehmas, M.; Teixeira, J.; Geandier, G.; Denis, S.; Sourmail, T.; Aeby-Gautier, E. *In situ* synchrotron X-ray diffraction and dilatometric sthdy of austenite formation in a multi-component steel: Influence of initial microstructure and heating rate. *Acta Mater.* **2014**, *80*, 118–131. [CrossRef]

12. Vyazovkin, S.; Burnham, A.K.; Criado, J.M.; Perez-Maqueda, L.A.; Popescu, C.; Sbirrazzuoli, N. ICTAC Kinetics Committee recommendations for performing kinetic computations on thermal analysis data. *Thermochim. Acta* **2011**, *520*, 1–19. [CrossRef]

13. Flynn, J.H.; Wall, L.A. General Treatment of the Thermogravimetry of Polymers. *J. Res. Natl. Bur. Stand.* **1966**, *70*, 487–523. [CrossRef]

14. Ozawa, T. A new method of quantitative differential thermal analysis. *Bull. Chem. Soc. Jpn.* **1966**, *39*, 2071–2085. [CrossRef]

15. Kissinger, H.E. Reaction Kinetics in Differential Thermal Analysis. *Anal. Chem.* **1957**, *29*, 1702–1706. [CrossRef]

16. Sunose, T.; Akahira, T. Method of determining activation deterioration constant of electrical insulating materials. *Res. Rep. Chiba Inst. Technol.* **1971**, *16*, 22–31.

17. Vyazovkin, S.; Dollimore, D. Linear and Nonlinear Procedures in Isoconversional Computations of the Activation Energy of Nonisothermal Reactions in Solids. *J. Chem. Inf. Comput. Sci.* **1996**, *36*, 42–45. [CrossRef]

18. Friedman, H.L. New methods for evaluating kinetic parameters from thermal analysis data. *J. Polym. Sci. B Polym. Lett.* **1969**, *7*, 41–46. [CrossRef]

19. Friedman, H.L. Kinetics of thermal degradation of char-forming plastics from thermogravimetry. Application to a phenolic plastic. *J. Polym. Sci. C Polym. Symp.* **1964**, *6*, 183–195. [CrossRef]

20. Luiggi, N.; Betancourt, M. On the non-isothermal precipitation of the β' and β phases in Al-12.6 mass% Mg alloys using dilatometric techniques. *J. Therm. Anal. Calorim.* **2003**, *74*, 883–894. [CrossRef]

21. Adorno, A.T.; Carvalho, T.M.; Magdalena, A.G.; dos Santos, C.M.A.; Silva, R.A.G. Activation energy for the reverse eutectoid reaction in hypo-eutectoid Cu–Al alloys. *Thermochim. Acta* **2012**, *531*, 35–41. [CrossRef]

22. Silva, R.A.G.; Gama, S.; Paganotti, A.; Adorno, A.T.; Carvalho, T.M.; Santos, C.M.A. Effect of Ag addition on phase transitions of the Cu-22.26 at. %Al-9.93 at. %Mn alloy. *Thermochim. Acta* **2013**, *554*, 71–75. [CrossRef]

23. Vyazovkin, S. Computational aspects of kinetic analysis. Part C. The ICTAC Kinetics Project—The light at the end of the tunnel? *Thermochim. Acta* **2000**, *355*, 155–163. [CrossRef]

24. Brown, M.E.; Dollimore, D.; Galwey, A.K. *Reactions in the Solid State, Comprehensive Chemical Kinetics*; Elsevier: Amsterdam, The Netherlands, 1980.

25. Brown, M.E. *Introduction to Thermal Analysis: Techniques and Applications*; Springer Science & Business Media: Lodon, UK, 2001; p. 264.
26. Brown, M.E.; Maciejewski, M.; Vyazovkin, S.; Nomen, R.; Sempere, J.; Burnham, A.; Opfermann, J.; Strey, R.; Anderson, H.L.; Kemmler, A.; *et al.* Computational aspects of kinetic analysis Part A: The ICTAC kinetics project-data, methods and results. *Thermochim. Acta* **2000**, *355*, 125–143. [CrossRef]
27. Marcilla, A.; Garcia-Quesada, J.C.; Ruiz-Femenia, R. Additional considerations to the paper entitled: "Computational aspects of kinetic analysis. Part B: The ICTAC Kinetics Project—The decomposition kinetics of calcium carbonate revisited, or some tips on survival in the kinetic minefield". *Thermochim. Acta* **2006**, *445*, 92–96. [CrossRef]
28. Chen, R.; Gu, J.; Han, L.; Pan, J. Austenization kinetics of 30Cr2Ni4MoV steel. *Trans. Mater. Heat Treat.* **2013**, *34*, 170–174.
29. Chiba, A. Isothermal transformation kinetics from ferrite to austenite in an Fe-8% Cr alloy. *Trans. Jpn. Inst. Met.* **1984**, *25*, 523–530. [CrossRef]
30. Burnham, A.K.; Dinh, L.N. A comparison of isoconversional and model-fitting approaches to kinetic parameter estimation and application predictions. *J. Therm. Anal. Calorim.* **2007**, *89*, 479–490. [CrossRef]

12

Microstructures and Mechanical Properties of Austempering SUS440 Steel Thin Plates

Cheng-Yi Chen, Fei-Yi Hung *, Truan-Sheng Lui and Li-Hui Chen

Department of Materials Science and Engineering, National Cheng Kung University, Tainan 701, Taiwan; n5897115@mail.ncku.edu.tw (C.-Y.C.); luits@mail.ncku.edu.tw (T.-S.L.); chenlh@mail.ncku.edu.tw (L.-H.C.)
* Correspondence: fyhung@mail.ncku.edu.tw

Academic Editor: Hugo F. Lopez

Abstract: SUS440 is a high-carbon stainless steel, and its martensite matrix has high heat resistance, high corrosion resistance, and high pressure resistance. It has been widely used in mechanical parts and critical materials. However, the SUS440 martempered matrix has reliability problems in thin plate applications and thus research uses different austempering heat treatments (tempering temperature: 200 °C–400 °C) to obtain a matrix containing bainite, retained austenite, martensite, and the M7C3 phase to investigate the relationships between the resulting microstructure and tensile mechanical properties. Experimental data showed that the austempering conditions of the specimen affected the volume fraction of phases and distribution of carbides. After austenitizing heat treatment (1080 °C for 30 min), the austempering of the SUS440 thin plates was carried out at a salt-bath temperature 300 °C for 120 min and water quenching was then used to obtain the bainite matrix with fine carbides, with the resulting material having a higher tensile fracture strength and average hardness (HRA 76) makes it suitable for use as a high-strength thin plate for industrial applications.

Keywords: stainless steel; austempering; bainite; mechanical properties

1. Introduction

SUS440 is a high-chrome and carbon martensite stainless steel with high strength and corrosion resistance properties and thus been widely used in machine parts, screws, and knives [1,2]. In general, martensite stainless steel almost always undergoes austenitizing heat treatment, followed by quenching and tempering. However, if this process is carried out improperly, then this will adversely affect the mechanical properties of the resulting materials [3,4]. This is because the quenching rate and temper-brittleness, and especially quench-tempering, can lead to reliability problems with regard to thin plates [5,6]. In the current study, SUS440 plate is made into a double-loop type thin plate specimen by a punch-shear process, in order to highlight the stress concentration and study the effects on brittleness. Austempering heat treatment can produce excellent mechanical properties in iron-based materials [7,8], although the characteristics of austempered SUS440 have still not been studied. Notably, the salt that was used in the current study can also meet the demand for a more environmentally-friendly process [9]. Therefore, this research controlled the heat treatment conditions (Ms temperature is about −40 °C [10]) to obtain SUS440 with a bainite matrix, retained austenite, and stable MC carbides, and then investigated the tensile strength and hardness of this material in order to obtain data for use in SUS440 thin plate applications [11].

2. Experimental Procedure

The chemical composition of SUS440 is given in Table 1, the carbon content is 0.75% by mass, the chrome content is 18% by mass, and it contains other alloying elements, such as Si, Mn, and V.

This research uses a double loop-type thin plate (t = 0.78 mm) specimen to examine the effects of the punch-shear process on brittleness. Figure 1 shows the dimensions of the specimen. Figure 2 shows a diagram of the austempering process. The heat treatment condition of the specimen was 1080 °C (vacuum) held for 30 min for austenitic treatment (the austenite conditions of all specimens are the same), and then the specimen was immediately moved into a salt-bath furnace for austempering. There were two different conditions for the salt-bath phase and thus two sets of specimens were prepared. The first set underwent the same salt-bath time (120 min) and different constant temperatures (200, 300, and 400 °C), and were then quenched in the water. The second set underwent the same salt-bath temperature (300 °C) and different time intervals (30, 60, 90, and 120 min), and were then quenched in the water. Each austempering condition was called x °C-y min, based on the salt-bath condition, such as 200 °C-120 min [7,8].

Table 1. The chemical composition of the SUS440 (mass %).

Element	C	Mn	Si	P	S	V	Cr	Mo	Fe
Content	0.75	1.00	1.00	0.04	0.04	0.15	18.20	0.30	Bal.

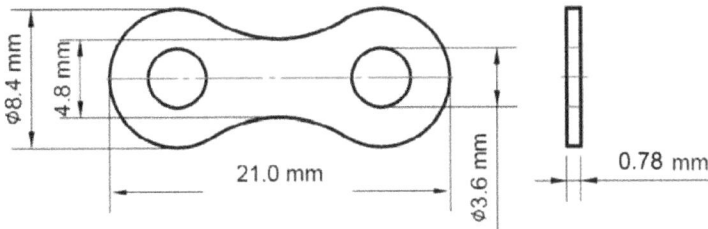

Figure 1. Configuration of a double-loop thin plate specimen.

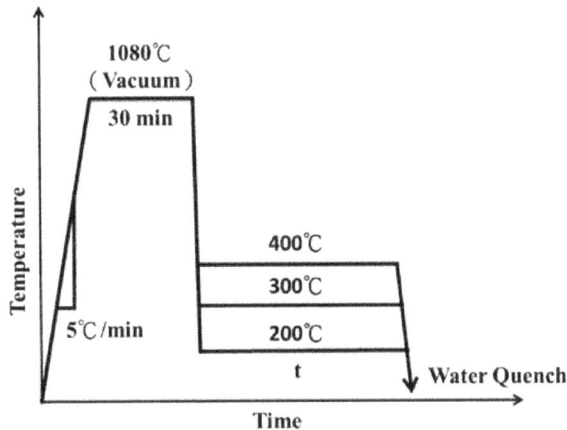

Figure 2. The tempering process with different salt-bath temperatures.

The characteristics of each austempered specimen were determined quantitatively by SEM (Hitachi SU8000, Hitachi, Tokyo, Japan) and an image analyzer (Oxford Instrument, Abingdon, UK). The hardness (HRA) and the tensile properties (tensile rate: 1 mm·min^{-1}) of each specimen were evaluated. The structural phases were identified by XRD (Bruker AXS, Karlsruhe, Germany). The short (30 min) and long (120 min) tempering affected the precipitation of the tempering carbides, and Electron Spectroscopy for Chemical Analysis (ESCA, PHI 5000 V, ULVAC-PHI, Inc., Kanagawa, Japan) was used to assess this. In addition, SEM and OM were used to observe the fracture surface and subsurface of the specimen to clarify the tensile failure characteristics of thin plate specimens [11].

3. Results and Discussion

Figure 3 shows the microstructures of the austempered bainite specimens (1080 °C-30 min) with different salt-bath conditions. The matrix is the feather bainite structure. Notably, the major phases were the retained austenite phases combined with martensite, and the particles were the carbides. In addition, the salt-bath specimen at 300 °C, became coarser was the austempering duration increased (Figure 4). A fully bainite matrix with finer primary carbides can be achieved with a longer austempering time [5,12–14]. However, if the tempering time was insufficient, then no feathery structure was seen in the matrix after etching.

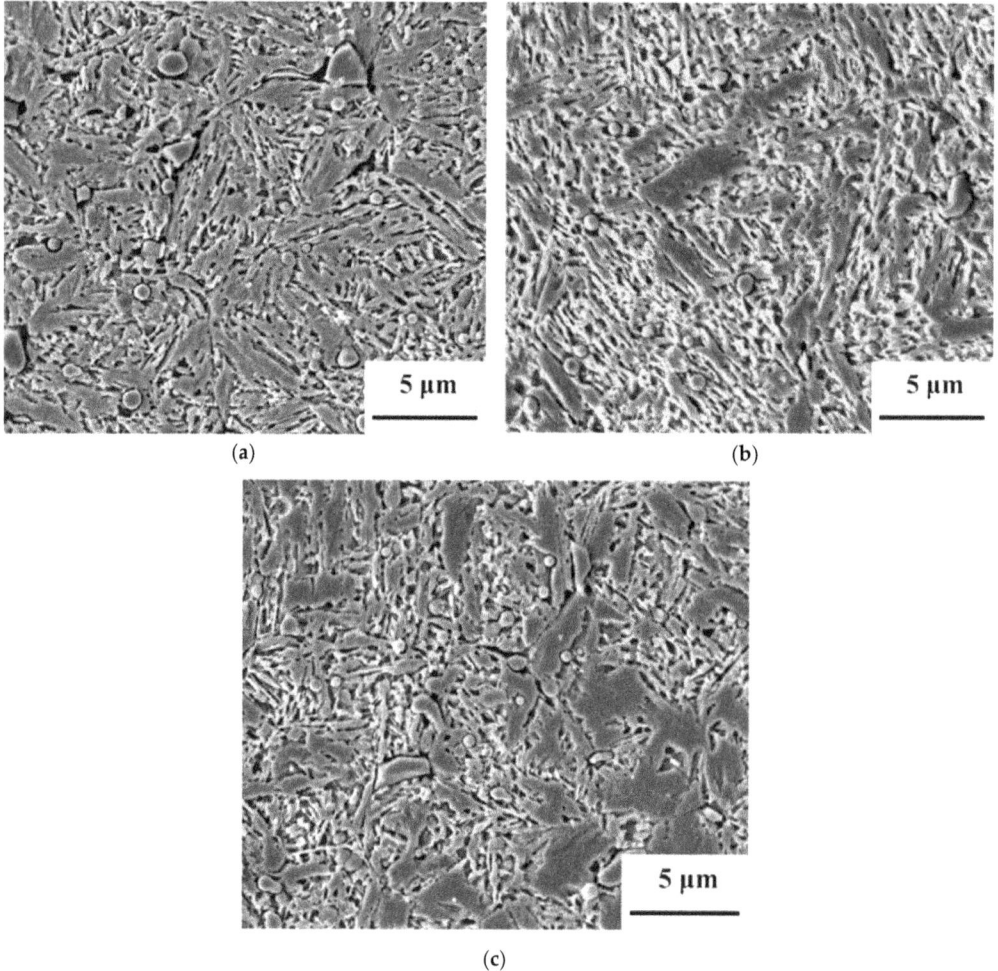

(a)   (b)

(c)

Figure 3. Microstructural characteristics of the austempered specimens (1080 °C-30 min) with a salt-bath time of 120 min and different salt-bath temperatures: (a) 200 °C, (b) 300 °C, and (c) 400 °C.

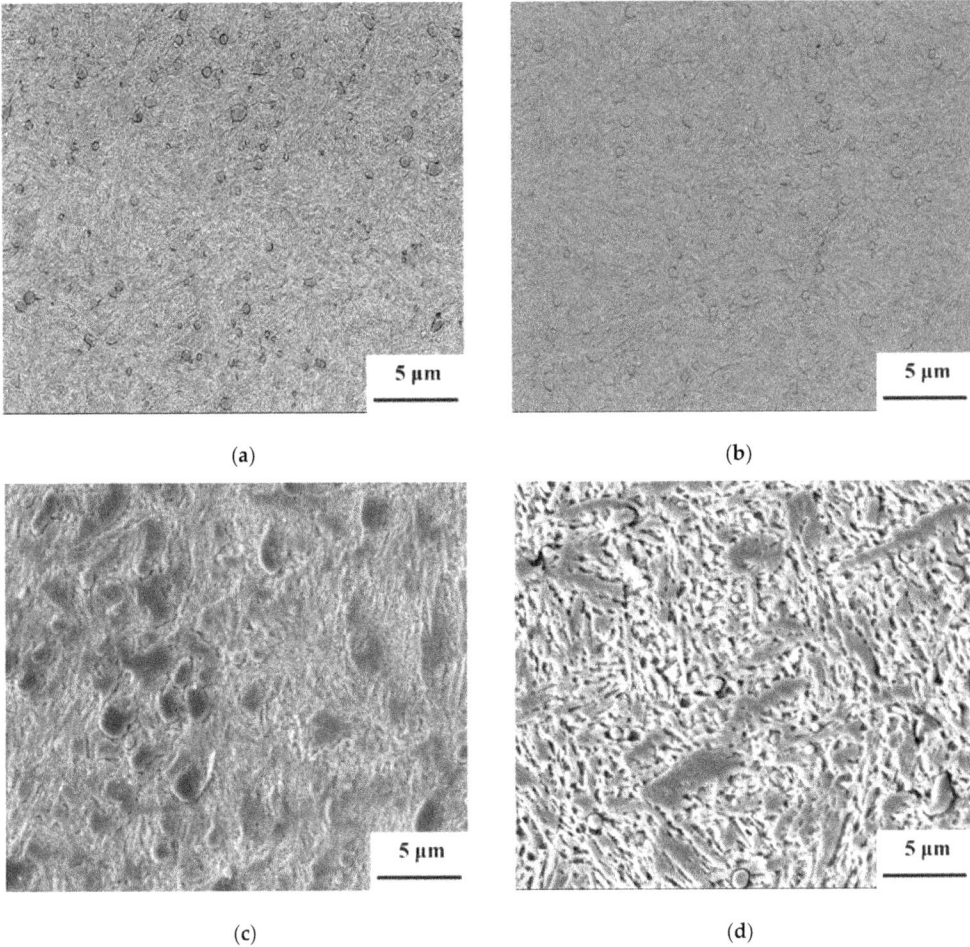

Figure 4. Microstructural characteristics of the austempered specimens (1080 °C-30 min) with a salt-bath temperature of 300 °C and different salt-bath times: (**a**) 30 min, (**b**) 60 min, (**c**) 90 min, and (**d**) 120 min.

The XRD patterns of the specimens produced under different salt-bath conditions after austempering at 1080 °C-30 min, are shown in Figures 5 and 6. All the specimens shown in the figure have a bainite matrix, M7C3 carbide, and retained austenite combining martensite, while the specimen produced at an austempering temperature of 200 °C had more martensite phases [8,15,16]. Figure 5 shows that the retained austenite content of the specimen produced at a salt-bath temperature of 400 °C is the highest, while Figure 6 shows that the austenite content of the specimen decreases as the austempering time increases.

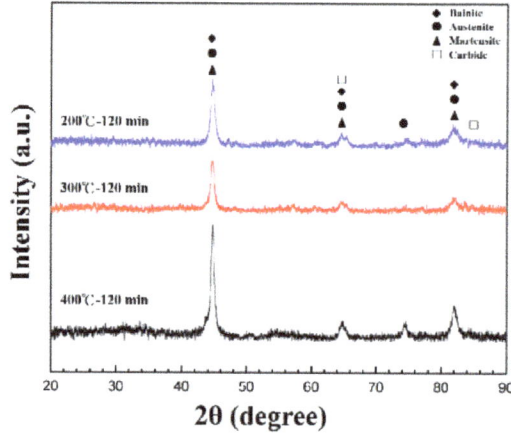

Figure 5. XRD of austempered specimens (1080 °C-30 min) with a salt-bath time of 120 min and different salt-bath temperatures.

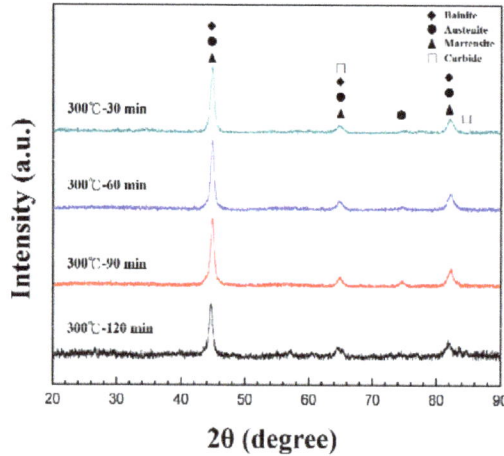

Figure 6. XRD of austempered specimen (1080 °C-30 min) with a salt-bath temperature of 300 °C and different salt-bath time.

Figure 7 is the schematic diagram of the tensile test method in this study we use two pins to insert into the double-loop type thin plate specimen and vertical. Figure 8 shows a comparison of the tensile properties obtained with different salt-bath temperatures. The tensile properties of the 300 °C specimen are significantly better than those of the other specimens. The main reason for this is that the austempering temperature of 200 °C is too low for complete austempering and thus the matrix also contains martensite, which increases the brittleness of the material [5,17,18]. The 400 °C specimen has more retained austenite, which results in lower tensile mechanical properties. In the hardness tests, the 200 °C specimen had greater hardness because of the martensite inside the matrix [2,19], while the other specimens had an HRA of about 76.

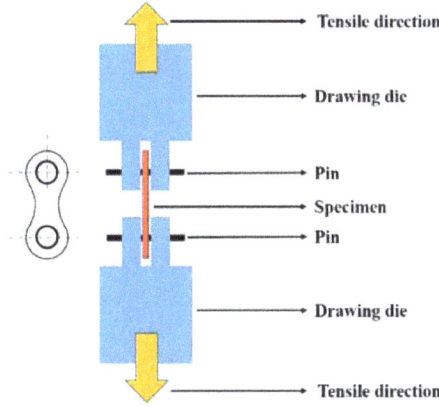

Figure 7. The schematic diagram of the tensile test method.

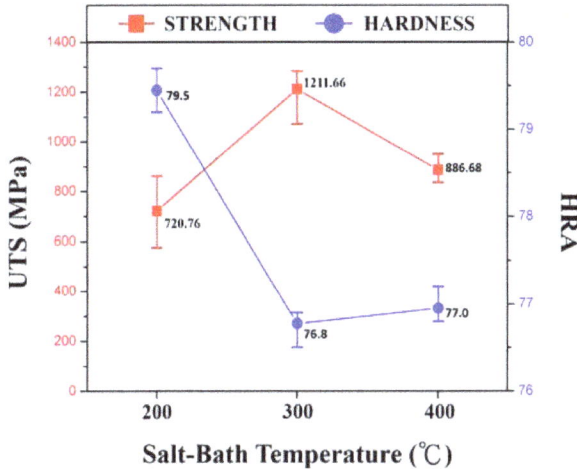

Figure 8. The mechanical properties of the austempered specimens (1080 °C-30 min) with a salt-bath time of 120 min and different salt-bath temperatures.

Figure 9 shows a comparison of the tensile properties for specimens obtained at different salt-bath times (30–120 min, fixed salt-bath temperature at 300 °C) after austenitizing at 1080 °C for 30 min. The average tensile strength of the 90 min specimen was greater than that of the others (>1100 MPa), and the hardness of all four heat treatment conditions was above HRA 76 [17,18]. Therefore, the condition of austenitizing at 1080 °C for 30 min, followed by austempering at salt-bath temperature of 300 °C for 90 min, led to the best tensile properties and hardness (and the composition of the resulting microstructure was 68.8% bainite + 12.6% martensite + 15.5% austenite + 3.1% carbides). The fracture characteristics of the various specimens were then compared using a tensile test.

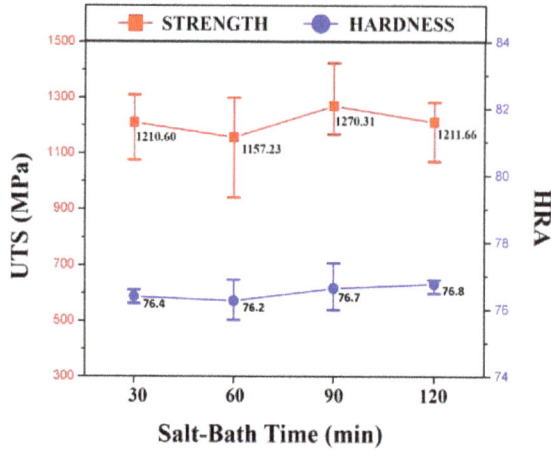

Figure 9. The mechanical properties of the austempered specimens (1080 °C-30 min) with a salt-bath temperature of 300 °C and different salt-bath times.

Figure 10 shows the strain-stress curve of the specimens produced under different salt-bath times (30–120 min, fixed salt-bath temperature at 300 °C) after austenitizing at 1080 °C-30 min. There is no necking characteristics that could be observed, so it is the brittle failure. Figure 11 shows the fracture surfaces of the specimens produced at 300 °C with different austempering durations. It can be seen that there are many peel-off structures, and this confirms the characteristics of ductile-brittle failure (Figure 11a–c). Figure 11d shows the peel-off microstructure, but some splitting behavior also occurred on the fractured surface, and this was due to the brittle fracture mode. Figure 12 shows the subsurface of the 300 °C specimens produced with salt-bath times of 30 min and 120 min. When the matrix was composed of a fine feather structure and martensite and subjected to a short duration (30 min) of austempering, the brittleness became significant and caused the subsurface to become irregular (Figure 12a) [12,20]. Figure 12b shows the flat fracture characteristic that was obtained with a longer austempering time (120 min). Since the austempering duration affected carbide precipitation and the tensile fracture behavior, electron spectroscopy chemical analysis (ESCA) was applied to the austempered SUS440 to measure the carbide contents (Figure 13). With the shorter austempering time (30 min), the main carbides were the $M_{23}C_6$ (M_7C_3) precipitate phases (Figure 13a). As the austempering time increased (Figure 13b), the content of $M_{23}C_6$ (M_7C_3) precipitate (particle-like) decreased, and the amount of stable M_7C_3 carbides increased [21].

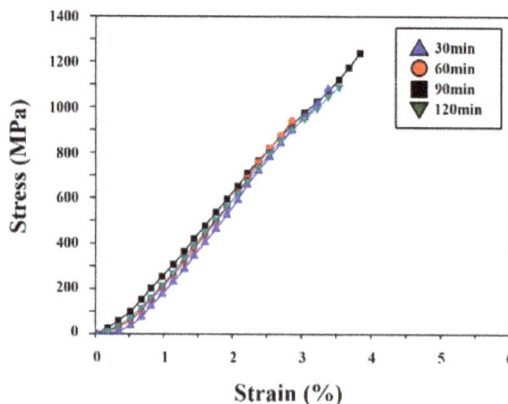

Figure 10. The strain-stress curve of the austempered specimens (1080 °C-30 min) with a salt-bath temperature of 300 °C and different salt-bath times.

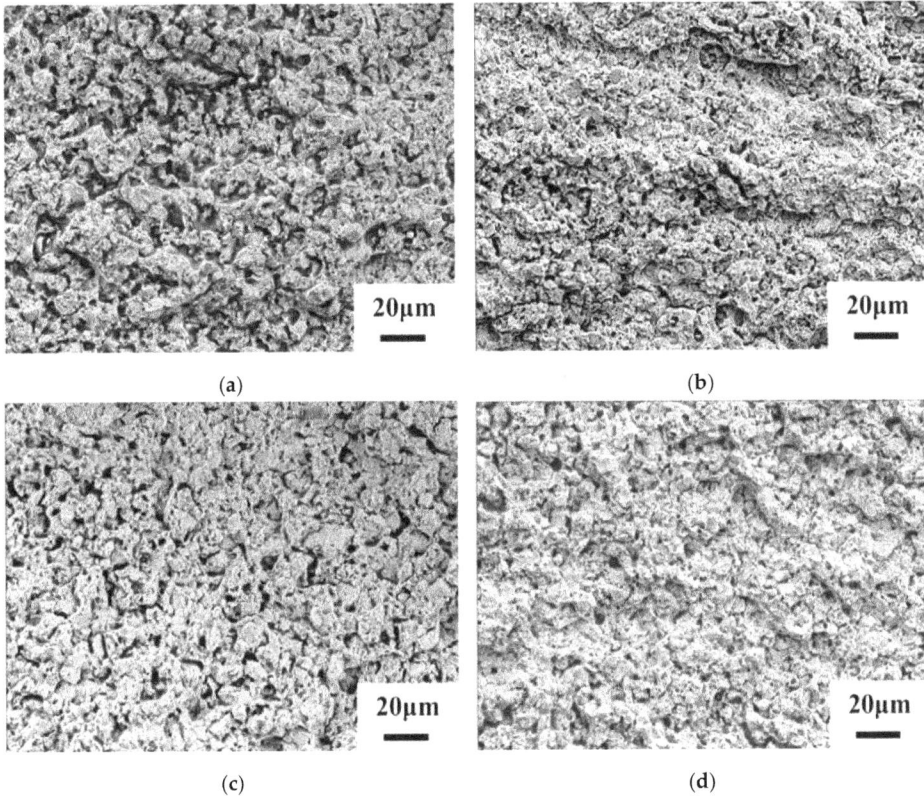

Figure 11. The fracture surfaces of the austempered specimens (1080 °C-30 min) with a salt-bath temperature of 300 °C and different salt-bath times: (**a**) 30 min, (**b**) 60 min, (**c**) 90 min, and (**d**) 120 min.

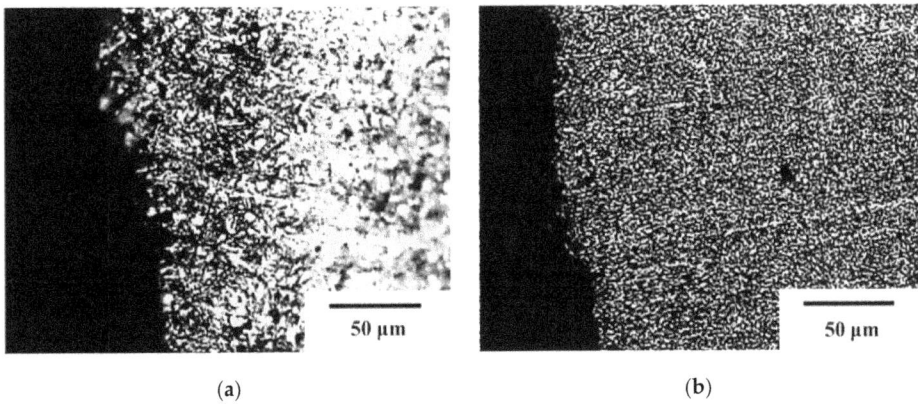

Figure 12. The subsurface of the austempered specimens (1080 °C-30 min) with a salt-bath temperature of 300 °C and different salt-bath times: (**a**) 30 min, and (**b**) 120 min.

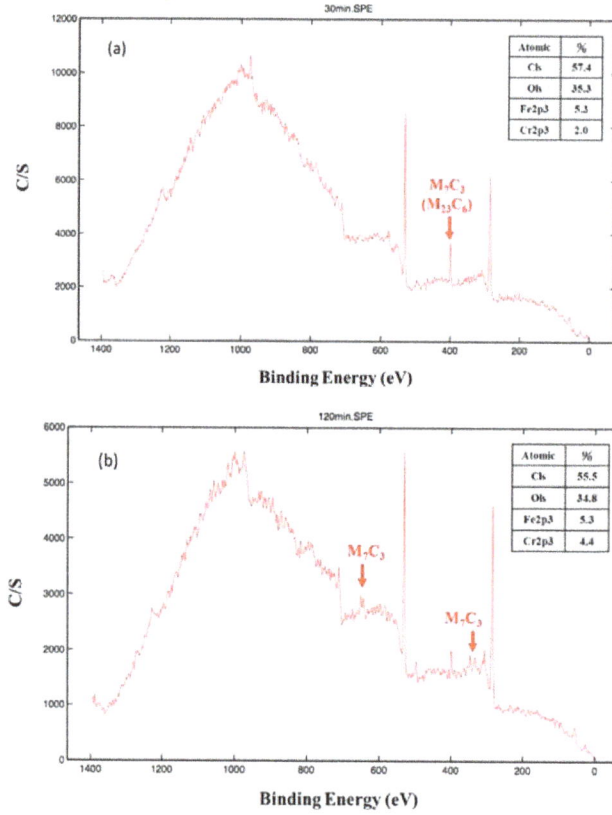

Figure 13. ESCA of carbides: (**a**) 30 min, and (**b**) 120min.

Figure 14 shows the mechanical properties of the corresponding microstructures at different salt-bath conditions. The bainite (high-retained austenite) with M7C3 coarsened as the salt-bath temperature increased. As the austempering time rose, the salt-bath condition of 300 °C-90 min, produced the specimen with the highest fracture strength. The results showed that an SUS440 thin plate with a suitable bainite matrix can have both lower brittleness and greater reliability with regard to the tensile mechanical properties.

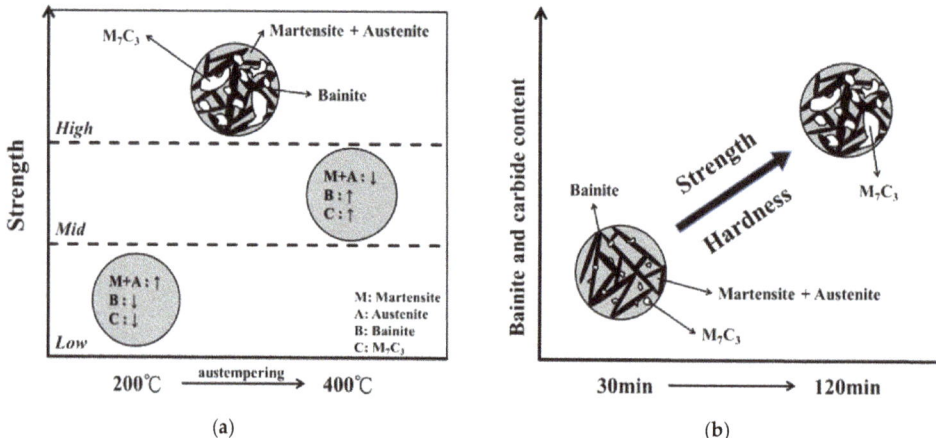

Figure 14. Diagram of tensile mechanical properties with different salt-bath conditions: (**a**) different salt-bath temperatures, and (**b**) different salt-bath times.

4. Conclusions

(1)   The brittleness of an SUS440 thin plate produced with a short austempering time will increase due to the martensite phases in the matrix. A longer austempering time reduces the strength of the SUS440 thin plate due to changes in the content of retained austenite and M7C3 precipitate. The 300 °C-90 min specimen had the best mechanical properties.

(2)   The martensite and retained austenite of the bainite matrix have a close relationship with the tensile brittleness of an SUS440 thin plate. The distribution of tempering M7C3 precipitate phases affected the failure mechanism.

Acknowledgments: The authors are grateful to The Instrument Center of National Cheng Kung University and 103-2221-E-006-066 for financial support for this research.

Author Contributions: C.-Y.C. and F.-Y.H. designed the research and wrote the manuscript with help from the other authors; C.-Y.C. performed the experiments and analyzed the data; F.-Y.H. C.-S.L. and L.-H.C. gave the support and comments.

Conflicts of Interest: The authors declare no conflict of interest.

References

1.   Bee, J.V.; Powell, G.L.F.; Bednarz, B. A substructure within the austenitic matrix of high chromium white irons. *Scr. Metall. Mater.* **1994**, *31*, 1735–1736. [CrossRef]
2.   Salleh, S.H.; Omar, M.Z.; Syarif, J. Carbide formation during precipitation hardening of SS440C steel. *Eur. J. Sci. Res.* **2009**, *34*, 83–91.
3.   Lin, Y.L.; Lin, C.; Tsai, T. Microstructure and mechanical properties of 0.63C-12.7Cr martensitic stainless steel during various tempering treatments. *Mater. Manuf. Proc.* **2010**, *25*, 246–248. [CrossRef]
4.   Salih, A.A.; Omar, M.Z.; Junaidi, S.; Sajuri, Z. Effect of different heat treatment on the SS440C martensitic stainless steel. *Aust. J. Basic Appl. Sci.* **2011**, *5*, 867–871.
5.   Bhadeshia, H. Martensite and bainite in steels: Transformation mechanism & mechanical properties. *J. Phys. IV* **1997**, *7*, 367–376.
6.   Lee, H.Y.; Yen, H.; Chang, H.; Yang, J. Substructures of martensite in Fe-1C-17Cr stainless steel. *Scr. Mater.* **2010**, *62*, 670–673. [CrossRef]
7.   Salleh, S.H. Investigation of microstructures and properties of 440C martensitic stainless steel. *Int. J. Mech. Mater. Eng.* **2009**, *4*, 123–126.
8.   Carpenter, S.D.; Carpenter, D. X-ray diffraction study of M7C3 carbide within a high chromium white iron. *Mater. Lett.* **2003**, *57*, 4456–4459. [CrossRef]
9.   Carpenter, S.D.; Carpenter, D.; Pearce, J.T.H. XRD and electron microscope study of an as-cast 26.6% chromium white iron microstructure. *Mater. Chem. Phys.* **2004**, *85*, 32–40. [CrossRef]
10. Liu, C.; Zhao, Z.; Northwood, D.O.; Liu, Y. A new empirical formula for the calculation of M_S temperatures in pure iron and super-low carbon alloy steels. *J. Mater. Proc. Technol.* **2001**, *113*, 556–562. [CrossRef]
11. Chen, C.Y.; Hung, F.; Lui, T.; Chen, L. Microstructures and mechanical properties of austempering Cr-Mo (SCM 435) alloy steel. *Mater. Trans.* **2013**, *54*, 56–60. [CrossRef]
12. Tsuzaki, K.; Maki, T. Some aspects of bainite transformation in Fe-based alloys. *J. Phys. IV* **1995**, *5*, 61–70. [CrossRef]
13. Funatani, K. Low-temperature salt bath nitriding of steels. *Met. Sci. Heat Treat.* **2004**, *46*, 277–281. [CrossRef]
14. Lee, B.-J. On the stability of Cr carbides. *Calphad* **1992**, *16*, 121–149. [CrossRef]
15. Kwok, C.T.; Lo, K.; Cheng, F.; Man, H. Effect of processing conditions on the corrosion performance of laser surface-melted AISI 440C martensitic stainless steel. *Surf. Coat. Technol.* **2003**, *166*, 221–230. [CrossRef]
16. Carpenter, S.D.; Carpenter, D.; Pearce, J.T.H. XRD and electron microscope study of a heat treated 26.6% chromium white iron microstructure. *Mater. Chem. Phys.* **2007**, *101*, 49–55. [CrossRef]
17. Avishan, B.; Yazdani, S.; Nedjad, S.H. Toughness variations in nanostructured bainitic steels. *Mater. Sci. Eng. A* **2012**, *548*, 106–111. [CrossRef]
18. Luo, Y.; Peng, J.; Wang, H.; Wu, X. Effect of tempering on microstructure and mechanical properties of a non-quenched bainitic steel. *Mater. Sci. Eng. A* **2010**, *527*, 3433–3437. [CrossRef]

19. Yang, J.R.; Yu, T.H.; Wang, C.H. Martensitic transformations in AISI 440C stainless steel. *Mater. Sci. Eng. A* **2006**, *438*, 276–280. [CrossRef]
20. Sajjadi, S.A.; Zebarjad, S.M. Isothermal transformation of austenite to bainite in high carbon steels. *J. Mater. Proc. Technol.* **2007**, *189*, 107–113. [CrossRef]
21. Isfahany, A.N.; Saghafian, H.; Borhani, G. The effect of heat treatment on mechanical properties and corrosion behavior of AISI420 martensitic stainless steel. *J. Alloys Compd.* **2011**, *509*, 3931–3936. [CrossRef]

13

Wear Behaviour of A356/TiAl₃ *in Situ* Composites Produced by Mechanical Alloying

Seda Çam, Vedat Demir and Dursun Özyürek *

Manufacturing Engineering of Technology Faculty, Karabuk University, 78100 Karabuk, Turkey;
msfe_0180@hotmail.com (S.C.); vdemir@karabuk.edu.tr (V.D.)
* Correspondence: dozyurek@karabuk.edu.tr

Academic Editor: Sherif D. El Wakil

Abstract: In this study, the effects of *in situ* TiAl₃ particles on dry sliding wear behavior of A356 aluminum alloy (added Ti) composites were investigated. The wear samples were prepared by adding different amounts of Ti (4%, 6%, and 8%) into A356 powder alloy by mechanical alloying. The mechanically alloyed powders were cold pressed at 600 MPa and sintered 530 °C for 1 h in argon atmosphere and cooled in the furnace. After the sintering process, the samples were characterized. The results show that AlTi and TiAl₃ intermetallic phases were formed and their amount increased depending on the amount of Ti added into A356 powder alloy. Out of the samples sintered with different titanium amounts (1 h at 530 °C), the highest hardness value and, accordingly, the lowest wear amount, were observed in the alloy containing 8% Ti.

Keywords: A356-Ti alloys; mechanical alloying; sintering; wear behavior

1. Introduction

Aluminum and its alloys are commonly used in structural applications in the automotive and aerospace industry, owing to their low densities and high specific resistances. However, these materials have poor tribological properties [1]. Magnesium is an important element which forms a significant solid solution in Al-Si alloys and increases the strengths of alloys with precipitation hardening. A356 alloys constitute the most important group of Al-Si-Mg alloys whose strength can be increased with Mg₂Si precipitates in Al matrix [2–5]. Another important method used to improve the mechanical properties of the alloy is to compose stable *in situ* intermetallic phases in the structure [6–8]. In the production of aluminum metal matrix composites (AMMC), which are defined as materials having superior tribological properties, the most common materials used as reinforcement phase are SiC [9–14], Al₂O₃ [15,16], TiB₂ [17,18] and TiC [19]. In the production of many AMMC materials, *ex situ* techniques are preferred (supplements made to the matrix *ex situ* in order to increase strength). The biggest contribution of *ex situ* addition of reinforcement materials to the matrix is that they protect the stable condition of the reinforcement phase in the structure at elevated temperatures. However, certain reinforcement phases are subject to phase transformation at elevated temperatures and, thus, AMMCs exhibit unstable structure. As to *in situ* composite materials, reinforcement phases are formed during solidification, sintering, or synthesizing [7,20–22]. Thus, *in situ* reinforcement phases make processes shorter and reduce the costs. In Al-Ti alloy system, TiAl₃ and AlTi intermetallic phases are formed *in situ* at the equilibrium conditions and these phases increase the hardness and strength of AMMC materials. Aluminum alloys reinforced with aluminates particles are preferred due to their high strengths and mechanical properties at elevated temperatures [23,24]. In comparison to other Al-rich (NiAl₃, FeAl₃, *etc.*) intermetallic, they are more attractive as they have lower densities (3.4 gr/cm³) and higher melting points (~1350 °C) [22,25–27]. This present study aims at developing new materials with improved wear properties by adding Ti to A356 alloy. The influence of intermetallic phases (TiAl$_3$ and AlTi) formed *in situ* during the sintering process on the wear behaviors were investigated.

2. Experimental Procedure

For the experimental studies, A356 (Al-Si-Mg) alloy powder having a mean particle size of <50 μm was used as the matrix material. The chemical composition of the alloy powder is given in Table 1. Titanium powder with 99.7% purity and 10 μm mean particle size was added into A356 alloy powder samples with three different percentages, namely 4%, 6%, and 8%, by weight. The samples were then mechanically alloyed for 45 min at 10:1 ball-to-powder ratio a SPEX SamplePreb unit (SPEX SamplePreb, Standmore, UK) [28]. The mechanically-alloyed A356-Ti powder samples were then cold pressed unidirectionally at 600 MPa to obtain 6 mm height small cylindrical samples having 10 mm diameter. The samples were sintered at 530 °C for 1 h at a 10 °C/min heating rate in an argon atmosphere and cooled in the furnace. After the sintering process had been completed, the samples were cleaned with alcohol and acetone and left to dry. Following standard metallographic processes, the samples were etched for 30–45 s with 2 mL (HF), 90 mL pure water, 5 mL (HNO$_3$), and 3 mL (HCl) for microstructural examinations. Archimedes' method was used for density measurements of the samples. Following the density measurements, the samples were characterized through X-ray diffraction method (XRD), scanning electron microscope (SEM) (FEI, Hilliboro, OR, USA), transmission electron microscope (TEM) (FEI, Hilliboro, OR, USA), and hardness measurements. For XRD examinations, an INEL unit (INEL, Artenay, France) was used, while a FEI Quanta 250 (30 kV) (FEI, Hilliboro, OR, USA) was used for SEM and EDS examinations. As for TEM examinations, an FEI Tecnai G2 F30 unit (200 kV) (FEI, Hilliboro, OR, USA) was used. Hardness values of the samples were determined by calculating the mean of five measurements obtained using an AFFRI hardness measurement device (HV2) (AFFRI, Olona, Italy). Wear tests were carried out on an AISI 4423 steel disc having 58 HRC and rotating continuously in a pin-on-disc wear device.

Table 1. Chemical composition of the A356 alloy.

Alloy	Si	Fe	Cu	Mn	Mg	Zn	Ti	Pb
A356	6.5	0.15	0.03	0.03	0.4	0.05	0.2	0.03

Prior to the wear tests, the surfaces of each sample were prepared with 1200 grade SiC abrasive to ensure a full contact between the abrasive steel disc and sample surfaces. All wear tests were carried out at room temperature and under dry sliding conditions. In these tests, a 30 N load, 1 m·s^{-1} sliding speed, and five different sliding distances, namely 400 m, 800 m, 1200 m, 1600 m, and 2000 m, were used. Following the wear tests, the worn surfaces of the samples were examined using SEM.

3. Results and Discussion

Figure 1 gives SEM images of the microstructure of A356-Ti alloy sintered at 530 °C for 1 h. It is clear in Figure 1 that Ti added into a A356 matrix at different ratios (4%, 6%, and 8%) showed different distribution patterns in the microstructure following the mechanical alloying process. The presence of large titanium grains within the A356 matrix may be explained in two different ways. First, since the size of the titanium particles added into the matrix was too small (10 μm), the powders might have become agglomerated. Secondly, this situation may have resulted from the characteristic feature of the mechanical alloying process. As is well known, mechanical alloying is a solid-state process that starts with cold welding of powder particles that are squeezed between two balls and/or between a ball and the container wall, as high-energy balls crash into one another and the container wall during the process. Then, powder particles undergo work hardening and in the last phase, particles are fragmented. Previous studies reported a slight increase in the powder dimension with the effect of cold welding taking place in the first stage of the mechanical alloying process [29–31]. Therefore, titanium powder particles in the matrix can be combined together during mechanical alloying and form coarse grains. Intermetallic compounds (TiAl$_3$), resulting from the reaction between Al matrix and Ti during sintering of cold-formed compacts, are also observed in the SEM images. It can also be seen that TiAl$_3$ intermetallic phase (Figure 1a) that seems to come out in the structure during sintering is surrounded by the matrix and is very small. In addition, a reaction takes place between the aluminum matrix and reinforcement phase interface in A356 alloy that has added titanium. This was also pointed out in a study carried out by Yu *et al.* [32].

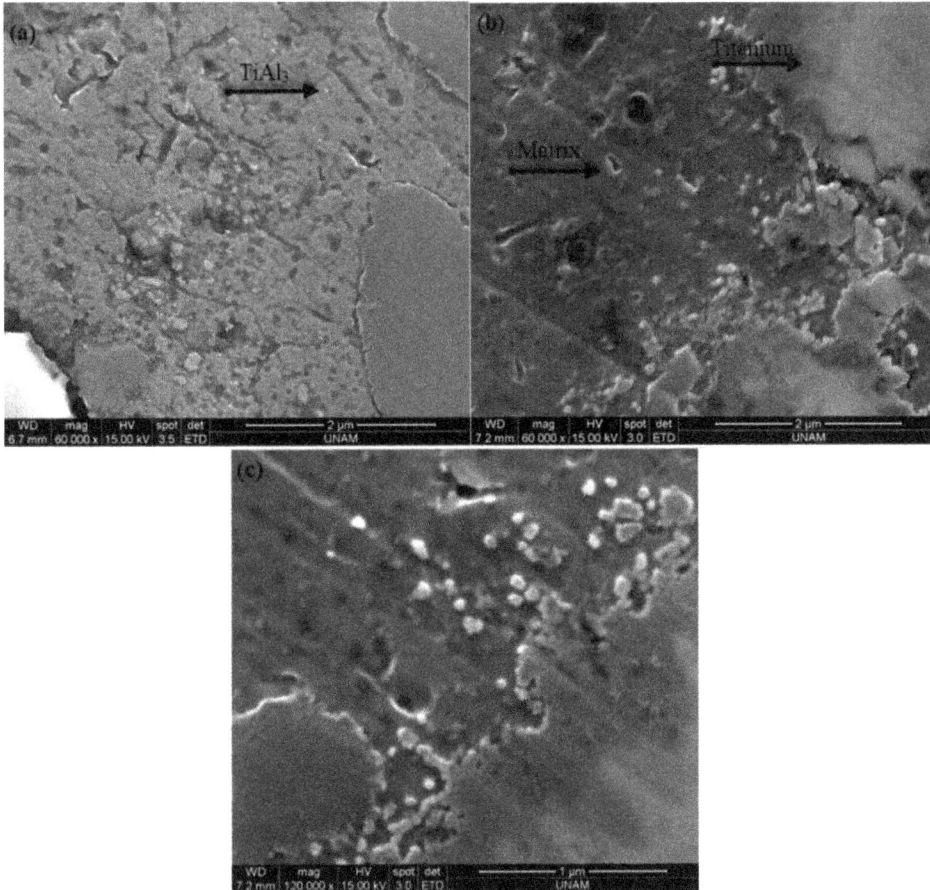

Figure 1. SEM images of Al-Ti alloys 4% Ti (**a**); 6% Ti (**b**); and 8% Ti (**c**) sintered at 530 °C for 1 h.

AlTi and TiAl$_3$ particles are formed *in situ* form within the matrix [8,33]. To better examine the dimensions of intermetallic compounds formed in the *in situ* structure, TEM examinations were carried out. Figure 2 gives an image of TEM bright-area and selected-area diffraction pattern, where it is seen that *in situ* titanium aluminate particles with a particle size of about 50 nm are formed in the structure during the sintering.

In the bright-area TEM image, TiAl$_3$ intermetallic phase is evident in A356 matrix depending on the sintering process. Additionally, a ringed structure is clearly seen in the diffraction pattern image of the selected area. The ringed appearance is associated with deformation of the crystal structure during the mechanical alloying process as well as the formation of a harmonic, wide-angle grain structure. An increase in the amount of deformation occurring in the powder particles during the mechanical alloying process reduces the recrystallization temperature. Thus, a larger grained structure is expected to form during the sintering following the deformation. However, as Ti is added into the A356 matrix it reacts and forms a TiAl$_3$ phase, it is clear that the grain growth of newly-formed grains in the structure is not excessive. This is because of the fact that certain phases added into the Al matrix, such as TiAl$_3$, TiB$_2$, TiC, and B, inhibit the grain growth [17,34]. X-ray diffraction results of A356-Ti alloy having different ratios of Ti are given in Figure 3.

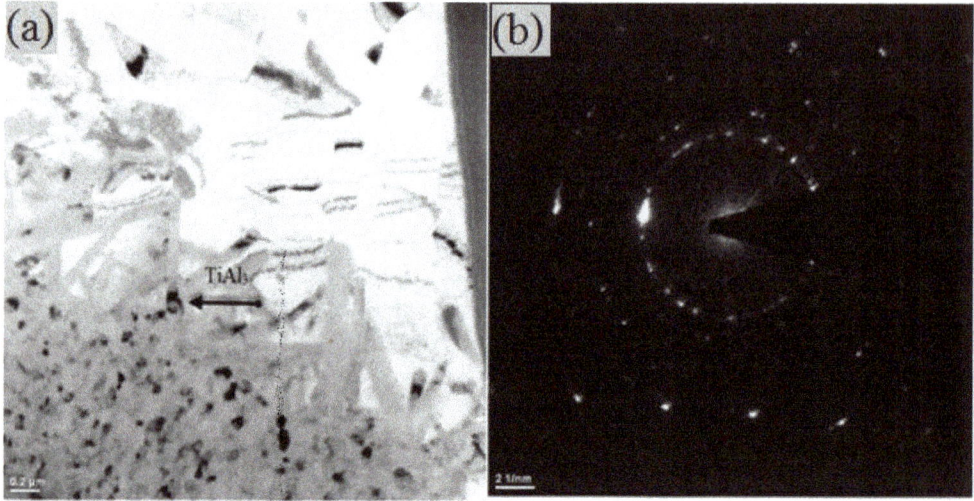

Figure 2. TEM images of 6% Ti added A356 alloy bright-area (**a**) and selected-area (**b**) diffraction pattern.

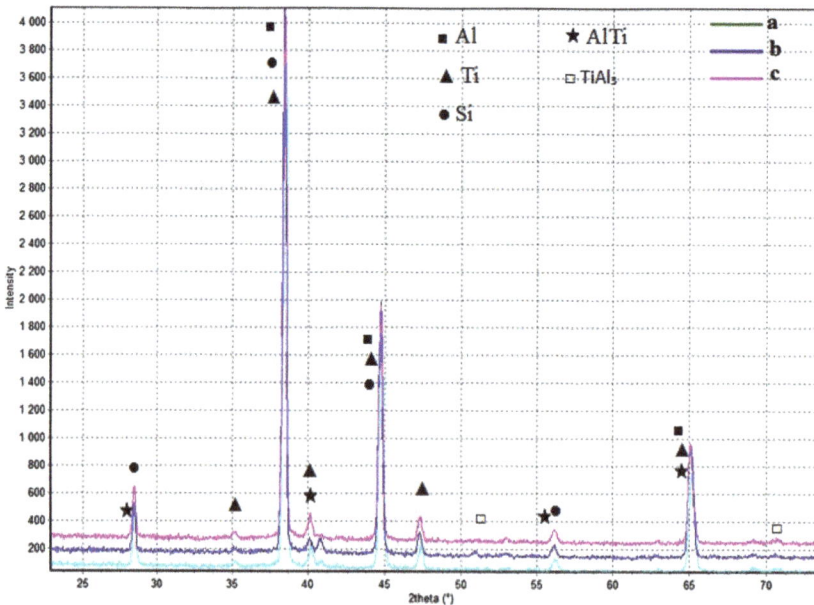

Figure 3. XRD results of A356-Ti alloy with 4% (**a**); 6% (**b**); and 8% (**c**) Ti.

It is known that different intermetallic compounds (α_2-Ti$_3$Al, γ-TiAl, TiAl$_3$, and TiAl$_2$) are formed in the Al-Ti phase diagram depending on the ratios of Al and Ti elements and alloying conditions [35]. In the present study, it was determined that TiAl$_3$ and AlTi reinforcement phases which were expected to be formed in the structure with X-ray diffraction were actually formed via *in situ* reactions. This *in situ* reaction is as follows:

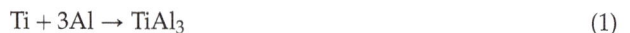

$$Ti + 3Al \rightarrow TiAl_3 \tag{1}$$

In situ titanium aluminate intermetallic compounds in the structure are intermetallic compounds whose crystal structures are different from metals. In the study conducted by Chianeh *et al.* [33], it was reported that only the TiAl$_3$ intermetallic phase was obtained in the samples sintered at 500 °C for 9 h and the samples sintered at 600 °C for 3 h. In this study, XRD results given in Figure 3 show that both AlTi and TiAl$_3$ phases can be formed following sintering at 530 °C for 1 h. XRD results also demonstrate that the TiAl$_3$ phase can be formed during the sintering lasting for 1 h and increases depending on the increase in the Ti amount added into the alloy. Post-sintering hardness values of A356-Ti powder alloy produced with mechanical alloying where different amounts of Ti are added are given in Figure 4.

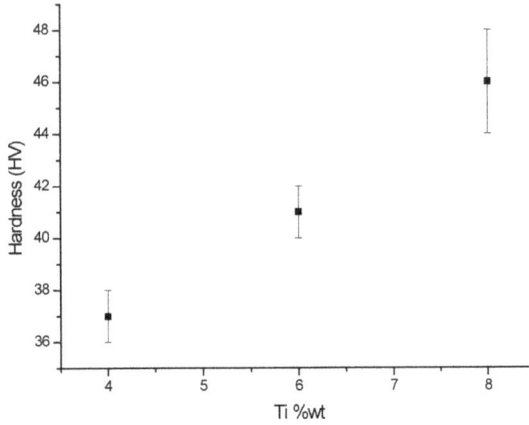

Figure 4. Hardness values of A356 different amounts of Ti added.

As can be seen in Figure 4, the hardness values of samples containing 4% Ti, 6% Ti, and 8% Ti are 37 HV, 41 HV, and 46 HV, respectively. The lowest hardness value was observed for the samples containing 4% Ti, while the highest hardness value was measured for the samples containing 8% Ti. Accordingly, as the Ti amount in the alloy increases, the hardness of Al-Ti alloy also increases. That increase in hardness is associated with the increase of the volume fraction of the TiAl$_3$ intermetallic phase. This finding is supported by a previous study [8]. Post-sintering densities of A356 alloy into which different amounts of Ti were added are given in Figure 5. Archimedes method was used for the density measurement.

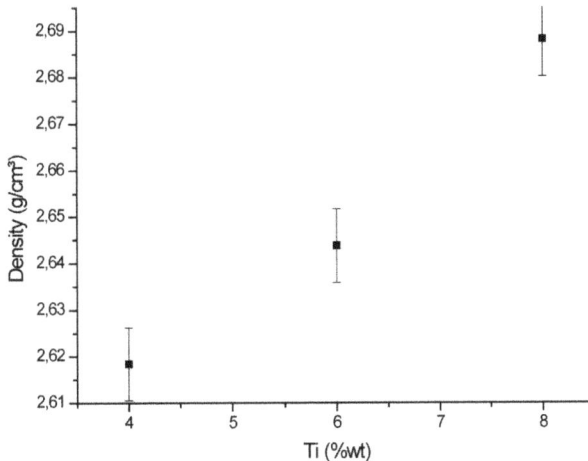

Figure 5. Densities of A356 alloy containing different amounts of Ti.

When post-sintering density values of A356 alloy into which different amounts of Ti were added are examined, the highest value is observed for the samples containing 8% Ti (2.68 g·cm^{-3}), while the lowest value is obtained for the samples containing 4% Ti (2.61 g·cm^{-3}). As the amount of % Ti in the alloy increases, density and hardness values of alloys also increase. The reason behind this hardness and density increase is the fact that the density of titanium is higher than the density of A356 alloy. Weight losses of A356-Ti alloys produced with mechanical alloying under 30 N load for different distances (400 m, 800 m, 1200 m, 1600 m, and 2000 m) are given in Figure 6.

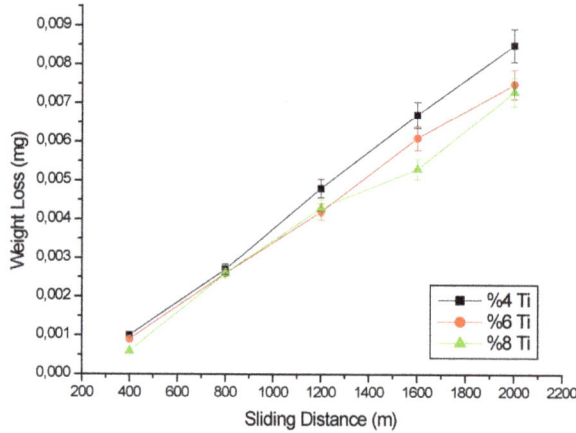

Figure 6. Weight losses of the samples depending on the sliding distances.

As evident in Figure 6, the highest weight loss occurs in the alloy containing 4% Ti, while the lowest weight loss is observed for the alloy containing 8% Ti. However, weight loss values of the alloys containing 6% and 8% Ti are extremely close to each other. The highest weight loss of the sample containing 4% Ti can be attributed to its lower hardness, as well as the lower amount of intermetallic phases. The hardness of the AMCs increased with increasing Ti amount. As it is known, the weight loss of a material in a wear test decreases with an increase in the hardness of the material [12,15]. Vencl et al. [36] pointed out that the rate of reinforcement should be increased to reduce plastic deformation and to increase hardness of the material. Average friction coefficients of Al-Ti alloys containing different amounts of titanium (at the end of 2000 m) are given in Figure 7.

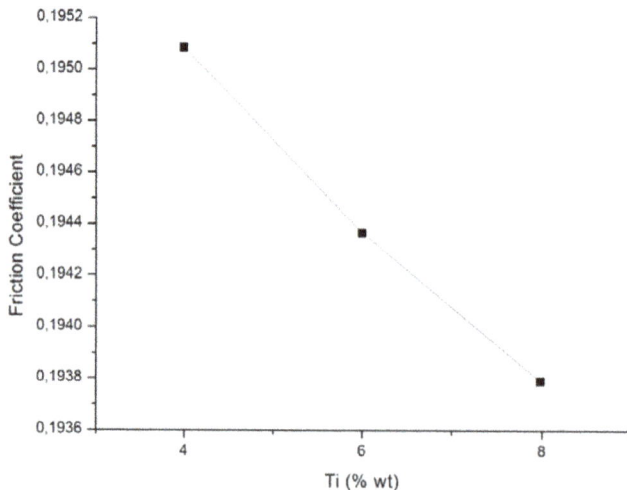

Figure 7. Average friction coefficients of Al-Ti alloys.

The friction coefficients given in the Figure 7 are average values obtained for each alloy at the end of 2000 m sliding distance. The lowest friction coefficient was obtained for the alloy containing 8% Ti, while the highest friction coefficient was obtained for the alloy containing 4% Ti. As there is a relationship between weight losses and friction coefficients, the friction coefficient decreases with the increase in the hardness of the alloy. It was determined that both the weight loss and friction coefficient of the alloy, which has the lowest hardness value and contains 4% Ti are higher than the others. SEM images taken from the worn surfaces of Al-Ti alloys are given in the Figure 8.

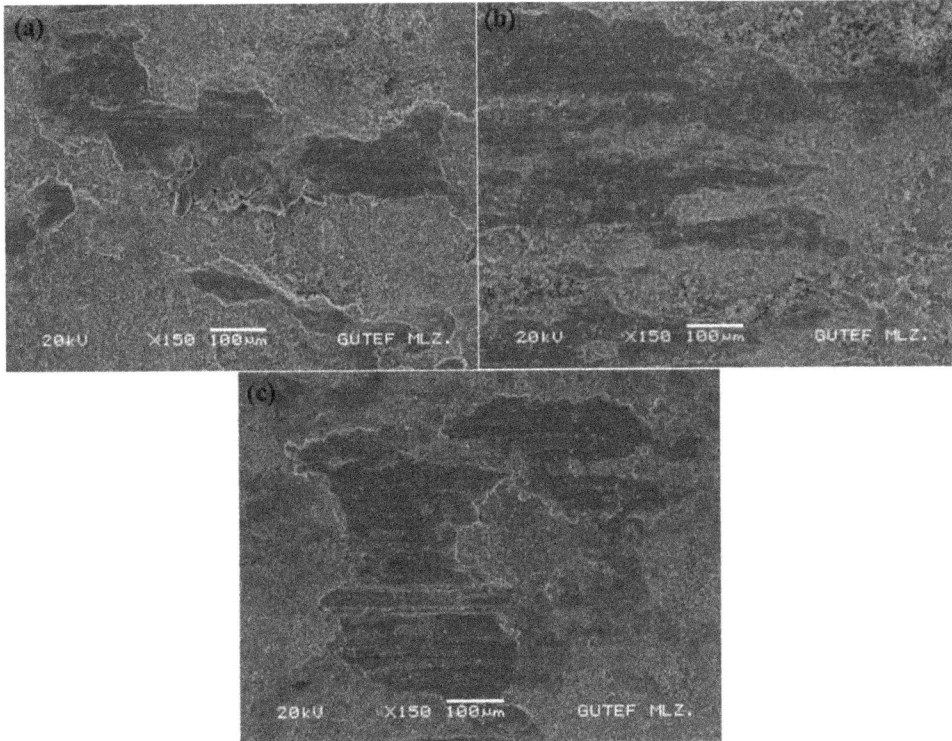

Figure 8. SEM images of the worn surface of Al-Ti alloys 4% Ti (**a**); 6% Ti (**b**); and 8% Ti (**c**).

When the SEM image of the worn surface of Al-4% Ti alloy that has the highest weight loss is examined (Figure 8a), it is seen that plastic deformation and adhesion occurred on the surface. That can be explained by the fact that the particles pulled off the surface during the wear test are moved and smeared onto the worn surface once more [37,38]. Smearing which occurs during the wear test is also observed in the worn surfaces of other alloys containing 6% Ti and 8% Ti. Another important point observed from the worn surfaces is the lack of deep marks on the surface. Additionally, it is seen that oxidation occurs on the worn surfaces of the samples into which 4%, 6%, and 8% titanium are added. As mentioned in a previous study, the oxidation on the worn surfaces results from the heat caused by the friction occurring during the wear tests [18]. That stable oxide structure forming on the worn surfaces acts as solid lubricant during dry sliding. When looking at the hardness values in Figure 4, it is clear that as the amount of titanium added into the alloy increases, hardness also increases. Another impact of hardness values (Figure 7) is that friction coefficients decrease depending on the titanium amount added into the alloy.

4. Conclusions

In the present study, Al-Ti alloy powders were produced by mechanical alloying to contain 4%, 6% and 8% titanium respectively. They were then compacted cold and sintered at 530 °C for 1 h. The effect of the percentage of titanium on the hardness and the wear properties of the various sintered samples were examined. The following could be concluded:

- As the amount of Ti in Al-Ti alloys increases, the hardness value increases.
- As the amount of added Ti increases in the composites, the density value also increases since the density of Ti is higher than that of Al.
- The XRD examinations show that the titanium aluminate phases ($TiAl_3$ and $AlTi$), which increase strength in the samples, are formed in the alloy following the sintering process.
- The amount of intermetallic phases in the structure during sintering increases depending on the increasing titanium addition and that affects the wear properties.
- Sintered samples with higher percentages of titanium would exhibit less weight loss and a lower coefficient of friction.

Author Contributions: Study design: D. Özyürek and V. Demir; Experimental work: S. Cam; Results analysis and Manuscript preparation: All authors; Manuscript proof and submission: D. Özyürek.

Conflicts of Interest: The authors declare no conflict of interest.

References

1. Kumar, S.; Balasubramanian, V. Developing a mathematical model to evaluate wear rate of $AA7075/SiC_p$ powder metallurgy. *Wear* **1999**, *264*, 1026–1034. [CrossRef]
2. Azadi, M.; Shirazabad, M.M. Heat treatment effect on thermo-mechanical fatigue and low cycle fatigue behaviors of A356.0 aluminum alloy. *Mater. Des.* **2013**, *45*, 279–285. [CrossRef]
3. Ragab, K.A.; Samuel, A.M.; Al-Ahmari, A.M.A.; Samuel, F.H.; Doty, H.W. Influence of fluidized sand bed heat treatment on the performance of Al-Si cast alloys. *Mater. Des.* **2011**, *32*, 1177–1193. [CrossRef]
4. Zhu, M.; Jian, Z.; Yang, G.; Zhou, Y. Effects of T6 heat treatment on the microstructure, tensile properties and fracture behavior of the modified A356 alloys. *Mater. Des.* **2012**, *36*, 243–249. [CrossRef]
5. Kliauga, A.M.; Vieira, E.A.; Ferrante, M. The influence of impurity level and tin addition on the aging heat treatment of the 356 class alloy. *Mater. Sci. Eng. A* **2008**, *480*, 5–16. [CrossRef]
6. Ahmed, S.; Haseeb, A.S.M.A.; Kurny, A.S.W. Study of wear behaviour of Al-4.5% Cu-3.4% Fe *in situ* composite: Effect of termal and mechanical processing. *J. Mater. Process. Tech.* **2007**, *182*, 327–332. [CrossRef]
7. Niu, L.; Zhang, J.; Yang, X. *In-situ* synthesis of Al_3Ti particles reinforced Al-Based composite coating. *Trans. Nonferrous Met. Soc. China* **2012**, *22*, 1387–1393. [CrossRef]
8. Özyürek, D.; Tekeli, S.; Tuncay, T.; Yılmaz, R. The effect of synthesis time on the wear behavior of Al-8% Ti alloy produced by mechanical alloying. *Powder Metall. Met. Ceram.* **2012**, *51*, 491–495.
9. Lu, L.; Lai, M.O.; Ng, C.W. Enhanced mechanical properties of an Al based metal matrix composite prepared using mechanical alloying. *Mater. Sci. Eng. A* **1998**, *252*, 203–211. [CrossRef]
10. Hu, Q.; McColl, I.R.; Harris, S.J.; Waterhouse, R.B. The role of debris in the fretting wear of a SiC reinforced aluminium alloy matris composite. *Wear* **2000**, *245*, 10–21. [CrossRef]
11. Zhao, N.; Nash, P.; Yang, X. The effect of mechanical alloying on SiC distribution and the properties of 6061 aluminum composite. *J. Mater. Process. Tech.* **2005**, *170*, 586–592. [CrossRef]
12. Özyürek, D.; Tekeli, S. An investigation on wear resistance of SiC_p-reinforced aluminium composites produced by mechanical alloying. *Sci. Eng. Compos. Mater.* **2010**, *17*, 31–38. [CrossRef]
13. Aztekin, H.; Özyürek, D.; Çetinkaya, K. Production of hypo-eutectic Al-Si alloy based metal matrix composite with thixomoulding processing. *High Temp. Mater. Process.* **2010**, *29*, 169–178. [CrossRef]
14. Walker, J.C.; Rainforth, W.M.; Jones, H. Lubricated sliding wear behaviour of aluminium alloy composites. *Wear* **2005**, *259*, 577–589. [CrossRef]
15. Özyürek, D.; Tekeli, S.; Güral, A.; Meyveci, A.; Gürü, M. Effect of Al_2O_3 amount on microstructure and wear properties of Al-Al_2O_3 metal matrix composites prepared using mechanical alloying method. *Powder Metall. Met. Ceram.* **2010**, *49*, 289–294. [CrossRef]
16. Daoud, A.; Reif, W. Influence of Al_2O_3 particulate on the aging response of A356 Al-based composites. *J. Mater. Process. Tech.* **2002**, *123*, 313–318. [CrossRef]

17. Mandal, A.; Murty, B.S.; Chakraborty, M. Wear behaviour of near eutectic Al-Si alloy reinforced with *in-situ* TiB$_2$ particles. *Mater. Sci. Eng. A* **2009**, *506*, 27–33. [CrossRef]

18. Özyürek, D.; Ciftci, I. An investigation into wear behaviour of TiB$_2$ particle reinforced aluminum composites produced by mechanical alloying. *Sci. Eng. Compos. Mater.* **2011**, *18*, 5–12. [CrossRef]

19. Ruiz-Navas, E.M.; Fogognolo, J.B.; Valesco, J.B.; Ruiz-Prieto, J.M.; Froyen, L. One step production of aluminium matrix composite powders by mechanical alloying. *Compos. A Appl. Sci. Manuf.* **2006**, *37*, 2114–2120. [CrossRef]

20. Clemens, H.; Wallgram, W.; Kremmer, S.; Ther, V.G.; Otto, A.; Bartels, A. Design of novel β-solidifying TiAl alloys with adjustable β/B2-phase fraction and excellent hot-workability. *Adv. Eng. Mater.* **2008**, *10*, 707–713. [CrossRef]

21. Tetsui, T.; Shindo, K.; Kaji, S.; Kobayashi, S.; Takeyama, M. Fabrication of TiAl components by means of hot forging and machining. *Intermetallics* **2005**, *13*, 971–978. [CrossRef]

22. Hsu, C.J.; Chang, C.Y.; Kao, P.W.; Ho, N.J.; Chang, C.P. Al-Al$_3$Ti nonocomposites produced *in-situ* by friction stir processing. *Acta Mater.* **2006**, *54*, 5241–5249. [CrossRef]

23. Yeh, C.L.; Su, S.H. *In situ* formation of TiAl-TiB$_2$ composite by SHS. *J. Alloy. Compd.* **2006**, *407*, 150–156. [CrossRef]

24. Liu, Y.; Xiu, Z.; Wu, G.; Jiang, L.; Gou, H.; Jiang, G. Microstructure evolution of Ti-Al-C system composite. *Rare Met. Mater. Eng.* **2010**, *39*, 1152–1156.

25. Shi, T.; Zhang, J.; Wang, L.; Jiang, W.; Chen, L. Fabrication, microstructure and mechanical properties of TiC/Ti$_2$AlC/TiAl$_3$ *in situ* composite. *J. Mater. Process. Tech.* **2011**, *27*, 239–244.

26. Wang, T.; Zhang, J. Thermoanalytical and metallographical investigations on the synthesis of TiAl$_3$ from elemantary powders. *Mater. Chem. Phys.* **2006**, *99*, 20–25. [CrossRef]

27. Prakash, U.; Sauthoff, G. Structure and properties of Fe-Al-Ti intermetallic alloys. *Intermetallics* **2001**, *9*, 107–112. [CrossRef]

28. Özyürek, D.; Tunçay, T.; Evlen, H.; Çiftçi, I. Synthesis, Characterization and Dry Sliding Wear Behavior of *In-situ* Formed TiAl$_3$ Precipitate Reinforced A356 Alloy Produced by Mechanical Alloying Method. *Mater. Res.* **2015**, *18*, 813–820.

29. Rafiei, M.; Khademzadeh, S.; Parvin, N. Characterization and formation mechanism of nanocrystalline W-Al alloy prepared by mechanical alloying. *J. Alloy. Compd.* **2010**, *489*, 224–227. [CrossRef]

30. Tang, H.; Cheng, Z.; Liu, J.; Ma, X. Preparation of a high strength Al-Cu-Mg alloy by mechanical alloying and press-forming. *Mater. Sci. Eng. A* **2012**, *550*, 51–54. [CrossRef]

31. Gu, D.; Zhang, G.; Dai, D.; Wang, H.; Shen, Y. Nanocrystalline tungsten-nickel heavy alloy reinforced by *in-situ* tungsten carbide: Mechanical alloying preparation and microstructural evolution. *Int. J. Refract. Met. Hard Mater.* **2013**, *37*, 45–51. [CrossRef]

32. Yu, P.; Zhang, L.C.; Zhang, W.Y.; Das, J.; Kim, K.B.; Eckert, J. Interfacial reaction during the fabrication of Ni60Nb40 metallic glass particles-reinforced Al based MMCs. *Mater. Sci. Eng. A* **2007**, *444*, 206–213. [CrossRef]

33. Chianeh, V.A.; Hossein, H.R.; Nofar, M. Microstructural features and mechanical properties of Al-Al$_3$Ti composite fabricated by *in-situ* powder metallurgy route. *J. Alloy. Compd.* **2009**, *473*, 127–132. [CrossRef]

34. Rao, A.K.P.; Das, K.; Murty, B.S.; Chakraborty, M. Microstructural and wear behavior of hypoeutectic Al-Si alloy (LM25) grain refined and modified with Al-Ti-C-Sr master alloy. *Wear* **2006**, *261*, 133–139.

35. Mishin, Y.; Herzig, C. Diffusion in the Ti-Al system. *Acta Mater.* **2000**, *48*, 589–623. [CrossRef]

36. Vencl, A.; Rac, A.; Bobic, I.; Miskovic, Z. Tribological properties of Al-Si alloy reinforced with Al$_2$O$_3$ particles. *Tribol. Ind.* **2006**, *28*, 27–31.

37. Sağlam, I.; Özyürek, D.; Çetinkaya, K. Effect of ageing treatment on wear properties and electrical conductivity of Cu-Cr-Zr alloy. *Bull. Mater. Sci.* **2011**, *34*, 1465–1470. [CrossRef]

38. Attar, H.; Prashanth, K.G.; Chaubey, A.K.; Calin, M.; Zhang, L.C.; Scudino, S.; Eckert, J. Comparison of wear properties of commercially pure titanium prepared by selective laser melting and casting processes. *Mater. Lett.* **2015**, *142*, 38–41. [CrossRef]

14

Monte Carlo Modelling of Single-Crystal Diffuse Scattering from Intermetallics

Darren J. Goossens

School of Physical, Environmental and Mathematical Sciences, University of New South Wales, Canberra ACT 2600, Australia; d.goossens@adfa.edu.au

Academic Editor: Klaus-Dieter Liss

Abstract: Single-crystal diffuse scattering (SCDS) reveals detailed structural insights into materials. In particular, it is sensitive to two-body correlations, whereas traditional Bragg peak-based methods are sensitive to single-body correlations. This means that diffuse scattering is sensitive to ordering that persists for just a few unit cells: nanoscale order, sometimes referred to as "local structure", which is often crucial for understanding a material and its function. Metals and alloys were early candidates for SCDS studies because of the availability of large single crystals. While great progress has been made in areas like *ab initio* modelling and molecular dynamics, a place remains for Monte Carlo modelling of model crystals because of its ability to model very large systems; important when correlations are relatively long (though still finite) in range. This paper briefly outlines, and gives examples of, some Monte Carlo methods appropriate for the modelling of SCDS from metallic compounds, and considers data collection as well as analysis. Even if the interest in the material is driven primarily by magnetism or transport behaviour, an understanding of the local structure can underpin such studies and give an indication of nanoscale inhomogeneity.

Keywords: diffuse scattering; single crystal; short-range order; CePdSb; Kondo

1. Introduction

Short-range order (SRO) is present in almost all families of crystalline compounds, from metals to proteins [1–8]. SRO can influence electrical, magnetic and most other physical properties, including ferroelectricity, superconductivity and multiferroic behaviour.

SRO manifests in the diffuse scattering, the coherent scattered intensity which is not localised on the reciprocal lattice; in other words, it is found throughout reciprocal space, not just on the Bragg reflections at integer hkl. Thus, to best investigate the diffuse scattering it is necessary to survey a large region (area or volume) of reciprocal space with low noise and high dynamic range. This is not a trivial exercise, and much effort has gone into data collection and reduction [6,9–11].

Data are typically presented as reciprocal space cuts or sections, which essentially plot diffracted intensity as a function of position in reciprocal space.

Metals were an early test-bed for ways of modelling SRO, in particular chemical SRO as modelled by, for example, Cowley SRO parameters [12–15]. Cowley realised that Fourier transforming the diffuse intensity could give atomic pair correlations when the scattering admitted a direct interpretation, for example when looking at a diffuse peak that would sharpen to a Bragg spot on going through a phase transition. Warren and co-workers showed how the atomic size effect (the dependence of interatomic spacing on species, most simply conceptualised as thinking about atoms as being of different radii) caused asymmetries in the scattering [16]. When the system is relatively simple, sometimes an analytical form can be found to yield the distribution of scattering.

If the underlying crystallography is simple, it may be possible to use an essentially analytic analysis, as for example can be obtained by expanding the diffraction equations [17] and using conditional probabilities to express the various terms. These probabilities can then be adjusted and the expected scattering calculated.

However, in more complex cases, in particular systems containing many atomic species and/or in which the atoms form into clusters with their own structure factors that then conflate with the scattering from the defects and the local ordering, it is often difficult or impossible to interpret the scattering directly or to meaningfully invert it to get the real space structures. These, and cases where we must allow for displacive relaxation around defects, require a more model-based approach. When contrast between scatterers is weak (atoms nearby on the periodic table will have very similar X-ray scattering factors), it may be necessary to use neutron and/or X-ray single-crystal diffuse scattering (SCDS) data. Neutron diffraction requires larger crystals, which may be difficult to obtain, so it may be that X-ray SCDS is coupled with neutron pair distribution function analysis (PDF; [18–20]), obtained from polycrystalline specimens.

A wide range of local structures have been observed in metallic compounds, from classic examples like chemical substitution and resulting clustering or anti-clustering in alloys, through to subtle phenomena related to the atomic size effect and even the rotation of large motifs, such as the cages of atoms seen in complex intermetallics [21]. For relatively simple systems, recent advances allow almost direct interpretation of the diffuse scattering, while developments in detailed calculation methods, like density functional theory and molecular dynamics, allow direct calculation of low energy short-ranged order configurations when not too many atoms are required [22–25].

However, when many atoms are involved and the correlation lengths encompass many unit cells, the number of atoms involved is beyond the scope of such methods. Then, the ability to model a crystal of $>10^5$ atoms becomes useful. Methods like 3D-ΔPDF [26] offer what are almost "direct methods" for such systems and are currently a fascinating field of development. The reverse Monte Carlo (RMC) approach [19,27,28] offers a means to directly fit the diffuse scattering data, but can be limited in the size of simulation that can be implemented because of the way in which a single atomic move must have a significant effect on the goodness of fit of the model.

Thus, at this time, the most flexible approach remains the forward Monte Carlo (MC), though it has its own weaknesses, in particular one must posit the nature of the disorder and then find a means of introducing that disordered structure into the model, before calculating the Fourier transform of the model and testing the theory. The process can be slow; models are difficult to optimise; and knowing what to include in the model (what forms of disorder and how to induce them) requires considerable insight. Further, since disorder can take on so many forms, it is often necessary to write bespoke computer code to tackle a given problem, something which is time consuming and not conducive to broad acceptance of the technique.

This paper aims to very briefly look at Monte Carlo analysis of diffuse scattering, particularly as it pertains to metallic materials, alloys and the like. The fascinating field of quasicrystals, many of which are metallic, will not be covered. This field has been surveyed in a range of detailed and high quality presentations, which need not be repeated here [29–31].

2. Data Collection

The experiments considered here use large slices of reciprocal space, rather than collecting intensity at a few key scattering vectors. This allows elucidation of SRO that is anisotropic or only affects small regions of reciprocal space. Similarly, the use of pair distribution function and powder diffraction is not discussed, though both are very important techniques [3,18,19,32].

The quality and quantity of data required depends, of course, on the experiment being undertaken. Ideally, the different scatterers will have well-differentiated cross-sections for the radiation being used. If the disorder is anisotropic, then data that extend in three dimensions are desirable. If local ordering is only significant in, say, the *ab* plane, then collection of the *hk*0 section of reciprocal space may be sufficient. If quantitative comparison of the calculated SCDS with the observed is desired [33], the observed data must show low noise, few artefacts, and a background that can be removed either by subtraction of "blank" runs or some other method, like fitting a function to it. For qualitative comparison with calculations, showing whether features are present or not, for example, noisier data may be acceptable, and the less quantitative results of electron diffraction are also useful. Analysis of SCDS is often limited by the data that can be obtained, but as long as features in the scattering can be identified as "real", then some insight can be gained.

2.1. X-ray

Assuming that the X-ray source is a constant wavelength, monochromated source, volumes of diffuse scattering are collected by rotating a sample in front of an area detector. Earlier work often made use of a line counter [34], but the modern prevalence of area detectors has rendered this approach largely redundant.

The main variation is in the choice of detectors. In particular, while much important data collection has made use of image plates [31,35–42], the use of electronic counters that can provide a high dynamic range has become possible [43–45]. These have a much improved duty-cycle. Experiments with image plates at synchrotrons, where beams are very intense, can follow an exposure of a few seconds, rarely more than 30 s, with a readout time of a minute or more, which is not good use of the intense and expensive beam.

Figure 1 presents a generic schematic diagram of a constant wavelength experiment. The main parameters include the sample to detector distance, the wavelength and whether the beam path is enclosed in a vacuum or He-filled vessel, which reduces noise, or is through air, which tends to result in intense forward scattering that requires careful correction and collections of "blank" runs, which can then be subtracted from the data. Other corrections may be required depending on the nature of the detector and the stability of the beam and the nature of the beam. If a laboratory source is being used, the compromise between intensity and quality of monochromation can result in the beam possessing a white component, which is much weaker than the characteristic radiation, but nevertheless results in a radial streak through the Bragg peaks, because of the long exposures required to reveal the diffuse scattering. Other artefacts that would not be apparent in an experiment using shorter exposure times may also be revealed. These include X-rays that pass through the image plate and scatter off components of the detector and re-enter the image plate from behind (this was discovered when the shadows of the image plate mounting screws were projected onto the detector(!)), as well as resolution streaks, discussed in Figure 15 of [46].

The high intensities at a synchrotron can cause problems when the area detector intercepts a Bragg reflection; depending on the design of the detector, a wire or a pixel can become saturated. In CCD devices, charge can spill over and contaminate surrounding pixels (deep depletion devices overcome this somewhat); in a wire detector, a bright spot anywhere on the wire may force the removal from the data set of all "pixels" measured by that wire [11].

Other issues include ghosting, when a pixel value on a measurement is partly influenced by the previous measurement. This can happen in image plates, where a very highly exposed pixel may not be fully "reset" by the readout, and thus, its value on the next exposure is not correct.

Traditionally, flat reciprocal space cuts have been reconstructed from the curved sections collected in an experiment such as that in Figure 1. Flat sections generally admit to easier visual interpretation, as the normal is everywhere the same and corresponds to a particular reciprocal space direction. However, from a computational point of view there is little difference between calculating the scattering in a flat or curved section. Further, at high X-ray energies the radius of the Ewald sphere is so large that each exposure is almost a flat section in reciprocal space anyway. In such cases, it is sensible to align the crystal carefully, such that useful data can be obtained with relatively few exposures. This leads to the ability to do parametric studies of diffuse scattering, which is an area under-exploited at this time.

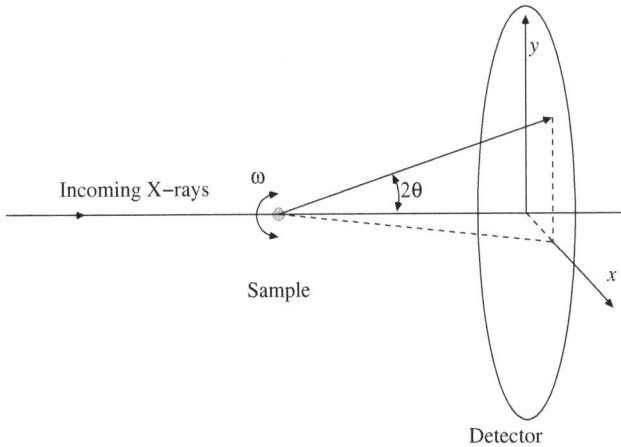

Figure 1. A schematic diagram of a diffuse scattering collection using a 2D detector. The sample angle is ω; incoming X-rays are of known wavelength, λ; and the scattering angle is as usual 2θ; but because we wish to transform the detector coordinates into hkl's, we work with x and y coordinates on the detector. During a single exposure, the sample is typically rocked through an angle $d\omega \sim 0.25°$, then ω is incremented by $d\omega$ and the measurement repeated. After $180/d\omega$ such exposures, enough data points have been collected to reconstruct most of reciprocal space [10] out of the the maximum value, which is given by the radius of the detector, the sample-detector distance, and λ.

It may be noted that static and dynamic displacements cannot be distinguished with an X-ray experiment because, compared to the high energies of the X-rays, all atomic motions are of very low energy (seem very "slow") and are seen as "static"; this is an area where neutrons may be preferable.

2.2. Neutron

Neutron diffraction comes in essentially two varieties: constant wavelength and time of flight. The latter is most commonly found at a spallation neutron source, while the former is found at reactor sources or a steady-state spallation source, like SINQ, the Swiss Spallation Neutron Source.

A constant wavelength experiment essentially uses the same configuration as for the X-ray case (Figure 1). A typical example is the Wombat instrument at the Bragg Institute at the Australian Nuclear Science and Technology Organisation (ANSTO) [47,48]. This instrument uses a two-dimensional detector to collect a sort of "cake slice" of diffraction space, such that data collected at multiple sample angles can be combined to give a volume from which sections can be extracted. Such an instrument does not select for neutron energy, so scattering from dynamic effects like phonons overlaps with that from static structures like chemical short-range order. This is much as for X-rays, except that the neutron energy is much lower, and inelastic effects may change the neutron wavelength substantially, which has the effect of "moving" the scattered beam around on the detector and, thus, shifting the inelastic scattered intensity to different positions in the reciprocal space map. Such effects can in some cases be interpreted usefully [49]. They do lead to a reduction of the symmetry of the pattern and may limit the ability to quantitatively model the scattering. If diffuse scattering is measured using an instrument that can select for neutron energy, for example a chopper spectrometer, then static can be separated from dynamic, although that depends on the energy resolution of the instrument; quasi-elastic scattering may be binned in with the "strictly elastic" scattering.

At a spallation neutron source, the time structure of the pulse collapses an entire diffraction pattern into a single pixel on a detector, meaning that such instruments, for example SXD (single crystal diffractometer) at ISIS [50,51] and TOPAZ at the Spallation Neutron Source [52], collect very large volumes of reciprocal space with a single sample setting. Rotating the sample leads to rapidly scanning a large volume, generally much larger than that accessible at a constant wavelength source. On the other hand, instrument resolution can vary dramatically from forward- to back-scattering detectors, and since the experiment is essentially imaging reciprocal space, this can affect the interpretability of some patterns. Further, such instruments are often "open" in geometry, without collimation between sample and detector. Thus, they effectively image the sample onto the detector, meaning that anisotropic sample shape can lead to odd-shaped features. This is not an issue when the feature is to be integrated up to get an intensity for conventional Bragg analysis, but when reciprocal space maps are being looked at, it can have an effect.

It is possible to use energy discrimination on spallation instruments [49,53], and again, this yields the possibility of separating dynamic from static effects.

Whether constant wavelength or spallation, polarisation analysis can be used to separate magnetic from structural diffuse scattering [54–57].

3. Basic Principles of Monte Carlo Modelling of SRO

This topic is dealt with in great detail elsewhere [17,58–60], so a simple outline will suffice; Figure 2 summarises the process.

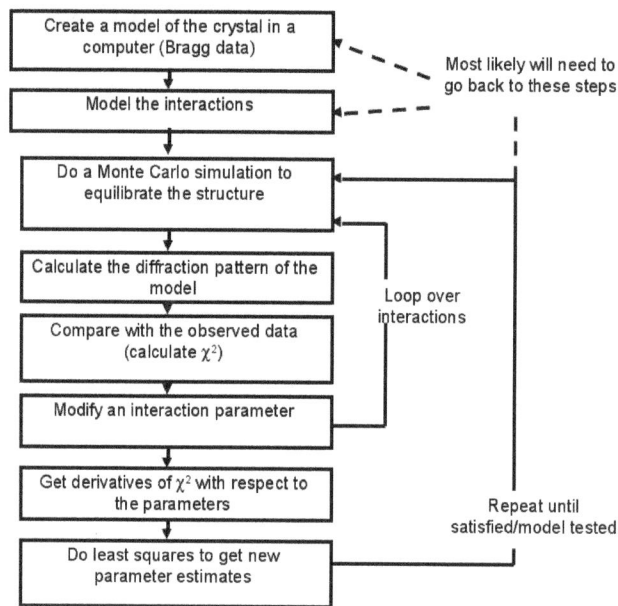

Figure 2. The overall MC modelling procedure. The flow chart illustrated in Figure 3 is an expansion of the box labelled "Do a Monte Carlo simulation to equilibrate the structure". This diagram assumes a least squares procedure based on calculating a χ^2 statistic for the model (or perhaps a kind of R-factor [61,62]); but often, the comparison will be done heuristically by the investigator, and the results will be more qualitative. The initial model is based on the average structure from Bragg data.

At its simplest, the type of MC modelling considered here has just a few steps.

- Decide on a starting configuration for the model. This usually means creating (in a computer) a $M \times N \times P$ array of unit cells, typically 32 on a side, and populating it with atoms based on the average structure determined by conventional studies.
- Choose some interactions between atoms. To set up chemical SRO when there are two species, a typical interaction is a Ising-like potential for the energy associated with the occupancy of site i, E_{occ}^i:

$$E_{occ}^i = -J_{NN} \sum_{NN} S_i S_j \qquad (1)$$

where j indexes nearest neighbours and the sign of J determines whether a positive or negative nearest neighbour occupancy correlation, C_{NN}, is energetically favourable. Further, such terms may be present for more distant neighbours. $S_j = \pm 1$.

If it is displacements that are of interest, the simplest choice is to connect atoms with Hooke's law springs The program ZMC [63] is designed to induce correlations amongst atomic and/or molecular displacements by causing the atoms to interact with surrounding atoms via Hooke's law springs of the form:

$$E_{\text{inter}} = \sum_{\text{cv}} F_i (d_i - d_{0i}(1 + \epsilon_i))^2 \qquad (2)$$

where d_i is the length of vector i connecting atoms, d_{0i} is its equilibrium length and F_i is its force constant. The sum is over all contact vectors (cv). ϵ_i is the "size-effect" term, which allows that the equilibrium length required for the calculation may not be the average length as determined from Bragg scattering; this is particularly likely to be the case in occupationally-disordered materials, where the Bragg-refined intermolecular distance is in fact an average over several different distances resulting from differing atomic or molecular species (or vacancies).

- The actual MC part happens as follows (summarised in Figure 3). An atom is chosen at random, and its energy is calculated. Its configuration is changed, and the energy calculation repeated. The new configuration is kept or rejected based on a simple criterion: if new energy is lower, it is kept, and it may be kept if new energy is higher, with some probability based on simulation "temperature".

- Note that the configuration may be changed by adding small random variations to an atom's variables (e.g., moving it slightly) or by swapping the variables of one site with those of another. Swapping is particularly useful as a means of maintaining an initial population of displacements or chemical species, while inducing correlations within that population.

- Once every site has been visited, on average, some large number of times, which could be ten, hundreds or thousands, depending on the needs of the simulation, the simulation is complete, and the atomic coordinates are read out.

- A Fourier transform program DIFFUSE [64] then calculates the diffuse scattering for comparison with the experiment.

- It is possible to embed this process within a procedure that automatically modifies the interaction parameters to try to improve the fit between calculated and observed diffuse scattering, although often useful results can be obtained by qualitative comparison, which can be used to reveal key aspects of the local order without comprehensive fitting.

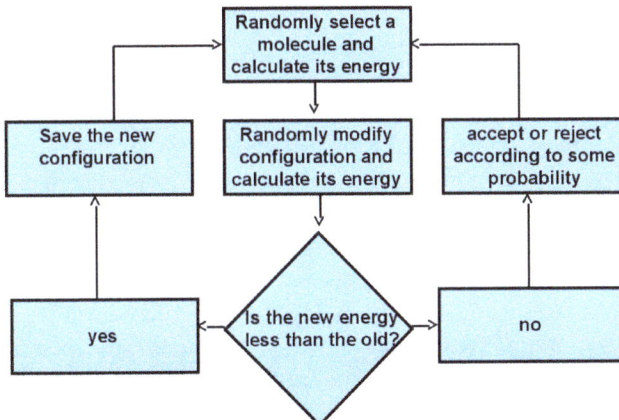

Figure 3. A simple representation of a single forward MC step; a molecule may be a single atom or a more complex motif.

The advantage of this approach is that the "energy" can be anything as long as it is quick to calculate. It may be relatively realistic or quite abstract, whatever suits the problem. However, knowing what disorder is present and then how to parameterise the interactions to induce it is not simple.

4. A Model System

In this section a model system, CePdSb, is considered from the point of view of inducing a range of local orderings and their resulting diffraction effects. No comparison with the observations is made, as we are looking simply to show how the disorder is modelled and some of the forms it can take.

CePdSb and related compounds form a family demonstrating a wide range of unusual magnetic phenomena, including the Kondo effect, heavy fermion behaviour and half-metal behaviour [65–71].

CePdSb itself shows a crystal structure in which the Pd and Sb lie on ordered sub-lattices at coordinates $(1/3, 2/3, 0.4684)$ and $(2/3, 1/3, 0.516)$ [66], with space-group $P6_3mc$ and lattice parameters approximately $a = 4.935\text{Å}$ and $c = 7.890\text{Å}$. This is different from an earlier structure in which the z coordinates of both Pd and Sb were taken as 0.5 [67] and the Pd and Sb were considered to be randomly mixed across the Pd/Sb sites.

The Ce atoms lie at $(0, 0, 1/3)$ 2a positions, forming chains along the c axis. The structure is represented in Figure 4. For the purposes of demonstrating various diffraction effects, we will explore what happens when the Ce2 site ($z = 2/3$) is occupied by approximately 67% Ce atoms and 33% vacancies.

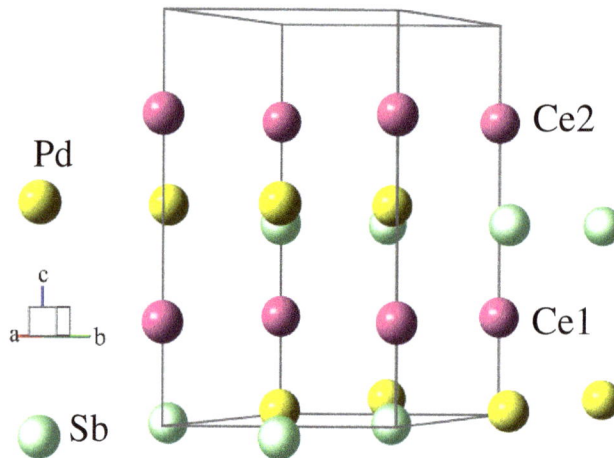

Figure 4. A schematic diagram of the structure of CePdSb, showing the Ce layers and the Pb/Sb layers, the latter of which are not flat, but "puckered" [72].

If we take the average structure of CePdSb [66] and calculate the diffuse scattering, we of course see nothing of interest, as there are no short-range correlations. However, we may, for example, connect atoms with Hooke's law springs (Equation (2)) and run a simulation. Figure 5 shows three sections through the diffuse scattering from CePdSb. The first row of images comes from a model in which there are no Ce vacancies and the atoms are connected by Hooke's law springs. The interactions induce streaks, most apparent in the $hk5.5$ layer. The second row shows the same cuts, but for a model in which there are 33% vacancies on the Ce2 layers, and they are forced to cluster. In the third row, the vacancies anti-cluster, and we can see in Figure 5i that this induces sharp spots in the half-layer, $hk5.5$, where previously, there were only streaks (the streaks are in fact sections through planes of scattering that can also be seen in the $hk5$ layer, though being less obvious due to the bright spots). We can also see that the clustering has little effect on the $hk5$ layer, while in $hk0$ it causes the spots that are present in hexagonal motifs around each Bragg peak (one hexagon is noted by white lines in Figure 5a) to extend closer to the origin. These spots actually come from the fact that the Pd and Sb atoms are not on idealised positions, such as $(1/3, 2/3, 1/2)$. When the vacancies cluster, we have large regions of the crystal where the scattering from the Ce2 layer is absent (effectively, these are like crystallites of composition $Ce_{0.5}PdSb$), giving different cancellation and allowing the spots to persist. When the vacancies anti-cluster (in the third row), the average scattering from Ce2 is preserved on the local scale, as well, and the cancellation is more like that seen in Figure 5a, though not identical.

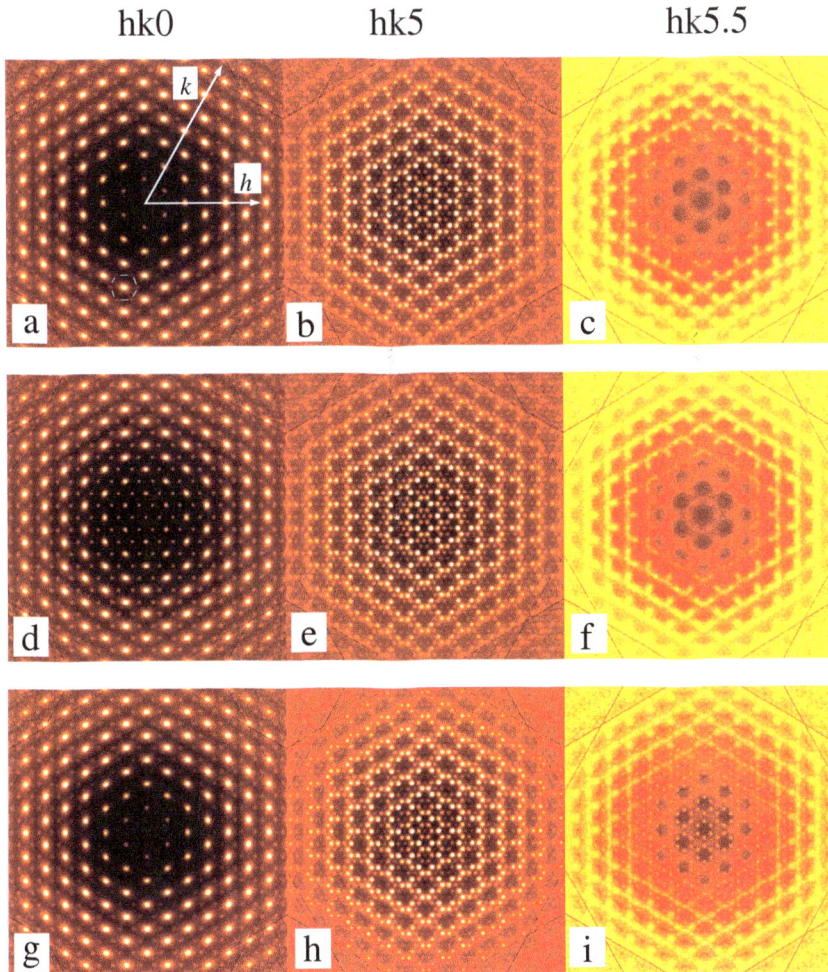

Figure 5. Slices of calculated diffuse scattering from different models of CePdSb. Row 1: no vacancies. Row 2: 33% vacancies on Ce2 site, clustering. Row 3: 33% vacancies on Ce2 site, anti-clustering. $hk5.5$ layers are normalised more brightly to bring out the details. For details, see the text. h and k axes noted on (**a**) to indicate directions.

In Figure 6, in rows 1 and 2, the displacive and occupancy effects are combined: the average distance atom-vacancy has been made 20% bigger than the average, while atom-atom is 10% smaller and vacancy-vacancy is 40% bigger. This is to mimic the effect sometimes seen where atoms move away from vacancies due to the lack of a bond, rather than moving into the gap. Row 1 is the model where the vacancies cluster; row 2 is where they avoid each other.

However, the third row of images in Figure 6 is the same as the second, but the size-effect signs have been reversed. Examining the two rectangles in Figure 6d,g shows how the brightness of consecutive spots is reversed by the change in size effect (white rectangles). Note however that other spots are relatively independent of this effect; this is one way in which this kind of modelling is useful, as it allows for the combined effects of the different structure factors (in a sense, each correlation has its own "structure factor") and form factors and how they interact.

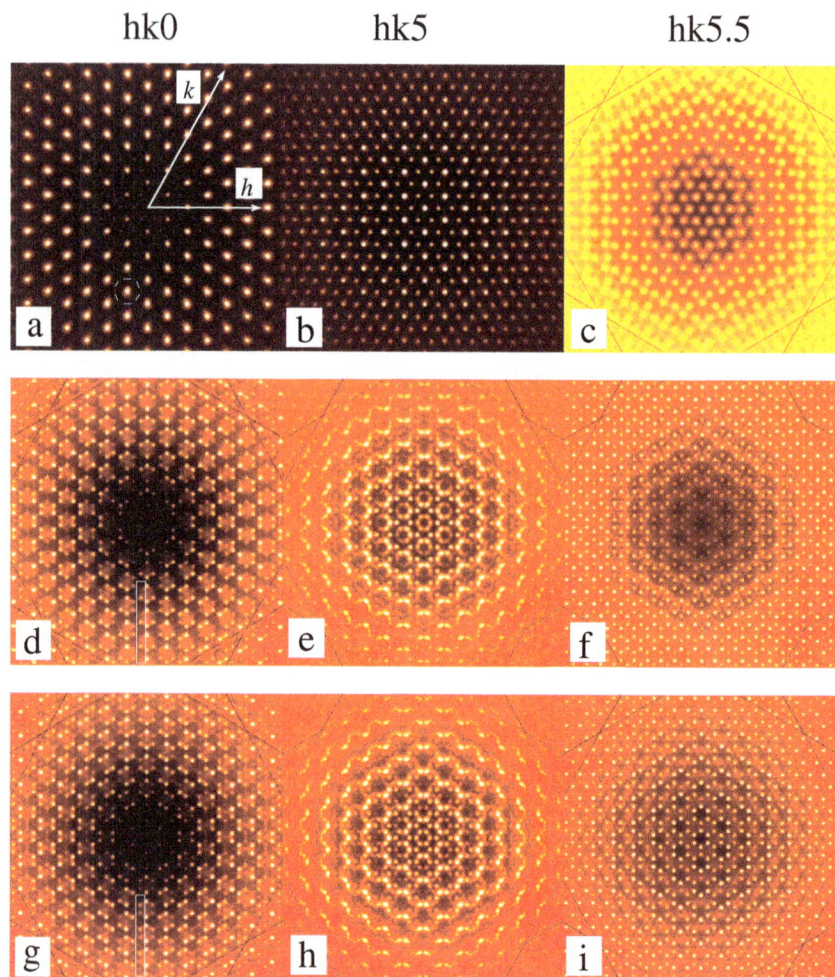

Figure 6. Slices of calculated diffuse scattering from different models of CePdSb, this time incorporating the atomic size effect. Row 1: Same as row 2 of Figure 5, but atoms move away from vacancies and vacancies away from each other. Row 2: Same as row 3 of Figure 5, but atoms move away from vacancies and vacancies away from each other. Row 3: Same as row 3 of Figure 5, but atoms move toward vacancies and vacancies toward each other. For details, see the text. h and k axes noted on one figure to indicate directions.

Note how the size effect is very different when applied to the clustering model (row 1) and the anti-clustering models (rows 2 and 3). Rows 2 and 3 of Figure 5 are different, but relatively subtly. Compare then rows 1 and 2 of Figure 6, which are the same two rows, now with the same kinds of size effects applied. Because the fraction of atom-vacancy bonds and atom-atom bonds is very different in the two models, the scattering is very different. This shows how strongly these effects can interact, something that can be difficult to disentangle without this kind of modelling to lean on.

Hence, even these relatively simple effects can have interesting and complex influences on the diffraction patterns of metallic systems. The MC model allows insight to be gained when the system is too complex to use direct inversion of the diffuse scattering to determine the correlations. In particular, exploring a range of representative models that look at various possible forms of SRO and their resulting diffraction is a useful guide to finding out what kinds of SRO are present in the real system.

5. Conclusions

Complex metallic systems, such as intermetallics, alloys, quasicrystals and Hume-Rothery phases, can all show detailed local ordering, which gives rise to highly structured and often very anisotropic single-crystal diffuse scattering. This paper reviews some of the issues associated with collecting and analysing such scattering and uses hypothetical calculation on the intermetallic CePdSb to illustrate some of the effects that may be observed in real systems.

Local order is important in determining many materials' properties and should not be ignored when trying to relate structure to function, especially when phenomena on the nanoscale are to be considered.

By qualitatively inducing various orderings in an MC model, the signatures of these orderings can be determined and compared to the observed data, providing guidance as to what structures are present in the real material.

Acknowledgements: Many thanks to T.R. Welberry and A.P. Heerdegen of the Australian National University for many useful discussions; they bear no responsibility for the opinions expressed herein. Thanks to Klaus-Dieter Liss for the invitation to write this article.

Conflicts of Interest: The author declares no conflict of interest.

References

1. Hukins, D.W.L. *X-ray Diffraction by Ordered and Disorderd Systems*; Pergamon Press: New York, NY, USA, 1981.
2. Krivoglaz, M.A. *Diffuse Scattering of X-rays and Neutrons by Fluctuations*; Springer-Verlag: Berlin, Germany, 1996.
3. Billinge, S.J.L.; Thorpe, M.F. *Local Structure from Diffraction*; Plenum: New York, NY, USA, 1998.
4. Welberry, T.R. Diffuse X-ray scattering and models of disorder. *Rep. Prog. Phys.* **1985**, *48*, 1543–1593.
5. Schweika, W. Disordered Alloys—Diffuse Scattering and Monte Carlo Simulation. In *Springer Tracts in Modern Physics*; Springer: Heidelberg, Germany, 1997; Volume 141.
6. Wall, M.; Adams, P.; Fraser, J.; Sauter, N. Diffuse X-ray Scattering to Model Protein Motions. *Structure* **2014**, *22*, 182–184.
7. Barabash, R.I.; Ice, G.E.; Turchi, P.E.A. *Diffuse Scattering and the Fundamental Properties of Materials*, 1st ed.; Momentum Press: New York, NY, USA, 2009.
8. Welberry, T.; Weber, T. One hundred years of diffuse scattering. *Crystallogr. Rev.* **2016**, *22*, 2–78.
9. Bürgi, H.B.; Weber, T. The structural complexity of a polar, molecular material brought to light by synchrotron radiation. *Mol. Cryst. Liq. Cryst.* **2003**, *390*, 1–4.
10. Estermann, M.A.; Steurer, W. Diffuse scattering data acquisition techniques. *Phase Transit.* **1998**, *67*, 165–195.
11. Welberry, T.R.; Goossens, D.J.; Heerdegen, A.P.; Lee, P.L. Problems in Measuring Diffuse X-ray Scattering. *Z. Krist.* **2005**, *220*, 1052–1058.
12. Cowley, J.M.; Gonnes, J. Diffuse scattering in electron diffraction. In *International Tables for Crystallography Volume B*; Springer: Dordrecht, The Netherlands, 1993; pp. 434–440.
13. Cowley, J.M. Kinematical Diffraction from Solid Solutions with Short Range Order and Size Effect. *Acta Crystallogr.* **1968**, *24*, 557–563.
14. Cowley, J.M. Short-Range Order and Long-Range Order Parameters. *Phys. Rev.* **1965**, doi:10.1103/PhysRev.138.A1384.
15. Cowley, J.M. Short- and Long-Range Order Parameters in Disordered Solid Solution. *Phys. Rev.* **1960**, *120*, 1648–1657.
16. Warren, B.E.; Averbach, B.L.; Roberts, B.W. Atomic Size Effect in the X-ray Scattering by Alloys. *J. Appl. Phys.* **1951**, *22*, 1493–1496.
17. Welberry, T.R. *Diffuse X-ray Scattering and Models of Disorder*; Oxford University Press: Oxford, UK, 2004.
18. Whitfield, R.E.; Goossens, D.J.; Welberry, T.R. Total scattering and pair distribution function analysis in modelling disorder in PZN ($PbZn_{1/3}Nb_{2/3}O_3$). *IUCrJ* **2016**, *3*, 20–31.

19. Neder, R.B.; Proffen, T. *Diffuse Scattering and Defect Structure Simulations: A Cook Book Using the Program DISCUS*; OUP: Oxford, UK, 2008.

20. Proffen, T.; Billinge, S.J.L. PDFFIT, a program for full profile structural refinement of the atomic pair distribution function. *J. Appl. Crystallogr.* **1999**, *32*, 572–575.

21. Henderson, R. A Cavalcade of Clusters: The Interplay Between Atomic and Electronic Structure in Complex Intermetallics. Ph.D. Thesis, Cornell University, Ithaca, NY, USA, January 2013.

22. Bosak, A.; Chernyshov, D. On model-free reconstruction of lattice dynamics from thermal diffuse scattering. *Acta Crystallogr. Sect. A* **2008**, *64*, 598–600.

23. Bosak, A.; Chernyshov, D.; Vakhrushev, S.; Krisch, M. Diffuse scattering in relaxor ferroelectrics: True three-dimensional mapping, experimental artefacts and modelling. *Acta Crystallogr. Sect. A* **2012**, *68*, 117–123.

24. Paściak, M.; Welberry, T.R. Diffuse scattering and local structure modeling in ferroelectrics. *Z. Krist.* **2011**, *226*, 113–125.

25. Maisel, S.B.; Schindzielorz, N.; Müller, S.; Reichert, H.; Bosak, A. An accidental visualization of the Brillouin zone in an Ni–W alloy via diffuse scattering. *J. Appl. Crystallogr.* **2013**, *46*, 1211–1215.

26. Simonov, A.; Weber, T.; Steurer, W. Yell: A computer program for diffuse scattering analysis *via* three-dimensional delta pair distribution function refinement. *J. Appl. Crystallogr.* **2014**, *47*, 1146–1152.

27. Nield, V.M.; Keen, D.A.; McGreevy, R.L. The interpretation of single-crystal diffuse scattering using reverse Monte Carlo modelling. *Acta Crystallogr. Sect. A* **1995**, *51*, 763–771.

28. Tucker, M.G.; Keen, D.A.; Dove, M.T.; Goodwin, A.L.; Hui, Q. RMCProfile: Reverse Monte Carlo for polycrystalline materials. *J. Phys. Condens. Matter* **2007**, *19*, 335218.

29. Steurer, W. Twenty years of structure research on quasicrystals. Part 1. Pentagonal, octagonal, decagonal and dodecagonal quasicrystals. *Z. Krist.* **2004**, *219*, 391–446.

30. Estermann, M.; Lemster, K.; Haibach, T.; Steurer, W. Towards the real structure of quasicrystals and approximants by analysing diffuse scattering and deconvolving the patterson. *Z. Krist.* **2000**, *215*, 584–596.

31. Estermann, M.; Steurer, W. Surveying the Entire Reciprocal Space of Quasicrystals with Imaging Plate Technology. In *Quasicrystals*; Janot, C., Mosseri, R., Eds.; World Scientific: Singapore, 1995.

32. Egami, T.; Billinge, S.J.L. *Underneath the Bragg Peaks, Structural Analysis of Complex Materials*; Pergamon: Oxford, UK, 2003.

33. Welberry, T.R. Diffuse X-ray Scattering and Disorder in *p*-methyl-*N*-(*p*-chlorobenzylidene)aniline $C_{14}H_{12}ClN$ (ClMe): Analysis *via* Automatic Refinement of a Monte Carlo Model. *Acta Crystallogr.* **2000**, *56*, 348–358.

34. Osborn, J.C.; Welberry, T.R. A Position-Sensitive Detector System for the Measurement of Diffuse X-ray Scattering. *J. Appl. Crystallogr.* **1990**, *23*, 476–484.

35. Templer, R.H.; Warrender, N.A.; Seddon, J.M.; Davis, J.M. The Intrinsic Resolution of X-ray Imaging Plates. *Nucl. Instrum. Methods* **1991**, *310*, 232–235.

36. Miyahara, J.; Takahashi, K.; Amemiya, Y.; Kamiya, N.; Satow, Y. A New Type of X-ray Area Detector Utilizing Laser Stimulated Luminescence. *Nucl. Instrum. Methods* **1986**, *246*, 572–578.

37. Gibaud, A.; Harlow, D.; Hastings, J.B.; Hill, J.P.; Chapman, D. A High-Energy Monochromatic Laue (MonoLaue) X-ray Diffuse Scattering Study of $KMnF_3$ Using an Image Plate. *J. Appl. Crystallogr.* **1997**, *30*, 16–20.

38. Amemiya, Y.; Matsushita, T.; Nakagawa, A.; Satow, Y.; Miyahara, J.; Chikawa, J. Design and Performance of an Imaging Plate System for X-ray Diffraction Study. *Nucl. Instrum. Methods* **1988**, *266*, 645–653.

39. Bourgeois, D.; Moy, J.P.; Svensson, S.O.; Kvick, A. The Point-Spread Function of X-ray Image-Intensifiers/CCD-Camera and Imaging-Plate Systems in Crystallography: Assessment and Consequences for the Dynamic Range. *J. Appl. Crystallogr.* **1994**, *27*, 868–877.

40. Iwasaki, H.; Matsuo, Y.; Ohshima, K.I.; Hashimoto, S. Time-Resolved Two-Dimensional Observation of the Change in X-ray Diffuse Scattering from an Alloy Single Crystal Using an Imaging Plate on a Synchrotron-Radiation Source. *J. Appl. Crystallogr.* **1990**, *23*, 509–514.

41. Thomas, L.H.; Welberry, T.R.; Goossens, D.J.; Heerdegen, A.P.; Gutmann, M.J.; Teat, S.J.; Wilson, C.C.; Lee, P.L.; Cole, J.M. Disorder in pentachloronitrobenzene, $C_6Cl_5NO_2$: A diffuse scattering study. *Acta Crystallogr. B* **2007**, *63*, 663–673.

42. Welberry, T.R.; Goossens, D.J.; Haeffner, D.R.; Lee, P.L.; Almer, J. High-energy diffuse scattering on the 1-ID beamline at the Advanced Photon Source. *J. Synchrotron Radiat.* **2003**, *10*, 284–286.

43. Arndt, U.W. X-ray Position-Sensitive Detectors. *J. Appl. Crystallogr.* **1986**, *19*, 145–163.

44. Henrich, B.; Bergamaschi, A.; Broennimann, C.; Dinapoli, R.; Eikenberry, E.; Johnson, I.; Kobas, M.; Kraft, P.; Mozzanica, A.; Schmitt, B. PILATUS: A single photon counting pixel detector for X-ray applications. *Nucl. Instrum. Methods Phys. Res. Sect. A* **2009**, *607*, 247–249.

45. Seeck, O.H.; Murphy, B. *X-ray Diffraction: Modern Experimental Techniques*, 1st ed.; CRC Press: Singapore, 2015.

46. Liss, K.D.; Bartels, A.; Schreyer, A.; Clemens, H. High-Energy X-rays: A tool for Advanced Bulk Investigations in Materials Science and Physics. *Textures Microstruct.* **2003**, *35*, 219–252.

47. Studer, A.J.; Hagen, M.E.; Noakes, T.J. Wombat: The high-intensity powder diffractometer at the OPAL reactor. *Phys. B Condens. Matter* **2006**, *385–386*, 1013–1015.

48. Whitfield, R.E.; Goossens, D.J.; Studer, A.J.; Forrester, J.S. Measuring Single-Crystal Diffuse Neutron Scattering on the Wombat High-Intensity Powder Diffractometer. *Metall. Mater. Trans. A* **2012**, *43A*, 1423–1428.

49. Welberry, T.R.; Goossens, D.J.; David, W.I.F.; Gutmann, M.J.; Bull, M.J.; Heerdegen, A.P. Diffuse neutron scattering in benzil, $C_{14}D_{10}O_2$, using the time-of-flight Laue technique. *J. Appl. Cryst.* **2003**, *36*, 1440–1447.

50. Keen, D.A.; Gutmann, M.J.; Wilson, C.C. SXD—The single-crystal diffractometer at the ISIS spallation neutron source. *J. Appl. Crystallogr.* **2006**, *39*, 714–722.

51. Welberry, T.R.; Gutmann, M.J.; Woo, H.; Goossens, D.J.; Xu, G.; Stock, C.; Chen, W.; Ye, Z.G. Single-crystal neutron diffuse scattering and Monte Carlo study of the relaxor ferroelectric $PbZn_{1/3}Nb_{2/3}O_3$ (PZN). *J. Appl. Crystallogr.* **2005**, *38*, 639–647.

52. Koetzle, T.F.; Bau, R.; Hoffmann, C.; Piccoli, P.M.B.; Schultz, A.J. Topaz: A single-crystal diffractometer for the spallation neutron source. *Acta Crystallogr. Sect. A* **2006**, *62*, s116.

53. Rosenkranz, S.; Osborn, R. Corelli: Efficient single crystal diffraction with elastic discrimination. *Pramana J. Phys.* **2008**, *71*, 705–711.

54. Schweika, W.; Böni, P. The instrument DNS: Polarization analysis for diffuse neutron scattering. *Physica B* **2001**, *297*, 155–159.

55. Ersez, T.; Kennedy, S.; Hicks, T.; Fei, Y.; Krist, T.; Miles, P. New features of the long-wavelength polarisation analysis spectrometer LONGPOL. *Phys. B Condens. Matter* **2003**, *335*, 183–187.

56. Stewart, J.R.; Deen, P.P.; Andersen, K.H.; Schober, H.; Barthélémy, J.F.; Hillier, J.M.; Murani, A.P.; Hayes, T.; Lindenau, B. Disordered materials studied using neutron polarization analysis on the multi-detector spectrometer, D7. *J. Appl. Crystallogr.* **2009**, *42*, 69–84.

57. Klose, F.; Constantine, P.; Kennedy, S.J.; Robinson, R.A. The Neutron Beam Expansion Program at the Bragg Institute. *J. Phys. Conf. Ser.* **2014**, *528*, 012026.

58. Welberry, T.R.; Goossens, D.J. The interpretation and analysis of diffuse scattering using Monte Carlo simulation methods. *Acta Crystallogr. Sect. A* **2008**, *64*, 23–32.

59. Schweika, W. *Disordered Alloys: Diffuse Scattering and Monte Carlo Simulations*; Springer: Berlin, Germany, 1998.

60. Binder, K. *Monte Carlo Methods in Statistical Physics*; Springer: Berlin, Germnay, 1979.

61. Chan, E.J.; Goossens, D.J. Study of the single-crystal X-ray diffuse scattering in paracetamol polymorphs. *Acta Cryst. B* **2012**, *B68*, 80–88.

62. Welberry, T.R.; Goossens, D.J.; Edwards, A.J.; David, W.I.F. Diffuse X-ray scattering from benzil, $C_{14}D_{10}O_2$: Analysis via automatic refinement of a Monte Carlo model. *Acta Cryst.* **2001**, *A57*, 101–109.

63. Goossens, D.J.; Heerdegen, A.P.; Chan, E.J.; Welberry, T.R. Monte Carlo Modelling of Diffuse Scattering from Single Crystals: The Program ZMC. *Metall. Mater. Trans. A* **2010**, *42A*, 23–31, doi:10.1007/s11661-010-0199-1.

64. Butler, B.D.; Welberry, T.R. Calculation of Diffuse Scattering from Simulated Crystals: A Comparison with Optical Transforms. *J. Appl. Crystallogr.* **1992**, *25*, 391–399.

65. Ślebarski, A. Half-metallic ferromagnetic ground state in CePdSb. *J. Alloy. Compd.* **2006**, *423*, 15–20.

66. Riedi, P.; Armitage, J.; Lord, J.; Adroja, D.; Rainford, B.; Fort, D. A ferromagnetic Kondo compound: CePdSb. *Phys. B Condens. Matter* **1994**, *199–200*, 558–560.

67. Malik, S.; Adroja, D. Magnetic behaviour of RPdSb (R = rare earth) compounds. *J. Magn. Magn. Mater.* **1991**, *102*, 42–46.

68. Katoh, K.; Ochiai, A.; Suzuki, T. Magnetic and transport properties of CePdAs and CePdSb. *Phys. B Condens. Matter* **1996**, *223–224*, 340–343.

69. Malik, S.K.; Adroja, D.T. CePdSb: A possible ferromagnetic Kondo-lattice system. *Phys. Rev. B* **1991**, *43*, 6295–6298.

70. Lord, J.S.; Tomka, G.J.; Riedi, P.C.; Thornton, M.J.; Rainford, B.D.; Adroja, D.T.; Fort, D. A nuclear magnetic resonance investigation of the ferromagnetic phase of CePdSb as a function of temperature and pressure. *J. Phys. Condens. Matter* **1996**, *8*, 5475.

71. Neville, A.; Rainford, B.; Adroja, D.; Schober, H. Anomalous spin dynamics of CePdSb. *Phys. B Condens. Matter* **1996**, *223–224*, 271–274.

72. Ozawa, T.C.; Kang, S.J. Balls & Sticks: Easy-to-use structure visualization and animation program. *J. Appl. Crystallogr.* **2004**, *37*, 679, doi:10.1107/S0021889804015456.

15

Formation and Dissolution of γ′ Precipitates in IN792 Superalloy at Elevated Temperatures

Pavel Strunz [1,*], Martin Petrenec [2,†,‡], Jaroslav Polák [2,3,†], Urs Gasser [4] and Gergely Farkas [5]

[1]   Nuclear Physics Institute ASCR, CZ-25068 Řež near Prague, Czech Republic
[2]   Institute of Physics of Materials of the ASCR, CZ-61662 Brno, Czech Republic; mpetrenec@gmail.com (M.P.); polak@ipm.cz (J.P.)
[3]   CEITEC Institute of Physics of Materials of the ASCR, CZ-61662 Brno, Czech Republic
[4]   Laboratory for Neutron Scattering, PSI, CH-5232 Villigen, Switzerland; urs.gasser@psi.ch
[5]   Department of Physics of Materials, Faculty of Mathematics and Physics, Charles University, Ke Karlovu 5, 121 16, Prague 2, Czech Republic; farkasgr@gmail.com
*   Correspondence: strunz@ujf.cas.cz
†   These authors contributed equally to this work.
‡   Present address: TESCAN, a.s., CZ-62300 Brno, Czech Republic.

Academic Editor: Hugo Lopez

Abstract: Precipitation of γ′ phase in nickel-base superalloy IN792-5A was studied using *in-situ* Small Angle Neutron Scattering (SANS). It was found that additional precipitates are formed after reheating above 600 °C when the material is previously fast cooled (100 K/min) from 900 °C. The size distribution and volume fraction of the additional γ′ precipitates as well as of the already present medium-size precipitates in dependence on temperature were evaluated. The small precipitates can influence mechanical properties of the alloy, which exhibits an anomaly in the temperature dependence of the yield stress. Volume fraction of all precipitate populations above 900 °C was estimated as well.

Keywords: metals; high temperature alloys; superalloy; precipitation; neutron scattering; *in-situ* neutron diffraction; small-angle neutron scattering

---

1. Introduction

Excellent strength of Ni-base superalloys comes from their microstructure composed of strengthening γ′-precipitates (L1$_2$ lattice) coherently embedded in γ solid solution (fcc) matrix [1,2]. Morphology of γ′ precipitates after standard heat treatment is spherical or cuboidal, depending on the lattice misfit, with dimensions in the range 1000–7000 Å [3].

Precipitate microstructure was examined in the past in IN738LC alloy both *ex situ* and *in situ* at elevated temperatures [4] using Small-Angle Neutron Scattering (SANS) technique [5]. It was found that additional precipitates in the channels between the large primary ones are formed either during slow cooling from high temperature or after reheating above 570 °C [4]. The new precipitates presumably affect mechanical properties as such small (tertiary) precipitates very significantly contribute to strengthening in polycrystalline superalloys [6,7].

Neutron diffraction offers a unique tool for *ex-* or *in-situ* bulk investigation of superalloy microstructure. While *ex-situ* SANS brings information on precipitate morphology, size and specific interface in superalloys (see e.g., [8,9] and references therein), *in-situ* SANS studies, moreover, are able to follow the evolution of the microstructure of superalloys directly at high temperature [10–14]. This approach has important benefits when compared with room temperature measurements, as the morphological changes occurring on cooling do not influence the results of the microstructural characterization. Moreover, SANS is an integral method that can extract information from a large amount of precipitates in bulk ($\approx 4 \times 10^{11}$ particles in the present experiment when counting only large precipitates). The results are thus not influenced by local inhomogeneities in the specimens, which could be the case when using microscopic methods. Therefore, SANS technique can be effectively used to map the temperature threshold where the formation of precipitates starts, to investigate the kinetics of their growth and to find the temperature limit for their dissolution. It can also bring information on volume fraction of precipitates, which is an important input parameter for strengthening modeling [6,15].

The present experiment was focused on *in-situ* investigation of precipitate formation and dissolution at elevated temperatures in another type of Inconel superalloy, IN792-5A. The initial *ex-situ* SANS tests carried out with IN792-5A alloy indicated that the formation of a new population of precipitates with slow kinetics occurs similarly to that in previously investigated IN738LC superalloy. The principal task of this study was similar as in the case of IN738LC alloy: to examine if and at which temperature the secondary precipitation occurs in IN792-5A and at which temperature the small precipitates disappear. The temperature intervals corresponding to the changes in the precipitate distribution can be correlated with the temperature domain where the anomaly of the mechanical properties appears. Simultaneously, the dissolution of large primary precipitates was studied in the present *in-situ* SANS experiment.

2. Material and Methods

2.1. Specimens and Thermo-Mechanical Treatment

IN792-5A is a cast polycrystalline Ni-base superalloy for turbine rim of small supplementary energy units in aircrafts. Its composition is reported in Table 1. The heat treatment of the superalloy was as follows: $(1120 \pm 5)\,^\circ\text{C}/4$ h air stream cooling, $(1080 \pm 5)\,^\circ\text{C}/4$ h air cooling, and $(845 \pm 5)\,^\circ\text{C}/24$ h air cooling.

Table 1. The chemical composition (wt. %) of the IN792-5A superalloy [16] and the approximate chemical composition of γ' precipitates in this superalloy taken from the reference [17] determined by energy dispersive spectroscopy. (Note: Hf reported in 0.5% to be present in the alloy used in [17] was not present in the alloy used in the present experiment. Proportional adjustment for the other elements to 100% was done). Scattering length densities (SLDs) at room temperature of the respective alloy (average) and of the γ' precipitates calculated using the given compositions are reported in the last column.

	Cr	Mo	C	Co	Fe	Zr	Nb	Al	B	Ti	Ta	W	Ni	SLD (10^9 cm^{-2})
alloy	12.28	1.81	0.078	8.87	0.16	0.031	0.1	3.36	0.015	3.98	4.12	4.1	rest	64.1
γ'	3.48			4.92				4.61		7.17	7.17	2.56	rest	66.8

Morphology of γ' precipitates after this standard heat treatment was bimodal, with large (mostly cuboidal) precipitates with size 630 nm and smaller spherical precipitates with dimension 190 nm [18]. Figure 1 displays Transmission Electron Microscopy (TEM) micrograph of typical precipitate microstructure in the material. 68% volume fraction was found using image analysis software (Adaptive Contrast Control) [18]. Table 1 also includes estimated composition of γ' precipitates taken from the EDS measurement reported by Dahl and Hald [17].

The specimens with diameter of 6 mm were subjected to cyclic loading (low-cycle fatigue test) [18] at various temperatures. The specimens were cyclically strained in a computer controlled electro-hydraulic MTS testing system with constant strain rate 2×10^{-3} s^{-1}. The total duration at the elevated temperature in the low-cycle fatigue test was roughly 6 h (approximately 4 h hold at the given temperature prior the cycling and 2 h cycling). The specimen was cycled at 900 °C with strain amplitude 0.28%. The saturated stress amplitude reached 375 MPa. The overall number of cycles was 3245. Nevertheless, similarly as in the case of IN738LC alloy [4], no influence of stress exposure on small precipitate formation and dissolution is expected.

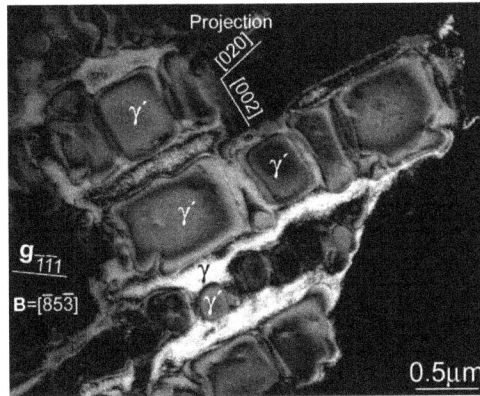

Figure 1. Transmission Electron Microscopy (dark field) micrograph of IN792-5A.

An interesting result for IN792-5A superalloy from an earlier study [19,20] was that it exhibits an anomaly in temperature dependence of the tensile properties in the temperature region 600–800 °C (Figure 2). Nevertheless, it is well known that strength of polycrystalline superalloys often increases between room temperature and 800 °C [6]. Such behavior cannot be explained by simple models of strengthening, and rather complex model considering all relevant strengthening mechanisms should be used [6].

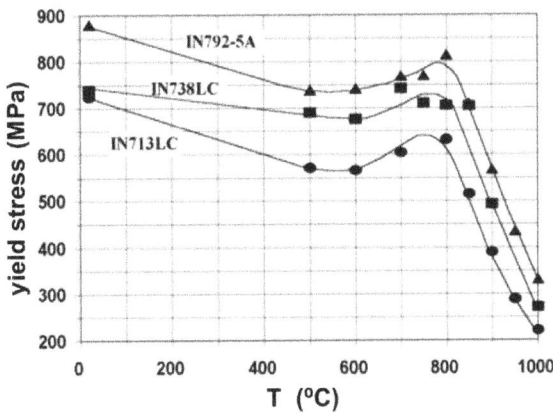

Figure 2. Temperature dependence of the yield stress for three Inconel superalloys. An anomaly (yield stress increase) was observed in the temperature region 600–800 °C [19].

2.2. SANS Technique

The IN792-5A sample was investigated by SANS *in situ* using vacuum furnace at temperatures up to 1120 °C. The pinhole SANS-II facility [21] at SINQ (PSI Villigen) was used for the measurement. Preliminary tests were performed using MAUD double-crystal SANS diffractometer (NPL lab of CANAM, NPI Řež, Czech Republic [22]).

Scattering length densities (SLDs) of the alloy and γ′ precipitates, necessary for the evaluation of the SANS data in an absolute scale, are reported in Table 1 (last column). When assuming volume fraction of precipitates to be 68% [18], then also the SLD of γ matrix can be calculated [4] with the result 58.3×10^9 cm^{-2}. The corresponding scattering contrast between γ′ precipitates and γ matrix is then 8.5×10^9 cm^{-2}.

Cylindrical samples were used for measurement. The neutron beam path was slightly less than 6 mm. The attenuation was still acceptable and multiple scattering did not influence significantly the scattering curves in the accessible region of scattering vector magnitude Q. The width of the slit was only 3.35 mm in order not to have an excessively broad distribution of thicknesses in the gauge volume. The average thickness (used in the raw-data treatment) was thus 5.62 mm. The slit height was 10 mm. Each sample installed into the furnace was adjusted to the beam using neutron sensitive camera. Expected thermal expansion of the stick, which was used for mounting the sample at high temperatures, was taken into account during the adjustment.

The scattering data were collected at several (reproducible) geometries during the hold at a particular temperature, *i.e.*, the measurements at various geometries were done with the same sample. The data acquisition time for one SANS pattern (*i.e.*, at one geometry) was 5–10 min, which enables to see possible microstructural change during hold at a temperature (the hold time > 50 min was always used).

The sample-to-detector distance was varied from 1.2 m to 6 m and the neutron wavelengths λ of 6.3 Å and 10.5 Å were used. The full covered range of Q ($Q = |\mathbf{Q}| = |\mathbf{k}-\mathbf{k}_0|$, where $\mathbf{k}_0$ and $\mathbf{k}$ are the wave vectors of the incident and of the scattered neutrons, respectively, and $|\mathbf{k}| = |\mathbf{k}_0| = 2\pi/\lambda$), was 4.0×10^{-3} Å^{-1}–0.13 Å^{-1} (*i.e.*, 4.0×10^{-2} nm$^{-1} < Q < 1.3$ nm^{-1}).

The *in-situ* SANS experiment was performed using tantalum furnace with maximum cooling rate approximately 100 K/min in the temperature region 900–600 °C. The uncertainty in temperature determination using a thermocouple was ±8 K due to the possible gradient of temperature in the relatively large gauge volume. The temperature profile during *in-situ* SANS measurement is plotted in Figure 2. It consists basically of three regions: temperature increase (step-by-step, the step equal to 25 K) from 550 up to 700 °C, which should provide information on formation of the small additional precipitates, then the further part where dissolution of these small precipitates between 700 and 900 °C was tested, and finally the temperature increase from 900 up to 1120 °C which was intended to provide information on dissolution of large primary precipitates in IN792-5A.

The part dealing with dissolution of the small precipitates between 700 and 900 °C consisted of series of thermal cycles 900 °C–400 °C–700 °C–T_D, hold at each of the given temperatures for 1 h. Temperature T_D has been increased in the subsequent cycles always by 25 K. The list of T_D is thus as follows: 725, 750, 775, 800, 825, 850, and 875 °C. The aim was to ensure—for each T_D—the same preconditions and thus the same microstructure of small precipitates at the beginning of the hold at T_D temperature. The temperature 900 °C served (similarly as in the case of previous study with IN738LC [4]) for dissolution of all populations of small- and medium-size precipitates from the previous heat treatment steps. The annealing at 900 °C was thus a kind of "reset" of the precipitate microstructure—only large precipitates remain, all the others were dissolved. Then, the sample was cooled to 400 °C and later heated to 700 °C (1 h hold) in order to produce a new population of small precipitates. Afterwards, the sample was heated to T_D in order to test the precipitate dissolution.

The last part of the *in-situ* experiment, *i.e.*, the temperature increase from 900 up to 1120 °C (step equal to 50 K) with one-hour hold at the particular temperature was carried out for assessment of the volume fraction of all precipitates (*i.e.*, not only small and medium, but also the large ones) in IN792-5A. Since the same sample was used, the evolution of the γ' volume fraction in the whole temperature range up to 1120 °C could be determined (see Evaluation and Discussion Section).

The measured raw SANS data were corrected for background scattering and calibrated to absolute scale using the measurement of the (attenuated) primary beam [23]. In this way, macroscopic differential cross section $d\Sigma/d\Omega$ (Q) was obtained. A correction for efficiency and solid angle of the individual pixels of the 2D detector was also performed. The scattered intensity is assumed to originate predominantly from the compositional variations in the superalloy, *i.e.*, due to the presence of γ' precipitates.

3. Results

The *in-situ* SANS experiment was mainly focused on the determination of the temperatures at which the formation of small precipitates starts and at which the γ′ precipitates are dissolved.

3.1. Formation of Precipitates

The temperature was increased step by step (see Figure 3) up to 700 °C and the sample was held at each temperature for a long time in order to observe if the formation of precipitates occurs at that particular temperature or not. This can be recognized from scattering intensity increase at the largest Q-values as the smallest precipitates after their formation contribute to the intensity just in that Q-range. A clear increase of the intensity occurs first at 625 °C during 2.5 h hold, as can be seen from the scattering curves in Figure 4.

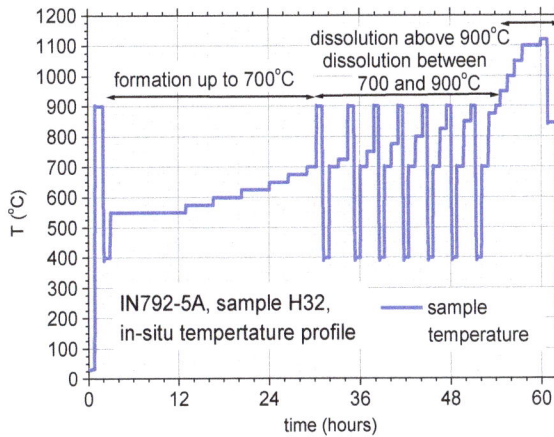

Figure 3. Thermal history for the *in situ* investigated sample. Different regions for the study of precipitate formation and dissolution by Small-Angle Neutron Scattering are marked.

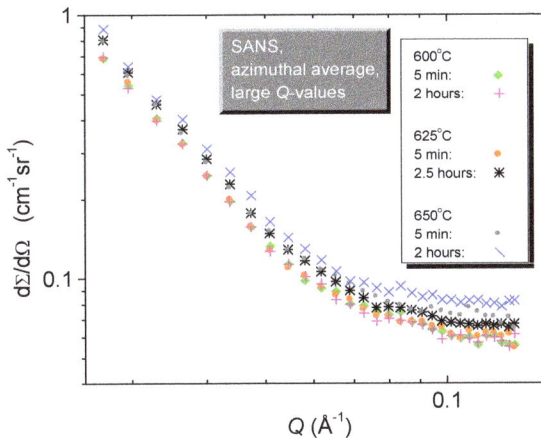

Figure 4. The large-Q part of the selected scattering curves (the scattering cross section dΣ/dΩ (Q)) taken at different temperatures in order to detect formation of the precipitates. Precipitates start to form only during 625 °C hold.

3.2. Evolution and Dissolution of Precipitates between 700 and 900 °C

Selected full scattering curves measured during *in-situ* heating are shown in Figure 5. From the scattering curves, it can be deduced that small precipitates grew with increasing temperature and also that they form a dense system for which the interparticle interference has to be taken into account when evaluating the SANS data.

The second part of the *in-situ* SANS experiment was focused on the determination of temperatures at which the small- and medium-size precipitates dissolve. The evolution of small precipitate morphology in the given temperature range can be deduced from the measured scattering curves. This temperature range was tested in a special sequence of temperatures described previously in the Experimental Section. The scattering curves measured in this temperature range can be seen in Figure 5.

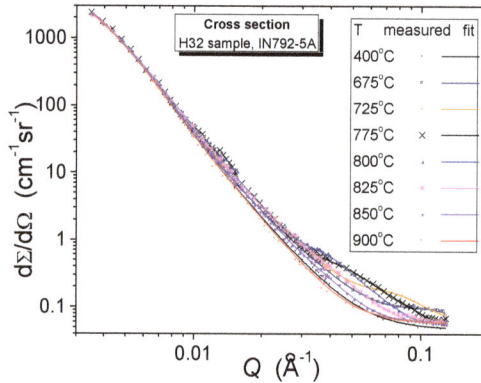

Figure 5. Selected scattering curves taken at different temperatures allowing detecting formation, evolution and dissolution of small precipitates. The solid lines are the fits of the model (discussed in "Evaluation and discussion" section) to the measured data.

3.3. Dissolution of Precipitates at and Above 900 °C

Primary large precipitates start to dissolve above 900 °C. A sequence of temperature steps and corresponding scattering curves at which this dissolution was tested by *in-situ* SANS can be seen in Figure 6. The intensity decreases with increasing temperature, but does not disappear completely at the highest temperature 1120 °C.

NOC program [24] was used for volume fraction estimation as this software takes into account multiple scattering from a thick sample, which is important for the large-precipitate scattering (unlike the scattering from the small and the medium-size populations).

Changes in the asymptotic part of the scattering intensity originating from the large precipitate population during the *in-situ* thermal exposure above 900 °C were then treated as coming from the relative decrease of their volume fraction. Proportionality was assumed.

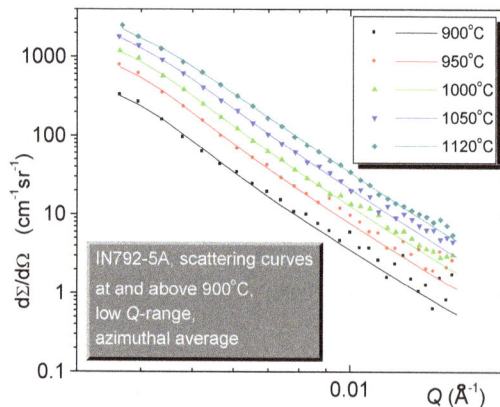

Figure 6. Scattering curves at low *Q*-values for the temperatures at and above 900 °C.

4. Evaluation and Discussion

4.1. Model for Evaluation

The SANS measurements at and below 900 °C were evaluated by the SASFIT program for SANS data treatment [25]. The analysis procedure is based on the simulation of a scattering profile generated from a set of size distributions of the particle system. In order to find the microstructural parameters which can be extracted from the measured data, the calculated SANS profile was matched with the experimental curve.

From the first qualitative analysis of SANS data and the observations using electron microscopy, a model was proposed which was used for the detailed analysis of the SANS data. The microstructural model was generally composed of three γ′ precipitate distributions:

(1) Small precipitates, mean sizes below 120 Å (radius < 60 Å), modeled by spherical-particle population with log-normal size distribution.

(2) Medium-size precipitates, mean sizes 70–600 Å, modeled by spherical-particle population with log-normal size distribution.

(3) Large precipitates with sizes larger than 600 Å (radius > 300 Å); these correspond predominantly to the secondary precipitates in the alloy. [Although shape of these precipitates is rather cuboidal, the spherical shape of particles was used in the evaluation also for these particles. It does not matter for evaluation as only the asymptotic part of the scattering is recorded from these particles at the used SANS facility and no details on shape and size of these particles are extracted from the measured data].

Pronounced interparticle interference effect for small particles (due to their relatively narrow size distribution as well as relatively dense arrangement of the small precipitates in the channels between the large precipitates) was observed in the initial period of new precipitate population formation. Therefore, hard-sphere model was used to approximate structure factor in SASFIT program. In all cases, this approximation was sufficient and the resulting model curves described very well the scattering data.

4.2. Volume Fraction

It has to be stressed that it is not possible to determine the total volume fraction of γ′ precipitates only from the SANS measurements because the size of the largest precipitates (size 6300 Å) is well above the detection limit for the used Q range. The scattering from these precipitates has nearly asymptotic character (except of the lowest Q-values) in the used Q range. At very low Q-values, due to the sample thickness, it is certain that also the multiple scattering takes place and influences the scattering curve.

Therefore, the estimate of the total volume fraction of precipitates in IN792-5A alloy at room temperature had to be taken from other techniques. The value 68% mentioned in the Experimental section of this paper was used.

As already mentioned above, the size of the large precipitates is well above the detection limit for the Q-range of the pin-hole SANS facility (SANS-II at SINQ) used for the *in-situ* measurements. The scattering from these precipitates is present but has nearly asymptotic character (*i.e.*, decreases as Q^{-4} with increasing Q) and the size cannot be determined from such shape of the scattering curve. Nevertheless, the scattering from the large precipitates has to be included in the model. Therefore, we modeled the scattering from large precipitates by scattering from the particles of their expected size (according to TEM results—see Figure 1) in order to get a realistic approximation of their part of the scattering. Naturally, no conclusion for size and for absolute volume fraction of large precipitates was drawn from the fit parameters for this particular part of the scattering. The volume fraction of large precipitates was adjusted in such a way, that the total volume fraction was equal to 68% at maximum.

On the other hand, the volume fraction of the medium-size and small precipitates can be determined from the SANS data. The magnitude of the scattering cross section can be employed for the evaluation of the volume fraction providing that the scattering contrast $\Delta\rho$ is known. We approximated $\Delta\rho$ by the value 8.5×10^9 cm^{-2} (see the Experimental Section).

4.3. Formation and Dissolution of the Precipitates

The evaluation of the *in-situ* SANS data resulted in a series of size distributions of small and medium precipitates at different elevated temperatures. The determined distributions are depicted in Figure 7. Further parameters were derived from these size distributions.

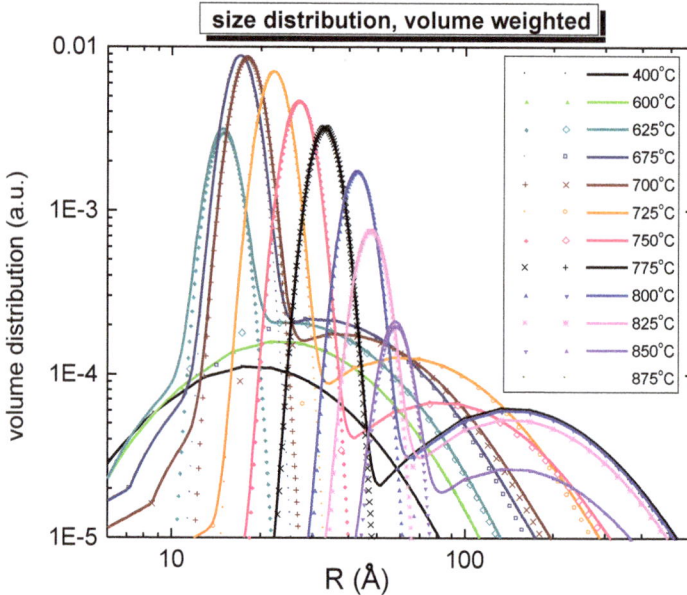

Figure 7. Size distributions determined for small- and medium-size precipitates at various temperatures. They corresponds to the fitted curves in Figure 6. Rather narow peak corresponds to the small precipitates, which is followed (towards larger radiuses) by a broader peak corresponding to the medium-size precipitates.

Figure 8 summarizes temperature dependence of the volume fraction during formation and dissolution. The volume fraction of small- and medium-size precipitates determined from the magnitude of the scattering cross section using the estimated scattering contrast is reported.

Precipitation of small particles—previously suppressed due to the fast cooling from high temperature (900 °C)—appears only after heating to temperatures above 600 °C. While the small precipitates form only on reheating, there was a certain amount of medium-size precipitates present already at 400 °C (*i.e.*, after the fast cooling from 900 °C).

The small- and medium-size precipitates are fully dissolved already around 875 °C. As can be seen from Figure 8, the dissolution process is size dependent: whereas the volume fraction of small precipitates decreases in the range 725–825 °C with increasing temperature, the volume fraction of the medium-size precipitates is constant or even slightly increases.

The volume fraction of small precipitates is at maximum in the temperature region 700–725 °C and is estimated to be around 0.045. On the other hand, there is less than 0.02 volume fraction of medium-size precipitates with maximum in the temperature region 775–800 °C. Figure 8 also shows total volume fraction of small- and medium-size precipitates with the maximum around 0.06 at 725 °C. Similarly as in [4], we estimate—thanks to the uncertainties in the scattering contrast determination—the error for the volume fraction determination of small- and medium-size precipitates to 30%.

The evolution of the size of the small precipitates after their formation could be determined from the *in-situ* SANS data. The size evolution for small- and medium-size precipitates in dependence on temperature and hold time is displayed in Figure 9. The small precipitates radii grew with time and temperature from 14 Å (at 625 °C) to 60 Å (at 850 °C, 50 min hold at the temperature).

The medium-size precipitates radii grew with time and temperature from 37 Å (at 400 °C) to 290 Å (at 775 °C, 50 min hold at the temperature). For the medium-size precipitates above 775 °C, it is not possible to determine their size due to the limited Q-range accessible at SANS-II and used for the measurement. Therefore, the size of medium-size precipitates was fixed for the evaluation of scattering curves above 775 °C, as shown in Figure 9.

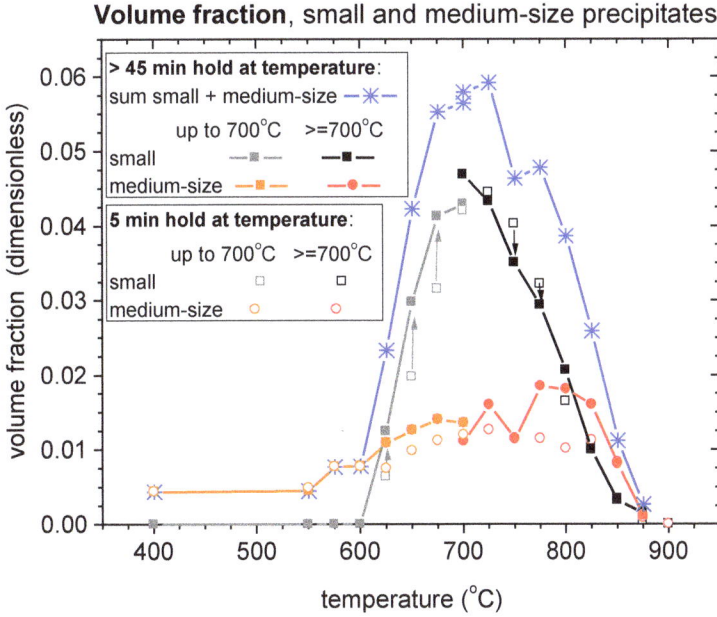

Figure 8. Volume fraction evolution of small- and medium-size precipitates in dependence on temperature and hold time.

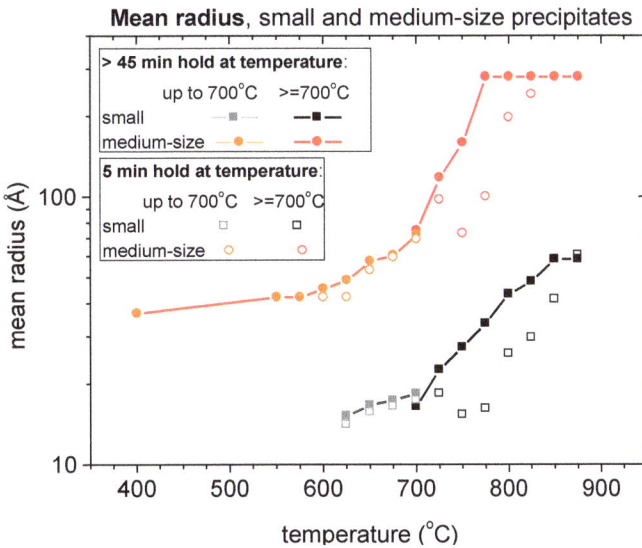

Figure 9. Size evolution of small- and medium-size precipitates in dependence on temperature and hold time.

In Figure 10, the volume fractions of individual precipitate populations are shown also for the temperatures, where large precipitates dissolve appreciably (particularly for temperatures 950, 1000, 1050 and 1120 °C). At 900 °C, both small- and medium-size precipitates are fully dissolved and only the large precipitates remain. Their volume fraction then gradually decreases with increasing temperature. Figure 10 also shows the total sum of all three precipitate populations.

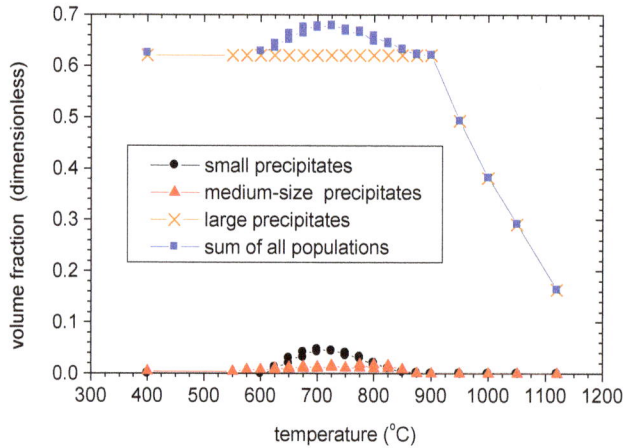

Figure 10. Estimation of total volume fraction dependence on temperature for IN792-5A previously fast cooled from 900 °C.

4.4. Strengthening by Tertiary Precipitates

As found by Kozar *et al.* for IN100 superalloy and its modification [6], the largest contribution to strengthening in multimodal superalloys are tertiary precipitates (they correspond in size to the small precipitates reported in the model used in this paper). Although there is several contributions for strengthening in polycrystalline superalloys (solid-solution strengthening, Hall–Petch mechanism, cross-slip pinning, strong pair-coupling, and weak pair-coupling), the main mechanism for the strengthening according to modeling [6] is—for alloy with large grains and significant amount of tertiary precipitates—weak pair-coupling due to tertiary (*i.e.*, small, with size less than 10–20 nm for IN100) precipitates followed by the strong pair-coupling mechanism due to the presence of larger (secondary) precipitates.

The earlier models of precipitate strengthening like the one of Kozar *et al.* [6] were improved recently to unify the weak pair-coupling and the strong pair-coupling mechanism into one comprehensive model [7]. Moreover, this model can be used for complex multimodal precipitate distributions as well. The new modeling approach was tested on several types of polycrystalline superalloys (RR1000, Udimet, IN100, KM4). Nevertheless, the qualitative output of the novel modeling approach [7] is for tertiary precipitates the same as found by Kozar *et al.*: the tertiary, *i.e.*, the small, precipitates provide the main part of strengthening.

Therefore, it is clear that small (tertiary) precipitates play a very significant role in strengthening. If they are formed at elevated temperatures in the channels between the larger (secondary) precipitates—as in our IN792-5A case—they certainly contribute to the strengthening of the superalloy and cause the increase of the yield strength with respect to lower temperatures. Moreover, the weak pair coupling strengthening, which is active up to the certain precipitate size, depends on the square root of precipitate size [6]. When the small precipitates grow, it thus additionally increases the strength of the alloy.

The last step of the standard heat treatment of the IN792-5A superalloy was the hold for 24 h at 845 °C, followed by air cooling. Such long hold resulted in microstructure with very low amount of tertiary precipitates (see Figure 8, small precipitate volume fraction at 845 °C) having, moreover, relatively large size. The relatively fast cooling (air cooling) from that temperature and slow kinetics hindered formation of new population of tertiary precipitates. Nevertheless, when heated again to elevated temperatures used for yield strength examination shown in Figure 2, new population of tertiary precipitates was formed during the tensile test in the temperature region 600–800 °C. It caused the observed (Figure 2) yield strength increase.

5. Conclusions

New precipitates in IN792-5A alloy arise at temperature (625 °C), higher than in IN738LC alloy (575 °C). The fast cooling from high temperature (900 °C) suppresses the precipitate formation due to the slow kinetics. The small precipitates (up to diameter of 120 Å) appear only after heating to temperatures higher than 600 °C. At temperatures above 875 °C, small precipitates are dissolved. The full dissolution of these small precipitates occurs in IN792-5A at a temperature that is approximately 25 K higher than in IN738LC alloy. The volume fraction of the small newly formed precipitates in IN792-5A is significantly lower than in IN738LC.

The size evolution and the volume fraction evolution of the small- and medium-size precipitates with temperature were determined. The maximum volume fraction of small precipitates at 725 °C is 4.5%. The full dissolution of the small- as well as the medium-size γ′ precipitates occurs around 875 °C. The presence of small- and medium-size precipitates in the temperature range 600–875 °C influences the yield strength of the IN792-5A alloy which exhibits an anomaly in this temperature range. Since the additional precipitation affects the mechanical properties of the alloy, these properties depend on the thermal history of the material. The observations in this study fit well with the modeling presented in earlier [6] or recent [7] studies highlighting significance of strengthening by tertiary precipitates in polycrystalline superalloys.

The volume fraction evolution of large primary γ′ precipitates at and above 900 °C was estimated. It was found that 1120 °C solution treatment (40 min) is not sufficient for their full dissolution.

Acknowledgments: The support by GACR project No. 14-36566G is gratefully acknowledged. The authors thank SINQ (PSI Villigen, Switzerland) and NPL (CANAM), NPI Řež, Czech Republic) for providing beamtime and support for the SANS measurements. Infrastructure projects support (NMI3-II EC project No. 283883 and CZ MSMT project No. LM2011019) are gratefully acknowledged as well.

Author Contributions: Martin Petrenec prepared the samples. Pavel Strunz, Urs Gasser and Gergely Farkas carried out the experiments. Pavel Strunz analyzed the SANS data. Pavel Strunz, with the help of Martin Petrenec and Jaroslav Polák, interpreted the results. Pavel Strunz and Jaroslav Polák prepared and revised the manuscript.

Conflicts of Interest: The authors declare no conflict of interest.

References

1. Ross, E.W.; Sims, C.T. Nickel-Base Alloys. In *Superalloys II: High-Temperature Materials for Aerospace and Industrial Power*, 2nd ed.; Sims, C.T., Stoloff, N.S., Hagel, W.C., Eds.; John Wiley and Sons: New York, NY, USA, 1987; Part 2, Chapter 4; pp. 97–134.

2. Reed, R.C. *The Superalloys, Fundamentals and Applications*; Cambridge University Press: New York, NY, USA, 2006.

3. Socrate, S.; Parks, D.M. Numerical determination of the elastic driving force for directional coarsening in Ni-superalloys. *Acta Metall.* **1993**, *41*, 2185–2209. [CrossRef]

4. Strunz, P.; Petrenec, M.; Gasser, U.; Tobiáš, J.; Polák, J.; Šaroun, J. Precipitate microstructure evolution in exposed IN738LC superalloy. *J. Alloy. Compd.* **2014**, *589*, 462–471. [CrossRef]

5. Kostorz, G. Small-Angle Scattering and Its Applications to Materials Science. In *Neutron Scattering: Treatise on Materials Science and Technology*; Kostorz, G., Ed.; Academic Press: New York, NY, USA, 1979; pp. 227–289.

6. Kozar, R.W.; Suzuki, A.; Milligan, W.W.; Schirra, J.J.; Savage, M.F.; Pollock, T.M. Strengthening mechanisms in polycrystalline multimodal nickel-base superalloys. *Metallurgical and Mater. Trans. A* **2009**, *40*, 1588–1603. [CrossRef]

7. Galindo-Nava, E.I.; Connor, L.D.; Rae, C.M.F. On the prediction of the yield stress of unimodal and multimodal gamma′ Nickel-base superalloys. *Acta Mater.* **2015**, *98*, 377–390. [CrossRef]

8.   Rogante, M.; Lebedev, V.T. Small angle neutron scattering comparative investigation of Udimet 520 and Udimet 720 samples submitted to different ageing treatments. *J. Alloy. Compd.* **2012**, *513*, 510–517. [CrossRef]

9.   Strunz, P.; Zrník, J.; Epishin, A.; Link, T.; Balog, S. Microstructure of creep-exposed single crystal nickel base superalloy CSMX4. *J. Phys.: Conf. Ser.* **2010**, *247*. [CrossRef]

10. Veron, M.; Bastie, P. Strain induced directional coarsening in nickel based superalloys: Investigation on kinetics using the small angle neutron scattering (SANS) technique. *Acta Mater.* **1997**, *45*, 3277–3282. [CrossRef]

11. Miller, R.J.R.; Messoloras, S.; Stewart, R.J.; Kostorz, G. Small-angle neutron-scattering study of temperature and stress dependence of microstructure of Nimonic alloys. *J. Appl. Cryst.* **1978**, *11*, 583–588. [CrossRef]

12. Mukherji, D.; Del Genovese, D.; Strunz, P.; Gilles, R.; Wiedenmann, A.; Rösler, J. Microstructural characterisation of a Ni-Fe-based superalloy by *in situ* small-angle neutron scattering measurements. *J. Phys.: Condens. Matter.* **2008**, *20*. [CrossRef]

13. Zickler, G.A.; Schnitzer, R.; Radis, R.; Hochfellner, R.; Schweins, R.; Stockinger, M.; Leitner, H. Microstructure and mechanical properties of the superalloy ATI Allvac®718Plus™. *Mater. Sci. Eng. A* **2009**, *523*, 295–303. [CrossRef]

14. Strunz, P.; Schumacher, G.; Klingelhöffer, H.; Wiedenmann, A.; Šaroun, J.; Keiderling, U. In situ observation of morphological changes of γ' precipitates in a pre-deformed single-crystal Ni-base superalloy. *J. Appl. Cryst.* **2011**, *44*, 935–944. [CrossRef]

15. Collins, D.M.; Stone, H.J. A modelling approach to yield strength optimisation in a nickel-base superalloy. *Int. J. Plast.* **2014**, *54*, 96–112. [CrossRef]

16. Petrenec, M.; Obrtlík, K.; Polák, J.; Kruml, T. Fatigue behaviour of cast nickel based superalloy Inconel 792–5A at 700 °C. *Mater. Technol.* **2006**, *40*, 175–178.

17. Dahl, K.V.; Hald, J. Identification of Precipitates in an IN792 Gas Turbine Blade after Service Exposure. *Pract. Metallogr.* **2013**, *50*, 432–450. [CrossRef]

18. Petrenec, M.; Obrtlík, K.; Polák, J.; Man, J. Dislocation structures in nickel based superalloy Inconel 792–5A fatigued at room temperature and 700 °C. *Mater. Sci. Forum* **2008**, *567–568*, 429–432. [CrossRef]

19. Podhorská, B.; Kudrman, J.; Hrbáček, K. Tepelné zpracování, mechanické vlastnosti a strukturní stabilita perspektivních litých niklových superslitin (in Czech). In METAL 2004, Proceedings of the 13th International Metallurgical & Material Conference, Hradec nad Moravicí, Czech Republic, May 2004; Tanger: Ostrava, Czech Republic, 2004; pp. 101–111.

20. Strunz, P.; Petrenec, M.; Gasser, U. Precipitate microstructure evolution in low-cycle fatigued Inconel superalloys. *Proceedings of the 15th International Small-Angle Scattering Conference*; McGillivary, D., Trewhella, J., Gilbert, E.P., Hanley, T.L., Eds.; Australian Nuclear Science and Technology Organization: Lucas Heights, New South Wales, Australia. Available online: http://trove.nla.gov.au/version/197519597 (accessed on 1 December 2015).

21. Strunz, P.; Mortensen, K.; Janssen, S. SANS-II at SINQ: Installation of the former Risø-SANS facility. *Phys. B* **2004**, *350*, e783–e785. [CrossRef]

22. Strunz, P.; Šaroun, J.; Mikula, P.; Lukáš, P.; Eichhorn, F. Double-Bent-Crystal Small-Angle Neutron Scattering setting and its applications. *J. Appl. Cryst.* **1997**, *30*, 844–848. [CrossRef]

23. Strunz, P.; Šaroun, J.; Keiderling, U.; Wiedenmann, A.; Przenioslo, R. General formula for determination of cross-section from measured SANS intensities. *J. Appl. Cryst.* **2000**, *33*, 829–833. [CrossRef]

24. Strunz, P.; Gilles, R.; Mukherji, D.; Wiedenmann, A. Evaluation of anisotropic small-angle neutron scattering data; a faster approach. *J. Appl. Cryst.* **2003**, *36*, 854–859. [CrossRef]

25. Kohlbrecher, J. SASfit: A Program for Fitting Simple Structural Models to Small Angle Scattering Data, 2014. Available online: https://kur.web.psi.ch/sans1/SANSSoft/sasfit.html, https://kur.web.psi.ch/sans1/SANSSoft/sasfit.pdf (accessed on 1 December 2015).

16

Continuous Casting of Incoloy800H Superalloy Billet under an Alternating Electromagnetic Field

Fei Wang, Lintao Zhang, Anyuan Deng, Xiujie Xu and Engang Wang *

Academic Editor: Johan Moverare

Key Laboratory of Electromagnetic Processing of Materials (Ministry of Education), Northeastern University, No. 3-11, Wenhua Road, Shenyang 110004, China; wangfei19860322@163.com (F.W.); l.zhang@swansea.ac.uk (L.Z.); dengay@epm.neu.edu.cn (A.D.); xuxj@epm.neu.edu.cn (X.X.)
* Correspondence: egwang@mail.neu.edu.cn

Abstract: We experimentally investigate the influence of an alternating electromagnetic field on the surface and internal qualities of Incoloy800H superalloy billets. The electromagnetic continuous casting experiments for Incoloy800H superalloy were successfully conducted and the billets (0.1 m × 0.1 m × 1.2 m) were obtained. We figure out that the high frequency (20.4 kHz) electromagnetic field which is applied in the mould region can improve the surface quality of Incoloy800H superalloy billet remarkably; the depth of oscillation mark decreases from 1.2 mm (without electromagnetic field) to 0.3 mm (with electromagnetic field). The internal quality of the billet was studied using a variety of characterization techniques. The low frequency (5 Hz) electromagnetic field which is applied in the second cooling region can improve the internal quality; the region of the equiaxed grain increases from 2.45% (without electromagnetic field) to 41.45% (with electromagnetic field). Furthermore, macro- and micro-segregation are suppressed and the TiN inclusion number is decreased as well.

Keywords: continuous casting; electromagnetic stirring; Incoloy800H superalloy; oscillation marks; segregation; TiN inclusion

1. Introduction

Incoloy800H superalloy (UNS N08810) is an austenitic Fe-Ni-Cr superalloy with excellent strength and resistance to oxidation and carburization, particularly under high-temperature conditions. The alloy is widely used in equipment that experiences long-term exposure to high temperatures and corrosive atmosphere, e.g., pigtails, radiant tubes and intermediate heat exchangers (IHX).

The corrosion behavior of Incoloy800H superalloy in supercritical water (374 °C, 22.1 MPa) was investigated [1] and the results showed that an oxidation process was observed as the primary corrosion behavior. Further research on this oxidation behavior, along with its mechanisms, in the vicinity of three different zones (containing the substrate, heat-effected zone and the melt zone) in dry and wet air conditions was conducted [2]. The results showed that the oxidation kinetics of three different zones followed the parabolic-rate law in dry and wet air. However, in wet air, the oxidation behavior experienced two different kinetics stages, consisting of an initial mass-gain stage followed by a second mass-loss stage. For the Incoloy800H superalloy welding parts which were processed by the method of laser shocking peening (LSP), the study of the microstructure and the residual stresses were carried out [3]. The mechanism of the grain refinement was due to twinning matrix formation and th strain-induced grain boundary cracking. Furthermore, the LSP process-welded joints exhibited high compressive and uniform distribution residual stress. The neutron diffraction method was also used to determine the residual stresses [4], e.g., for an 8 mm Incoloy800H weld.

The optimization of the diffusion welding process for Incoloy800H alloy was studied by numerical simulation [5]. Proper welding conditions were suggested, namely a temperature of 1423 K for an hour with an applied pressure of 5 MPa. Compression tests were conducted to investigate the hot deformation behavior of Incoloy800H superalloy at 1000 °C [6] and 750 °C [7], respectively. For temperatures above 1000 °C, a distinct change in slope of the linear fit curve regarding flow stress and temperature can be observed. For a temperature of 750 °C, the influence of the aging pre-treatment (for 0 h, 5 h, 10 h, 20 h and 50 h) is discussed in detail. The results revealed that the peak stress decreased whilst the aging time was increased. A further test regarding the relationship between fatigue crack growth (FCGR) and the external load ratios was conducted on Incoloy800H alloy at 750 °C [8]. As the load ratio increased, the crack growth increased as well. Through the thermo-mechanical process (TMP), the microstructure evolution, specifically of grain boundaries of Incoloy800H superalloy, was studied [9]. The results showed that the thermo-mechanical process can also introduce nanoscale precipitates in the matrix. Further research on the effect of TMP on the grain boundary character distribution (GBCD) in Incoloy800H was carried out [10]. The coincidence-site lattice boundaries increased with the pre-deformation level of the samples. For a highly deformed Incoloy800H superalloy, the nucleation and growth behavior were discussed [11]. The results showed that the oriented nucleation played a significant role in determining the final annealing texture. The dynamic strain aging (DSA) behavior in Incoloy800H superalloy was characterized under the condition of the strain rate in the range of 10^{-4} to 10^{-7} and temperatures between 295 K to 673 K [12]. It showed that, at temperatures above 473 K, the load serrations in Incoloy800H superalloy occurred over a wide range of strain rates. The tensile and creep-fatigue properties of Incoloy800H superalloy at a high temperature were investigated [13]. It was observed that the combined creep fatigue damage at 800 °C decreased with the decreasing total strain.

As seen in the short literature review above, most of the recent research focuses on the treatment and the post-processing of Incoloy800H superalloy, for better mechanical properties. The mechanical properties are determined by the surface and internal qualities of the product. Therefore, the manufacturing process of Incoloy800H superalloy billet is vital to obtaining an Incoloy800H alloy product with better mechanical properties. Electroslag remelting (ESR) is a secondary metal processing route for controlling the microstructure and chemical refining of the nickel-based superalloy [14]. For a higher production rate and lower production cost, continuous casting (CC) is an ideal technique for producing Incoloy800H superalloy billet. Continuous casting can remarkably increase the manufacturing efficiency; however, due to the special chemical composition of the alloy, the presence of defects, such as longitudinal cracks [15,16], can severely influence the quality of the whole Incoloy800H product. Electromagnetic casting (EMC) has been successfully applied to a large amount of metals to improve the surface and internal quality of the billet, even for some magnesium alloys and superalloy; e.g., Mg-Al-Zn-Ca alloy [17], IN100 superalloy [18] and Incoloy800H superalloy [19]. This is mainly because of the effect of electromagnetic force during the solidification process of the alloy. However, the application of the electromagnetic continuous casting technique to Inocoly800H superalloy is rare to see. Clearly, this shall be the task which will be completed in this paper. Our purpose is first to produce the Incoloy800H superalloy billet using the technique of electromagnetic continuous casting (EMCC). Secondly, we shall try to determine the influence of the alternating electromagnetic field on the surface and the internal quality of Incoloy800H superalloy. The influence of the electromagnetic field is represented by the two Incoloy800H superalloy continuous casting billets obtained through the contrast experiments: without (Set I) and with electromagnetic field (Set II), respectively.

The outline of the present paper is as follows. The experimental procedure is introduced in Section 2. In Section 3.1, the surface quality of the Incoloy800H continuous casting billet is discussed. In Section 3.2, the internal quality of the billet, including the macro and microstructure of the specimen (in Section 3.2.1), the macro and microsegregation (in Section 3.2.2) and the inclusion of TiN (in Section 3.2.3) are discussed, respectively. The main conclusions are summarized in Section 4.

2. Experimental Procedure

The Incoloy800H superalloy is selected as a working liquid metal in the present research. The reason for this selection had been discussed in the Section 1. Table 1 shows the chemical composition of the superalloy we used in the present research. Figure 1 is the schematic representation of the electromagnetic continuous casting system. Due to its symmetrical nature, only half of the whole system was drawn. In Figure 1, (3) and (5) denote the electromagnetic continuous casting coil and the stirring coils, respectively. The theories of electromagnetic continuous casting and electromagnetic stirring (EMS) were thoroughly discussed by Vives [20] and Moffatt [21]. Therefore, we shall not discuss those theories in detail here. The alternating currents for EMCC and EMS are 1310 A and 350 A, respectively. The frequencies are 20.4 kHz and 5 Hz, respectively.

Table 1. Chemical composition for the Incoloy800H superalloy (in wt. %).

Element	Carbon (C)	Silicon (Si)	Manganese (Mn)	Phosphorus (P)	Chromium (Cr)	Nickel (Ni)	Titanium (Ti)	Aluminum (Al)	Iron (Fe)
wt. %	0.08	0.52	0.8	0.023	20.13	30.88	0.52	0.28	46.76

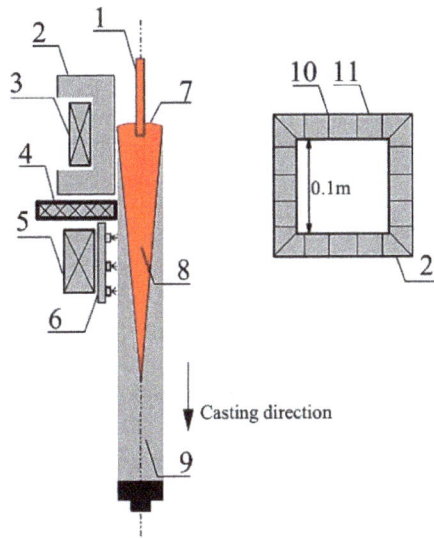

Figure 1. Schematic representation of the EMCC system (not scaled). (1) Nozzle, (2) mould, (3) high frequency induction coil for electromagnetic continuous casting (EMCC), (4) working platform, (5) electromagnetic stirring system (EMS), (6) water spray system, (7) meniscus, (8) liquid core, (9) EMCC billet (10) mould slit, and (11) mould segment.

The Incoloy800H superalloy (melting point: 1658 K) is first melted by a medium frequency furnace to 1700 K with a protective argon atmosphere. The temperature is captured by a B-type platinum rhodium thermocouple. The molten alloy is then poured into an EMCC copper mould (inner dimension: 0.1 m × 0.1 m). The EMCC mould has 20 slits in order to allow the high frequency electromagnetic field to permeate into the centre of the mould and act on the molten alloy. An ISP-200 kW super-sonic frequency power supply, with a frequency range from 10 kHz to 50 kHz, is used. Once the molten alloy free surface reaches the pre-set location, a casting speed is applied to the continuous caster and the experiment starts. Two sets of experiments are carried out: continuous casting without (Set I) and with (Set II) electromagnetic fields, respectively. The experimental conditions and operating parameters for both sets are shown in Table 2. In order to evaluate the influence of the electromagnetic field, the casting temperature, speed, cooling flows rate, mould oscillation frequency and amplitude are identical in Set I and II. For the Set II, the EMS is applied throughout the experiment. The magnetic flux density, which is due to the EMS system, is around 0.081 T with EMS current and frequency at 350 A and 5 Hz, respectively. However, the EMCC is only applied in a later stage of the experiment. The reason for this is that it is better for us to exclude the influence of the EMCC on the internal quality whilst we carry out the comparison study between Set I and II. Two billets with dimensions of 0.1 m × 0.1 m × 1.2 m are obtained from Set I and II, respectively.

Table 2. Experimental conditions and operating parameters for Set I and II.

Parameters	Set I	Set II
EMCC current, A	0	1310
EMCC frequency, kHz	0	20.4
EMS current, A	0	350
EMS frequency, Hz	0	5
Casting speed, m/min	0.6	0.6
Mould oscillation frequency, cpm	30	30
Mould oscillation amplitude, m	0.012	0.012
Mould cooling water flow rate, m^3/h	6.4	6.4
Secondary cooling water flow rate, m^3/h	3.0	3.0
Mould dimension, m	$0.1 \times 0.1 \times 0.4$	$0.1 \times 0.1 \times 0.4$
Mould wall thickness, m	0.03	0.03
Slit number, -	20	20
Slit height, m	0.4	0.4
Slit width, m	0.0005	0.0005

The depth of the oscillation mark is adopted to represent the surface quality of Incoloy800H superalloy billets. For the internal quality, the macrostructure of the cross-section of the billets is obtained after the samples were etched with the mixture etching agent. The microstructure morphology is investigated by using the OLYMPUS SZX16 microscope (OLYMPUS, Tokyo, Japan) and the grain size and the equiaxed grain fraction are estimated according to the ASTM standard [22]. The macrosegregation of carbon composition is captured by Leco CS844 infrared carbon-sulfur analyzer (Leco, San Jose, MO, USA) and the microsegregation of all chemical compositions was studied by Electron Probe Micro-analyzer (JXA-8530F EPMA, Shimadzu, Kyoto, Japan). The morphology of TiN was observed by Field-Emission Scanning Electron Microscope (JSM-7100F SEM, ZEISS, Jena, Germany) with an Energy Dispersive Spectrometer (EDS, ZEISS, Jena, Germany), and the number of TiN inclusion was captured by the inverted research microscope (Leica DMI5000M, Lecia, Barnack, Germany).

3. Results and Discussion

We first focus on the surface quality of the electromagnetic continuous casting billets of the Incoloy800H superalloy in Section 3.1 and then turn our attention to their internal quality in Section 3.2.

3.1. Surface Quality

Figure 2a,b shows the surface topographies of the billets obtained from Set I and II, respectively. It is clearly shown that the surface quality is highly improved by applying the high frequency electromagnetic field: the average depths of the oscillation marks decreased from 1.2 mm to 0.3 mm.

The mechanism for this improvement of the surface quality is similar to that for the other metals. The alternating current generates the alternating magnetic field with the same frequency as the initial current. The alternating magnetic field then generates the induced current in both the molten alloy and the mould segments. With the interaction of the magnetic field and the induced current in the molten alloy, the Lorentz force is generated with the direction pointing to the centre of the EMCC mould. The Lorentz force then reduces the contact pressure on the inner surface of the mould so that decreases the friction force between the molten alloy and the mould. This behavior improves the lubrication conditions dramatically so that the oscillation marks are obviously suppressed.

The improvement of the surface quality of the billet indicates that the EMCC technique also works on the Incoloy800H superalloy, which was not reported in the previous research.

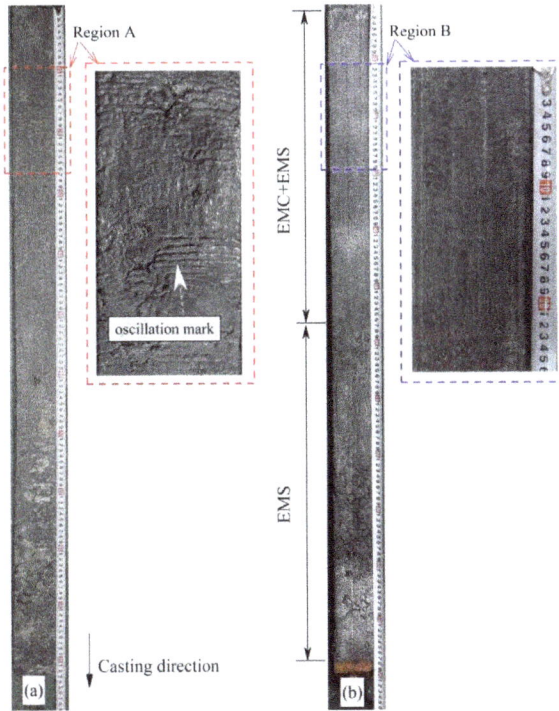

Figure 2. The surface topographies of Incoloy800H superalloy billets. Billets obtained from (**a**) Set I and (**b**) Set II. The depths of the oscillation marks decreased from 1.2 mm to 0.3 mm due to the effect of the alternating electromagnetic field.

3.2. Internal Quality

The internal quality of the billet is evaluated by the investigation of the macro and microstructure (Section 3.2.1), the segregation (Section 3.2.2) and the TiN inclusion (Section 3.2.3).

3.2.1. EMS on the Macro- and Microstructure

Figure 3 shows the cross-section macrostructure morphologies of the billets obtained from Set I (a) and II (b). With the aim of investigating the influence of EMS on the macrostructure of the Incoloy800H superalloy, for Set II, the aiming specimen for investigating should be obtained at the location when the EMCC current is at rest. The results show that the application of EMS can enlarge the region of the equiaxed grain from 2.45% to 41.45%, and can refine the equiaxed grain size significantly, from 10.83 mm to 1.28 mm.

A further study of EMS effect on Incoloy800H superalloy is conducted by investigating the microstructure. Three locations, edge, 1/4 width and 1/2 width (centre) of the billet cross-section, are selected for both specimens. The microstructure is obtained by using a DW/T-400 stero microscope and the microstructure morphologies at the selected locations are shown in Figure 4. For the specimen obtained from Set I, at the beginning of the solidification stage, a thin layer is formed by large numbers of tiny equiaxed dendrites (Figure 4a). The reason for the existence of the tiny equiaxed dendrites region is due to the large temperature gradient between the mould and the solidification front. This temperature gradient enforces a heat transfer process and results in a quick solidification. The tiny equiaxed dendrites grow rapidly with random orientation. For the 1/4 width region, some dendrites with a growing direction parallel to the temperature gradient (perpendicular to the mould wall) develop into the unidirectional columnar dendrites towards the centre of the billet, as shown in Figure 4b. In this region, coarse columnar dendrites are found. These unidirectional columnar dendrites continue to develop until they reach the centre of the specimen. Due to the symmetric features, the unidirectional columnar dendrites originated from all directions meet in the vicinity of the 1/2 width (centre of the specimen) region. The coarse equiaxed dendrites are observed, compared to the tiny equiaxed, as shown in Figure 4c.

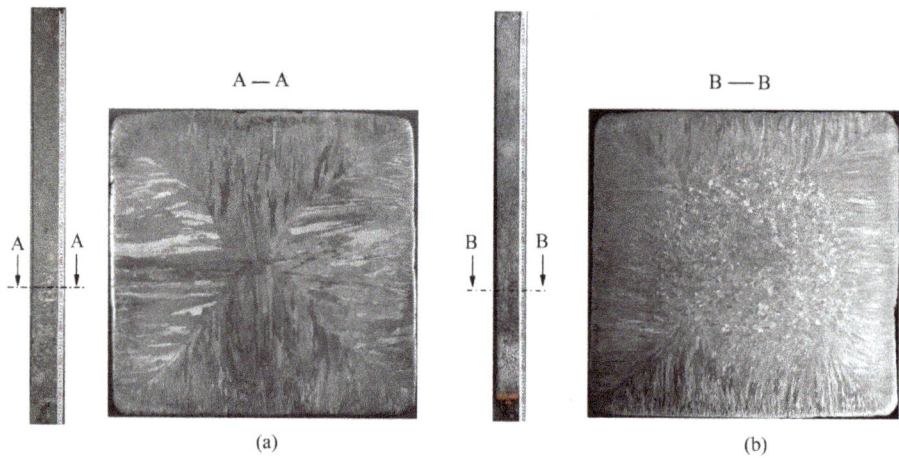

(a)   (b)

Figure 3. Macrostructure morphology of the billet cross-section: (**a**) Set I and (**b**) II.

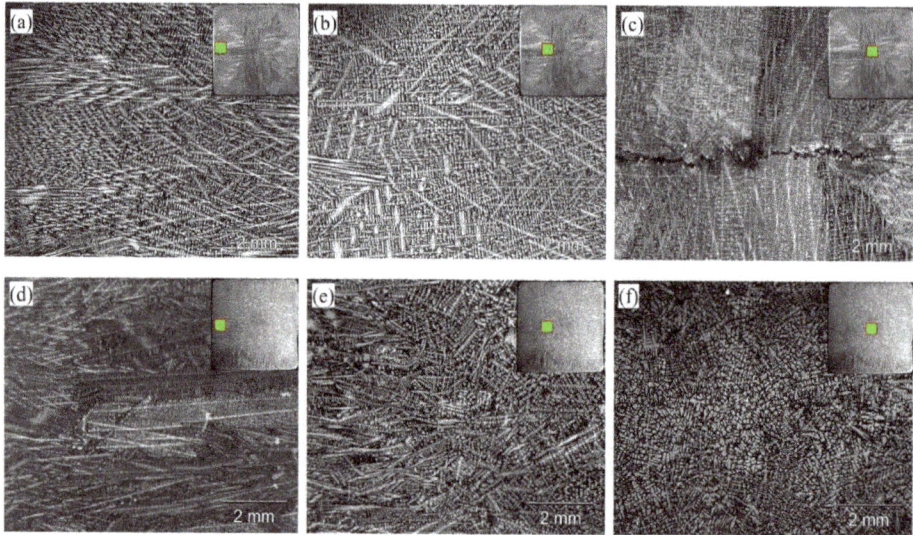

Figure 4. Microstructure of the billet: (**a–c**) represent the location of edge, 1/4 width and 1/2 width of the specimen obtained from Set I, respectively; (**d–f**) represent the location of edge, 1/4 width and 1/2 width of the specimen obtained from Set II, respectively.

Once the EMS is applied, the solidification behavior proceeds in the same manner as that of Set I near the edge region. The tiny equiaxed dendrites region appears (Figure 4d). However, the development of the coarse columnar dendrites which are parallel to the temperature gradient direction is interrupted by a forced convection which is generated due to EMS. The interruption behavior can be represented by the appearance of the dendrite arms at the solidification front. The dendrite arms are remelted or broken into fragments, and the length of primary dendrite arm decreases. This results in the appearance the random growing direction of the coarse columnar dendrites, as shown in Figure 4e. In the centre of the billet, a large number of fine equiaxed dendrites are formed (Figure 4f).

For the specimens obtained in Set I and II, the variation of the secondary dendrite arm spacing (SDAS) is also investigated, as shown in Figure 5. It is clear shown that the SDAS is remarkably decreased when the EMS is applied. For the Incoloy800H superalloy, this phenomenon can be understood as follows: the Lorentz force originated from EMS results in the movement of the molten alloy. This movement is accompanied by a heat/mass transfer. The dendrite arms are remelted. The fragments which depart from the original dendrite can become an effective nucleus region and increase the nucleation rate. Furthermore, the movement of the fluid flow can render the temperature field uniform and decrease the temperature gradient through this forced convection behavior. This can increase the heat transfer process from the liquid core of the billet to the mould. These factors result in the decrease of the SDAS and grain size and enhance the transformation from columnar grain to equiaxed grains.

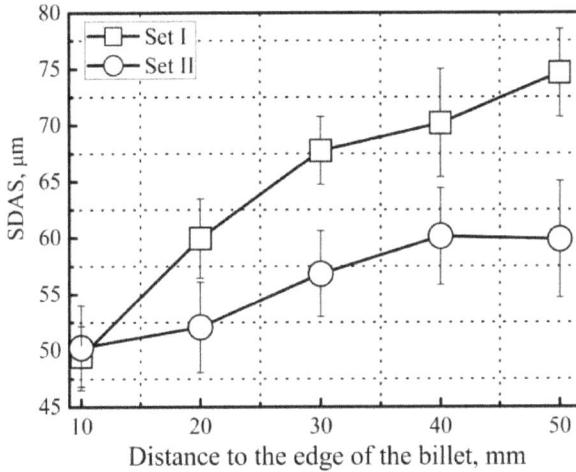

Figure 5. Variations of the secondary dendrite arm spacing (SDAS) with the distance from the edge to the centre of specimen.

3.2.2. EMS on the Segregation

Another major occurrence in solidification stage is the change in melt chemistry. Segregation is important in that it alters physical and mechanical properties: the segregation of the carbon composition can result in the potential cracks. For the macrosegregation, the two specimens without and with EMS obtained in Set I and II, are drilled 15 holes (diameter $\varphi = 5$ mm) with an internal of 6.25 mm from one edge to the other, respectively. The chips are collected for elemental analysis and the carbon segregation index ρ_i is defined as follows:

$$\rho_i = \frac{C_i}{\overline{\overline{C}}} \tag{1}$$

where C_i the carbon content at point i, $i \in [1-15]$, and $\overline{C}$ is the average carbon content of the 15 testing points. The level of microsegregation is evaluated by the segregation ratio, S_R, which is defined as follows:

$$S_R = \frac{C_{ID}}{C_{DC}} \tag{2}$$

where C_{ID} and C_{DC} are the concentrations in the interdendritic region and dendrite core, respectively. In the present research, 5 locations at the interdendritic region and dendrite core were selected, respectively. Therefore, 5 segregation ratios were obtained in the vicinity of the testing region. The average segregation ratio $\overline{S_R}$ is defined as follows:

$$\overline{S_R} = \sum_j^5 S_R \tag{3}$$

and was adopted. The positive and negative macrosegregation respectively: microsegregation) are defined when $\rho_i > 1$ (respectively: $\overline{S_R} > 1$) and $\rho i < 1$ (respectively: $\overline{S_R} < 1$), respectively.

The distribution of ρi along the middle line of the specimens (Set I and II) and the average microsegregation ratio $\overline{S_R}$ for each elements are shown in Figure 6 (left) and (right), respectively. For the macrosegregation, as shown in Figure 6 (left), the positive segregation of C elements reaches the maximum level of $\rho_i = 1.49$ in the vicinity of the centre region for the specimen obtained without EMS (Set I). For the specimen obtained from Set II (with EMS), the maximum ρ_i is decreased to 1.16. For the microsegregation of the specimen obtained in Set I, as shown in Figure 6 (right), most elements exhibit positive segregation behavior and indicate that these elements are enriched in the interdendritic region. It is clearly shown that the EMS reduces the microsegregation level for all the compositions. For Incoloy800H superalloy, the Ti element is the most serious and it reaches the peak value of $\overline{S_R}$ is 3.67 without EMS and decreases to 2.75 when EMS is applied.

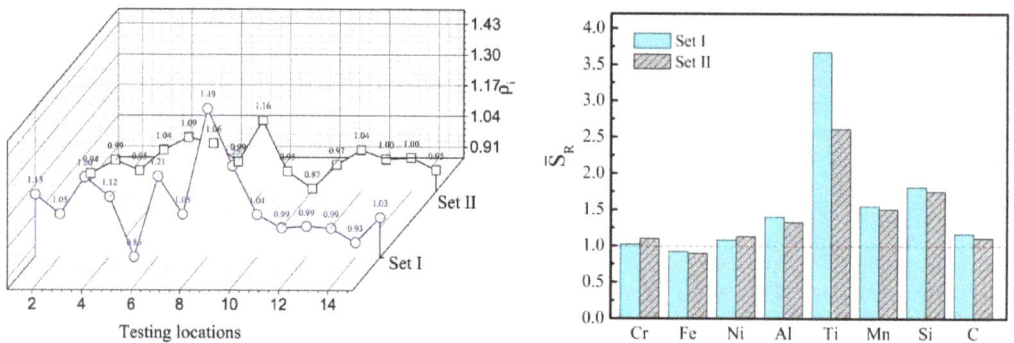

Figure 6. (Left) Variations of the content of carbon segregation index ρ_i for different testing locations; **(Right)** Microsegregation ratio distribution without (set I) and with (set II) EMS.

The reason for the central segregation of carbon composition without EMS has been thoroughly discussed. Here, it would more interesting to focus on the mechanism of how the EMS decreases the central segregation of Incoloy800H superalloy in continuous casting. The solidification structure of the molten alloy is formed by the development of the dendrites. In the mushy zone, the density of the solute, e.g., C, increases due to the selective crystallization feature. The solute concentration obtains a higher value in the vicinity of the dendrite core compared to that near the interdendritic regions because of the different solubility of solute in the solid and liquid. It may be worth mention that the mushy zone is assumed to be a saturated porous medium that offers frictional resistance toward fluid flow. Permeability in the mushy zone can be defined as following [23]:

$$K = \frac{d^2 \cdot g_l^3}{180 \times (1 - g_l)^2} \tag{4}$$

where K is the permeability, d is the representative size in the dendritric structure and g_l is the volume of fraction of the liquid. When EMS is not applied, the columnar dendrite grows towards to the centre of the billet. The higher value of the K causes the enrichment of the solute in the centre cross-section of the specimen. When EMS is applied, the permeability K is decreased due to the forced convection which is caused by the Lorentz force. This can be proofed by the decreasing of the SDAS values, as we discussed in Figure 5. The decreasing of the K uniforms the solute redistribution process and results in the decreasing of the segregation.

For the microsegregation, it is recognized as a result of a partition behavior of solute from the liquid phase to the solid phase. The solute partition coefficient k is defined as:

$$k_i = \frac{C_s^i}{C_l^i} \tag{5}$$

where C_s^i and C_l^i are the concentration of solute i in solid and liquid, respectively. The effective partitioning coefficient expressed as [24]:

$$k_e = \frac{k}{k + (1 - k)^{-R\frac{\delta}{D_L}}} \tag{6}$$

where here D_L is solute diffusion coefficient in liquid, R is growth rate, k is the equilibrium partitioning coefficient, and δ is the boundary layer thickness ahead of liquid/solid interface, respectively. When EMS is applied, the solidification time and the diffusion rate are decreased due to the increasing of the growing rate of nucleation, the forced convection originated from EMS constantly scours the solute-rich liquid and accelerates the homogenization of the solute elements in the remaining liquid. The increasing R results in the k_e becoming closer to 1, which indicates the decrease of the microsegregation between the dendrite core and the interdendritic regions.

3.2.3. TiN Inclusion

As we discussed in Section 3.2.2, the microsegregation of Ti element cannot be ignored because it plays an important role in the formation and growth of TiN. The inclusion of TiN will significantly affect the mechanical properties. Incoloy800H superalloy is more susceptible to the formation of TiN inclusions because of its relatively high level of Ti content. Therefore, the elimination of TiN inclusion is considered to be an important issue for improving the internal quality of Incoloy800H superalloy billet.

Figure 7 shows the TiN inclusion for the specimen obtained by Set I. For the billet of Incoloy800H superalloy obtained from Set I, the TiN inclusion has a typically cubic or rectangular-prim morphology, and some TiN inclusions have a black nucleus. In the figure, the existence of the crack in the vicinity of TiN inclusion is observed. The crack is because of the large degree of misfit between TiN inclusion and matix. The results also shows that the nucleus identified by SEM-EDS is $MgAl_2O_4$ spinel which was produced by the addition of the refractory material MgO and Al_2O_3 and can serve as a nucleation site for the formation of TiN. The range of TiN size is 1.85 to 10.94 μm.

The TiN inclusion number is estimated at the six locations with an internal of 10 mm from the edge to the centre of the cross-section. Each location is observed within thirty random fields of view at a magnification of $500\times$ by the optical microscope (Leica DMI5000M, Lecia, Barnack, Germany).

Figure 8 shows the variations of the number of the TiN inclusion along the radius direction of Incoloy800H billets. The results show that, for the specimens obtained from Set I and II, the TiN inclusion number shows no major differences in the vicinity of edge of the cross-section. This is because the initial solidification shell (5 mm to 10 mm) is already formed when the billet enters the EMS system covered region. The Lorentz force cannot handle the solidified shell. As the value of the distance from the specimen edge is increased, the number of TiN inclusion is significantly reduced for the specimen obtained in Set II compared to that in Set I, especially near the centre of the cross-section region, as shown in Figure 8.

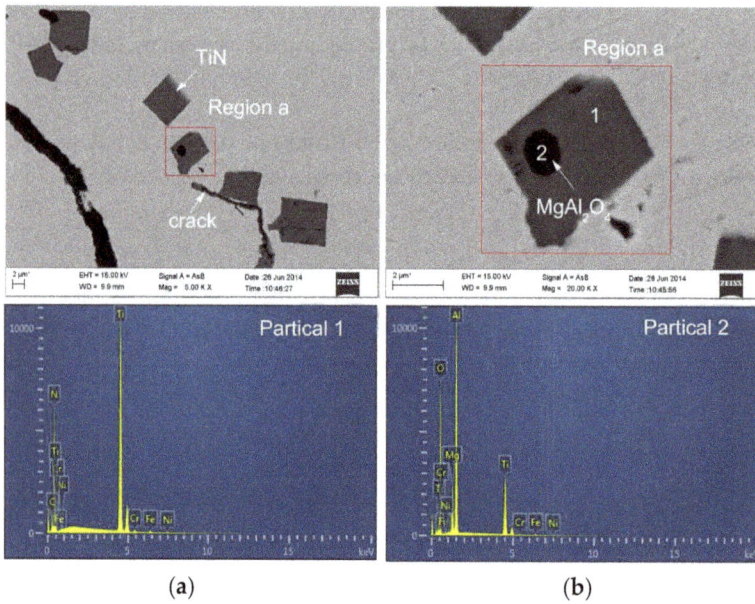

(a)   (b)

Figure 7. SEM images of an Incoloy800Hsuperalloy sample obtained in Set II. (a) TiN with and without nucleus; (b) TiN with nucleus.

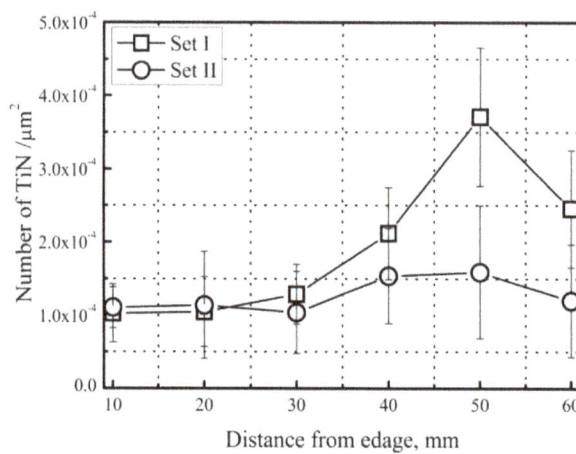

Figure 8. Variations of the TiN inclusion number along the radius direction of Incoloy800H superalloy billet in Set I and II.

The main reasons for the decrease of TiN inclusion by applying EMS can be understood as follows: firstly, as we discussed in Section 3.2.2, EMS can suppress the microsegregation of most compositions. Therefore, this obviously decreases the chance for the composition Ti and N to form TiN inclusion. Secondly, the rotating molten alloy produces turbulent vortex which can carry the TiN inclusions to the centre of the specimen. At the centre region, forced convection caused by EMS will promote the collision and coagulation of these TiN inclusions for forming TiN clusters. Due to the fact that the density of TiN is less than that of the surrounding molten alloy, TiN inclusions will float toward the upper part of the billet due to the buoyancy force, which escaped the solidifying matrix. Furthermore, the TiN clusters will move vertically upward more quickly than the fine TiN inclusions. These results cause a decrease in the TiN inclusion number when EMS is applied.

4. Conclusions

We successfully carried out an electromagnetic continuous casting experiment for Incoloy800H superalloy. With the aim of comparing the influence of the alternating electromagnetic field on the surface and internal qualities of the billet, a continuous casting experiment for Inconloy800H without an electromagnetic field was conducted as well. For both billets, the surface and internal qualities, which were represented by the depths of oscillation marks and macro- and microstructure segregations and inclusions, respectively, were investigated. The results can be broadly summarised as follows.

The electromagnetic continuous casting experiment was carefully designed and successfully carried out. An Incoloy800H superalloy billet with dimensions of $0.1 \times 0.1 \times 1.2$ m was obtained. For the surface quality of the billet, the depth of the oscillation mark significantly reduced from 1.2 mm to 0.3 mm while the high frequency (20.4 kHz) electromagnetic field was applied. For the internal quality, the macrostructure is refined with the low frequency (5 Hz) of the electromagnetic field. The fraction of the equiaxed grain increased from 2.45% to 41.4%, and the equiaxed grain size decreased from 10.83 mm to 1.28 mm. The segregation and the number of the TiN inclusions were suppressed as well.

Acknowledgments: This work was financially supported by the National Nature Science Foundation of China (No. 50834009), the Key Project of the Ministry of Education of China (No. 311014) and the Programme of Introducing Talents of Discipline to Universities (the 111 Project of China, No. B07015). The authors would also like to thank the referees for their work, which has contributed to this paper.

Author Contributions: Fei Wang and Engang Wang conceived and designed the experiments; Fei Wang, Anyuan Deng and Xiujie Xu performed the experiments; Fei Wang and Engang Wang analyzed the data; Anyuan Deng and Engang Wang contributed reagents/materials/analysis tools; Fei Wang, Lintao Zhang and Engang Wang contributed to writing and editing of the manuscript.

Conflicts of Interest: The authors declare no conflict of interest.

References

1. Tan, L.; Allen, T.T.; Yang, Y. Corrosion behavior of alloy 800H (Fe-21Cr-32Ni) in supercritical water. *Corros. Sci.* **2011**, *53*, 703–711. [CrossRef]

2. Chen, W.S.; Kai, W.; Tsay, L.W.; Kai, J.J. The oxidation behavior of three different zones of welded Incoloy 800H alloy. *Nucl. Eng. Design* **2014**, *272*, 92–98. [CrossRef]

3. Chen, X.; Wang, J.; Fang, Y.; Madigan, B.; Xu, G. Investigation of microstructures and residual stresses in laser peened Incoloy800H weldments. *Opt. Laser Technol.* **2014**, *57*, 159–164. [CrossRef]

4. Chen, X.; Zhang, S.; Wang, J.; Kelleher, J. Residual stresses determination in a 8 mm Incoloy800H weld via neutron diffraction. *Mater. Design* **2015**, *76*, 26–31. [CrossRef]

5. Mizia, R.E.; Clark, D.E.; Glazoff, M.V.; Lister, T.E.; Trowbridge, T.L. Optimizing the diffusion welding process for alloy 800H: Thermodynamic, diffusion modeling, and experimental work. *Metall. Mater. Trans. A* **2013**, *44*, 154–161. [CrossRef]

6. Cao, Y.; Di, H. Research on the hot deformation behavior of a Fe-Ni-Cr alloy (800H) at temperatures above 1000 °C. *J. Nucl. Mater.* **2015**, *465*, 104–115. [CrossRef]

7. Cao, Y.; Di, H.; Misra, R.D.K. The impact of aging pre-treatment on the hot deformation behavior of alloy 800H at 750 °C. *J. Nucl. Mater.* **2014**, *452*, 77–86. [CrossRef]

8. Kim, D.J.; Seo, D.Y.; Tsang, J.; Yang, J.H.; Lee, J.H.; Saari, H.; Seok, C.S. The crack growth behavior of Incoloy 800H under fatigue and dwell-fatigue conditions at elevated temperature. *J. Mech. Sci. Technol.* **2015**, *26*, 2023–2027. [CrossRef]

9. Tan, L.; Rakotojaona, L.; Allen, T.R.; Nanstad, R.K.; Busby, J.T. Microstructure optimization of austenitic alloy 800H (Fe-21Cr-32Ni). *Mater. Sci. Eng. A* **2011**, *528*, 2755–2761. [CrossRef]

10. Akhiani, H.; Nezakat, M.; Sanayei, M.; Szpunar, J. The effect of thermo-mechanical processing on grain boundary character distribution in Incoloy 800H/HT. *Mater. Sci. Eng. A* **2015**, *626*, 51–60. [CrossRef]

11. Akhiani, H.; Nezakat, M.; Sonboli, A.; Szpunar, J. The origin of annealing texture in a cold-rolled Incoloy 800H/HT after different strain paths. *Mater. Sci. Eng. A* **2014**, *626*, 334–344. [CrossRef]

12. Moss, T.E.; Was, G.S. Dynamic strain aging of Nickel-base alloys 800H and 690. *Metall. Mater. Trans. A* **2013**, *43A*, 3428–3431. [CrossRef]

13. Kolluri, M.; Pierick, P.T.; Bakker, T. Characterization of high temperature tensile and creep-fatigue properties of alloy 800H for intermediate heat exchanger components of (V)HTRs. *Nucl. Eng. Design* **2015**, *284*, 38–49. [CrossRef]

14. Busch, J.D.; Debaraadillo, J.J.; Krane, M.J.M. Flux entrapment and Titanium Nitride defects in electroslag remelting of INCOLOY alloys 800 and 825. *Metall. Mater. Trans. A* **2013**, *44*, 5295–5303. [CrossRef]

15. Kanbe, Y.; Ishii, T.; Todoroki, H.; Mizuno, K. Prevention of longitudinal cracks in a continuously cast slab of Fe-Cr-Ni superalloy containing Al and Ti. *Int. J. Cast Met. Res.* **2009**, *22*, 143–148. [CrossRef]

16. Todoroki, H.; Ishii, T.; Mizuno, K.; Hongo, A. Effect of crystallization behavior of mold flux on slab surface quality of a Ti-bearing Fe-Cr-Ni super alloy cast by means of continuous casing process. *Mater. Sci. Eng.* **2005**, *413–414*, 121–128. [CrossRef]

17. Park, J.P.; Kim, M.G.; Yoon, U.S.; Kim, W.J. Microstructures and mechanical properties of Ma-Al-Zn-Ca alloys fabricated by high frequency electromagnetic casting method. *J. Mater Sci.* **2009**, *44*, 47–54. [CrossRef]

18. Jin, W.; Li, T.; Yin, G. Research on vacuum-electromagnetic casting of IN100 superalloy ingots. *Sci. Technol. Adv. Mater.* **2007**, *8*, 1–4. [CrossRef]

19. Jiang, E.; Wang, E.; Deng, A. Experimental research on solidification structure of alloy 800H by linear electromagnetic stirring. *China Foundry* **2014**, *11*, 475–480.

20. Vives, C. Electromagnetic refining of aluminum alloy by the CREM process: Part I. working principle and metallurgical results. *Metall. Mater. Trans. B* **1988**, *20B*, 623–629. [CrossRef]

21. Moffatt, H.K. Electromagnetic stirring. *Phys. Fluids* **1991**, *3*, 1336–1343. [CrossRef]

22. Shepherd, B.F. The P-F Characteristic of Steel. *Trans. Am. Soc. Met.* **1934**, *22*, 979–1016.

23. Scheidegger, A.E. *The Physics of Flow Through Porous Media*, 3rd ed.; University of Toronto Press: Toronto, ON, Canada, 1974.

24. Poirier, D.R. Permeability for flow of interdendritic liquid in columnar-dendritic alloys. *Metall. Mater. Trans. B* **1987**, *8*, 245–255. [CrossRef]

17

Effects of Basicity and MgO in Slag on the Behaviors of Smelting Vanadium Titanomagnetite in the Direct Reduction-Electric Furnace Process

Tao Jiang, Shuai Wang, Yufeng Guo *, Feng Chen and Fuqiang Zheng

School of Minerals Processing and Bioengineering, Central South University, Changsha 410083, China; jiangtao@csu.edu.cn (T.J.); ws_csu@126.com (S.W.); csucf@126.com (F.C.); zfqcsu@126.com (F.Z.)
* Correspondence: guoyufengcsu@163.com

Academic Editor: Corby G. Anderson

Abstract: The effects of basicity and MgO content on reduction behavior and separation of iron and slag during smelting vanadium titanomagnetite by electric furnace were investigated. The reduction behaviors affect the separation of iron and slag in the direct reduction-electric furnace process. The recovery rates of Fe, V, and Ti grades in iron were analyzed to determine the effects of basicity and MgO content on the reduction of iron oxides, vanadium oxides, and titanium oxides. The chemical compositions of vanadium-bearing iron and main phases of titanium slag were detected by XRF and XRD, respectively. The results show that the higher level of basicity is beneficial to the reduction of iron oxides and vanadium oxides, and titanium content dropped in molten iron with the increasing basicity. As the content of MgO increased, the recovery rate of Fe increased slightly but the recovery rate of V increased considerably. The grades of Ti in molten iron were at a low level without significant change when MgO content was below 11%, but increased as MgO content increased to 12.75%. The optimum conditions for smelting vanadium titanomagnetite were about 11.38% content of MgO and quaternary basicity was about 1.10. The product, vanadium-bearing iron, can be applied in the converter steelmaking process, and titanium slag containing 50.34% TiO_2 can be used by the acid leaching method.

Keywords: basicity; MgO; vanadium titanomagnetite; direct reduction-electric furnace; iron-slag separation

1. Introduction

Vanadium titanomagnetite ore is a valuable resource which contains rich titanium, vanadium, and other metal elements. The Panxi region of China is rich in vanadium titanomagnetite resources with reserves of about 9.66 billion tons [1]. The proven deposits of titanium and vanadium account for 35.17% and 11.6% of world total reserves, respectively [2]. The blast furnace process has been used to utilize vanadium titanomagnetite resources since the 1970s in China [3], but the byproduct of the blast furnace, titanium slag which contains 22%–25% TiO_2, is difficult to be recovered on a large scale. Obviously, the titanium resource is wasted, to be stockpiled without effective utilization, and simultaneously causes environmental pollution. Actually, the recovery rates of Fe, V, and Ti of vanadium titanomagnetite under the main blast furnace process are about 70%, 47%, and 10%–20%, respectively [4]. Therefore, the key to raise up the utilization level of vanadium titanomagnetite is to increase the recovery rate of Ti. There are many methods [2,5–8] developed, or under development, including direct reduction-electric furnace smelting, sodium salt roasting-direct reduction-electric furnace smelting, and direct reduction-magnetic separation, and so on.

Compared with the other methods, which are still under investigation due to technological problems, the direct reduction-electric furnace smelting process has many advantages and has been commercialized in South Africa and New Zealand successfully [9]. However, reductions of vanadium oxides and titanium oxides in vanadium titanomagnetite and the separation of titanium slag from iron are difficult to control in this process. The separation of iron and slag would be difficult if titanium oxides were over-reduced. High basicity slag was used to solve this problem in electric furnace smelting in South Africa and New Zealand. As a result, the titanium slag contained about 30%–33% TiO_2 is obtained, but without effective recovery methods.

Previous research indicated that titanium resources could be recovered by acid leaching methods when the titanium slag contained about 50% TiO_2 [1]. Thus, the optimum composition of titanium slag which contained about 50% TiO_2 should be found. Furthermore, low melting temperature and low viscosity of titanium slag are needed to make the separation of titanium slag from iron successful. Fluxes and additives are commonly added to ensure the electric furnace smelting completes successfully [10]. CaO and MgO are important additives for the electric furnace process as well as blast furnace process [11,12]. Generally, basicity and MgO are important factors that can affect a slag viscosity and melting temperature, and have been widely studied for various slag systems [13–17]. He [18] studied for effect of MgO on viscosity of blast furnace slag and showed that the slag viscosity decreased with the increasing of MgO content in a definite range, and it decreased obviously with the increasing of slag basicity. Fan's experiment showed that the melting temperature of slag decreased with the increasing of MgO content in slag [16]. However, effects of basicity and MgO in many previous researches are of blast furnace slag [14,19]. The effects of basicity and MgO for vanadium titanomagnetite ore in the electric furnace smelting process are not fully understood, especially the effect of MgO in titanium slag.

In the present study, effects of basicity and MgO content of titanium slag on reduction behaviors of vanadium oxides and titanium oxides, as well as the separation behaviors of titanium slag from iron in smelting vanadium titanomagnetite ore by the direct reduction-electric furnace process were investigated. Different contents of CaO and MgO were added when smelting vanadium titanomagnetite by electric furnace. The vanadium-bearing iron and titanium slag were analyzed and detected by XRF and XRD. The recovery rates of Fe, V were calculated and Ti grades in iron were detected to display the effects of basicity and MgO on reduction of iron oxides, vanadium oxides and titanium oxides. In addition, separation behaviors of titanium slag from iron were analyzed through the pictures of iron and titanium slag and yield of metallized iron in magnetic separation. At last, these findings will provide the technological basis for commercial utilization of vanadium titanomagnetite ore by the direct reduction-electric furnace process in China.

2. Materials and Experimental

2.1. Materials

The vanadium titanomagnetite concentrate used in this investigation was provided by the Chongqing Iron and Steel Company (Sichuan, China). The direct reduction metallized pellet was produced from vanadium titanomagnetite concentrate pellet by coal-based direct reduction in a rotary kiln ($\varnothing$1000 mm × 550 mm, produced by Central South University, Changsha, China).

The main chemical composition of metallized pellet (diameter range 10–15 mm) is listed in Table 1. It shows the metallization rate (=MFe/TFe × 100%) of pellet is 90.45%. The metallized pellet contained 15.53 wt. % of titanium dioxide and 0.82 wt. % of vanadium oxide. Graphite was used in this research as the reductant. All chemicals and reagents used in this study were of analytical grade.

Table 1. Chemical compositions of metallized pellet (wt. %).

TFe	MFe	CaO	SiO_2	MgO	Al_2O_3	V_2O_5	TiO_2	C	S
71.38	64.56	1.01	3.71	2.58	3.83	0.82	15.53	0.036	0.027

The X-ray diffraction (XRD) pattern of metallized pellets is shown in Figure 1. The major phases in the metallized pellet are metallic iron (Fe), ferrous-pseudobrookite ($Fe_{1-x}Mg_xTi_2O_5$, $0 \leqslant x \leqslant 1$) and magnesia-alumina spinel ($Mg(AlSi_3O_8)$), respectively.

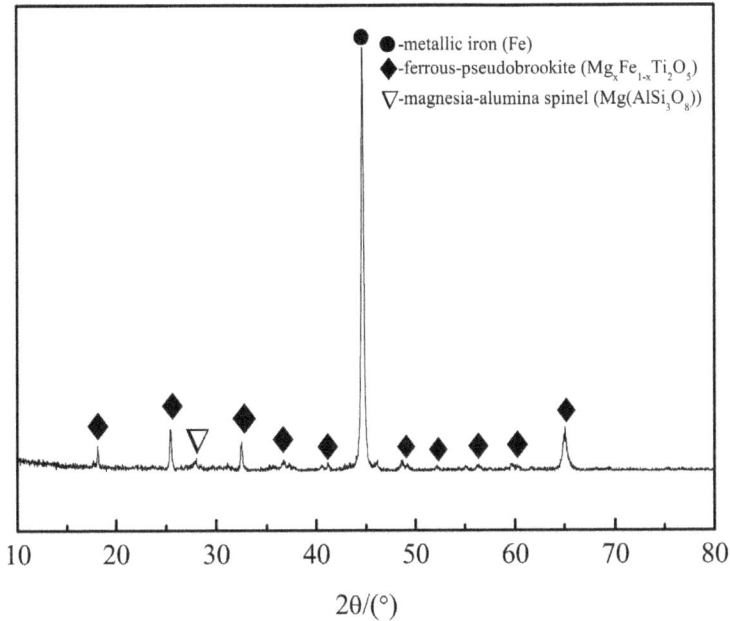

Figure 1. XRD pattern of metallized pellet.

2.2. Thermodynamics Analysis Methods

FactSage 7.0 (Thermfact/CRCT, Montreal, QC, Canada; GTT-Technologies, Herzogenrath, Germany) was used to draw the phase diagram and calculate the liquid temperature and viscosity of simulated titanium slag in this study [20]. The used modules were "Reaction", "Equlib", "Phase Diagram", "Viscosity", and the corresponding database was "FToxid".

2.3. Experimental Procedure

The high-temperature electric furnace (model: SG-160-MoSi$_2$, The Great Wall Furnace, Changsha, China) was used to smelt vanadium titanomagnetite ore (Figure 2). The metallized pellets were crushed at first (<1 mm), then mixed with additives (CaO, MgO) and graphite together. The well-mixed samples (200 grams) were placed in the graphite crucible ($\varnothing$50 × 100 mm) after the target temperature was reached and they were reduced in the high-temperature laboratory furnace at various temperatures with the protection of Ar. After smelting time was up, the crucible was taken out and cooled in neutral condition with Ar. Finally, the vanadium-bearing iron and titanium slag were crushed and detected by XRF and XRD.

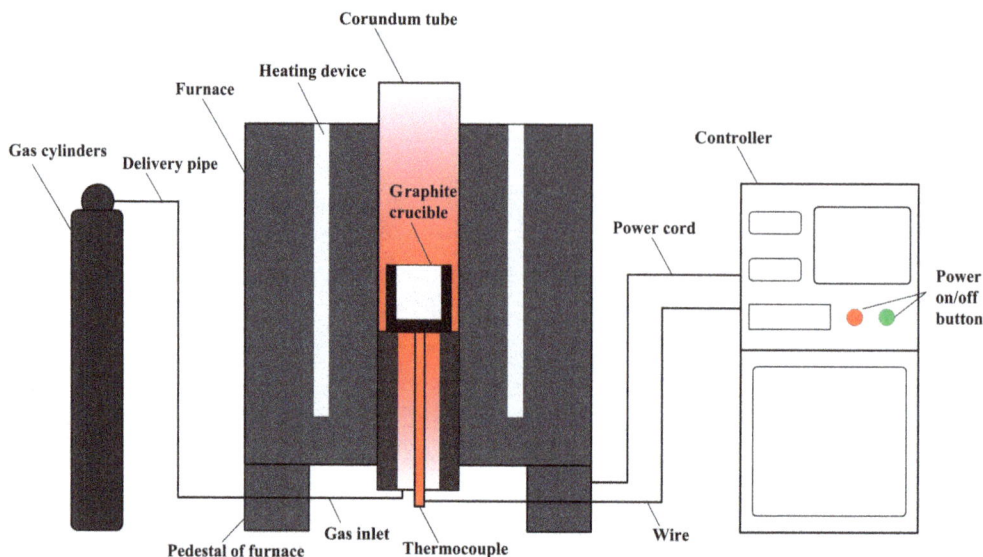

Figure 2. The SG-160-MoSi$_2$ furnace in the laboratory.

2.4. Definition of Parameters

2.4.1. Basicity

The basicity in this paper was quaternary basicity, which is represented as Equation (1):

$$R = \frac{\text{wt(CaO)} + \text{wt(MgO)}}{\text{wt(Al}_2\text{O}_3) + \text{wt(SiO}_2)} \tag{1}$$

2.4.2. Recovery Rates

The recovery rates of Fe and V were calculated by Equation (2):

$$\varphi(\text{Fe/V}) = \frac{\text{wt(Fe/V)}_{\text{iron}} \times \text{mass(iron)}}{\text{wt(Fe/V)}_{\text{pellet}} \times \text{mass(pellet)}} \times 100\% \tag{2}$$

The reduction behaviors of iron oxides and vanadium oxides were evaluated by the recovery rates of Fe and V, while the reduction behaviors of titanium oxides were evaluated by the Ti grade in iron.

The separation behaviors of titanium slag from iron were evaluated by observing the macroscopic pictures of iron and titanium slag. As residual metallized iron is the only magnetic material in titanium slag after the reduction, the residual metallized iron in titanium slag were separated by magnetic separation with the magnetic field intensity of 0.1 T. Thus, the yield of residual metallized iron in slag was another evaluation index of the separation of iron and titanium slag.

2.5. Analysis of Samples

The chemical composition of the vanadium-bearing iron and titanium slag were measured by X-ray fluorescence spectroscopy (XRF) (PANalytical Axios mAX, PANalytical B.V., Almelo, The Netherlands).

The chemical composition of metallized pellet was analyzed by Chinese standard YB/T 4170-2008.

The main phases of metallized pellet and titanium slag were analyzed by X-ray diffraction (XRD) (Cu Kα radiation, λ = 1.54056 Å, D/Max2200, Rigaku, Japan).

3. Experimental Fundamental Thermodynamics

Iron and vanadium are reduced into molten iron and titanium is kept as titanium slag in the direct reduction-electric furnace process. Therefore, the reductions of iron and vanadium were improved and the reduction of titanium was inhibited during the reduction in order to control the smelting direction of Fe, V, and Ti.

Equation (3) shows that the reduction of vanadium starts at a low temperature and is easier at high temperature due to the small value of ΔG_T^{θ} [21]:

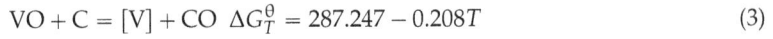

$$VO + C = [V] + CO \quad \Delta G_T^{\theta} = 287.247 - 0.208T \tag{3}$$

In addition, reduction of titanium oxides was controlled to limit the amount of titanium carbides in slag. The development of titanium carbides in slag increases the viscosity of slag and prevent separation of iron from titanium slag [22]. Thus, titanium oxides which are difficult to reduce to TiC were found in order to decrease the appearance of TiC in slag.

The reduction reactions of possible titanium oxides in smelting process according to the composition of metallized pellet were calculated by FactSage 7.0. Equations (4)–(7) shows that the reduction temperatures to TiC of $MgTi_2O_5$, Al_2TiO_5, $CaTiO_3$, and $CaSiTiO_5$ are 1629.77 K, 1590.26 K, 1863.16 K, 1709.16 K, respectively. It indicates that $MgTi_2O_5$, $CaTiO_3$, and $CaSiTiO_5$ have high reaction temperatures, however, Al_2TiO_5 is easier to be reduced to TiC due to its lower reduction temperature.

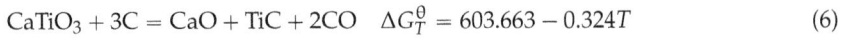

$$MgTi_2O_5 + 6C = MgO + 2TiC + 4CO \quad \Delta G_T^{\theta} = 1059.35 - 0.650T \tag{4}$$

$$Al_2TiO_5 + 3C = Al_2O_3 + TiC + 2CO \quad \Delta G_T^{\theta} = 478.668 - 0.301T \tag{5}$$

$$CaTiO_3 + 3C = CaO + TiC + 2CO \quad \Delta G_T^{\theta} = 603.663 - 0.324T \tag{6}$$

$$CaSiTiO_5 + 3C = CaSiO_3 + TiC + 2CO \quad \Delta G_T^{\theta} = 500.784 - 0.293T \tag{7}$$

Furthermore, the TiO_2 in titanium slag should increase to about 50% to ensure the recovery of titanium from slag by the acid leaching method. Therefore, the calculated TiO_2 content of titanium oxides, in the order from high to low, is shown:

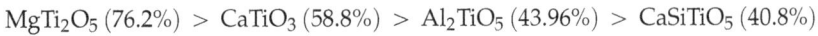

$$MgTi_2O_5 \ (76.2\%) \ > \ CaTiO_3 \ (58.8\%) \ > \ Al_2TiO_5 \ (43.96\%) \ > \ CaSiTiO_5 \ (40.8\%)$$

It is clearly seen that, even though $CaSiTiO_5$ has a higher reduction temperature, but it should not as the target mineral due to its low titanium content. The Al_2TiO_5 was inappropriate to be chosen due to the same reason, and the content of SiO_2 was kept constant due to the addition of SiO_2 decreased TiO_2 content of titanium slag. As a result, the $MgTi_2O_5$ and $CaTiO_3$ were suitable minerals in titanium slag, and the effects of CaO and MgO in slag needed to be investigated.

The separation of iron and titanium slag during the electric furnace process requires both low-viscosity slag and low melting temperature conditions. In this study, related phase diagrams were analyzed to identify the optimum compositions of slag suited to low melting temperature. The main chemical compositions of the metallized pellet are TiO_2, CaO, SiO_2, MgO, and Al_2O_3; the phase diagram of CaO-SiO_2-MgO-TiO_2-Al_2O_3 (Figure 3) was accordingly drawn and analyzed, with special focus on keeping the two compositions in the phase diagram constant. The content of TiO_2 was maintained at 50% constant, because analysis indicated that this amount of TiO_2 titanium slag can be used to recover Ti via the sulfuric acid process; Al_2O_3 content was kept constant at 13%. The optimum range of MgO content and suitable basicity in the titanium slag were identified according to the isothermal section of phase diagram (Figure 3) at 1400 °C of CaO-SiO_2-MgO-TiO_2 (50%)-Al_2O_3 (13%) drawn in FactSage 7.0. The red area in this phase diagram, *i.e.*, the melting temperature below 1400 °C.

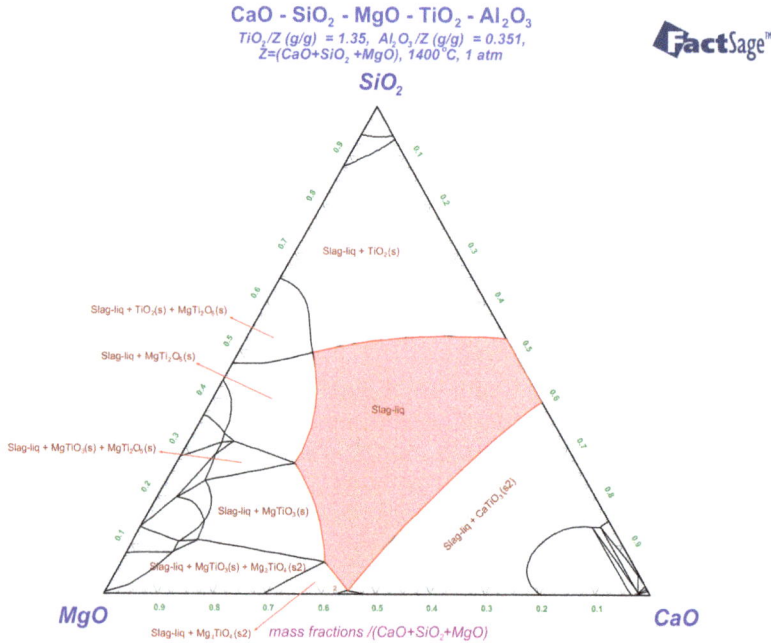

Figure 3. Isothermal section of phase diagram of CaO-SiO$_2$-MgO-TiO$_2$ (50%)-Al$_2$O$_3$ (13%) at 1400 °C.

Optimum contents of CaO, SiO$_2$, and MgO in titanium slag were determined as 8%–20%, 8%–18%, and 6%–14%, respectively, through careful analysis of the phase diagram. To investigate the effects of basicity and MgO content on melting temperatures, different binary basicity levels, MgO contents and the corresponding melting temperatures were also calculated by FactSage 7.0.

Figure 4 shows the changes of melting temperatures with different binary basicity and MgO contents, where we found that the melting temperature decreased as MgO content increased and decreased at first, then gradually increased with the increasing binary basicity. The suitable binary basicity range was 0.8–1.2 and MgO content was approximately 8%–12% within the appropriate temperature range.

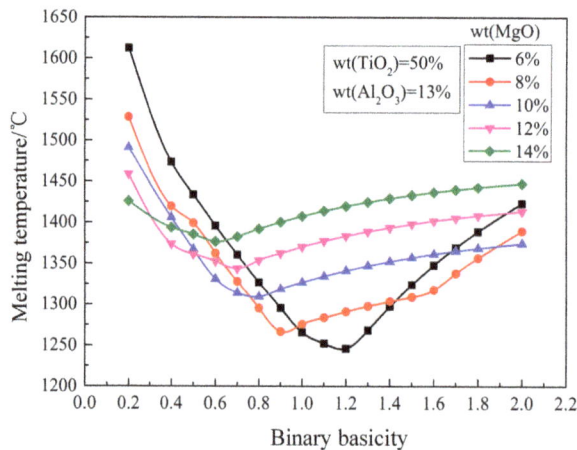

Figure 4. Effects of binary basicity and MgO content on melting temperature (FactSage 7.0).

Viscosity is one of the important physiochemical properties of slags in high-temperature extractive metallurgy [13]; generally, low slag viscosity promotes effective separation. The viscosity of titanium slag with different quaternary basicity was investigated by adjusting different CaO and MgO contents in FactSage 7.0 as shown in Figures 5 and 6.

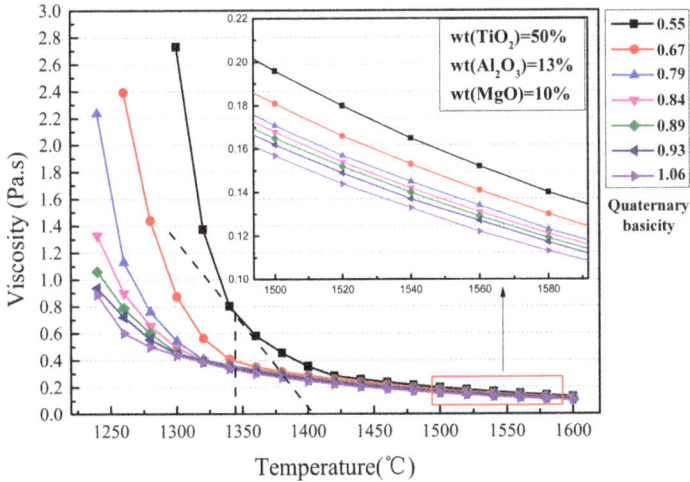

Figure 5. Effects of quaternary basicity by addition of CaO on titanium slag viscosity (FactSage 7.0).

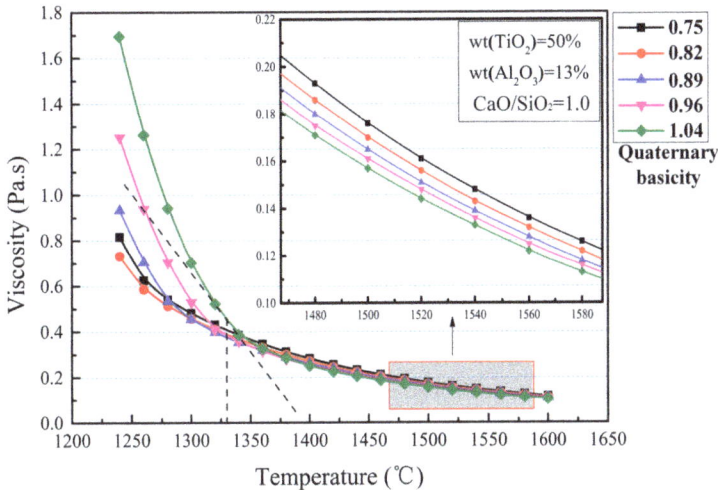

Figure 6. Effects of quaternary basicity by addition of MgO on titanium slag viscosity (FactSage 7.0).

The quaternary basicity was changed by adjusting content of CaO with constant MgO content in Figure 5, in contrast, MgO content was changed with constant CaO/SiO_2 to change the quaternary basicity in Figure 6. Figure 5 shows where the viscosity of the titanium slag decreases as its quaternary basicity increases, and the expanded image in Figure 6 indicates that the viscosity of titanium slag decreases as MgO content increases. Previous studies have also reported that the addition of MgO decreases slag viscosity [14,18]. In effect, addition of CaO and MgO to increase quaternary basicity can decrease the viscosity of titanium slag and promote the optimal separation of iron and slag.

Free running temperature is also a critical factor affecting slag properties. It is defined as the lowest temperature at which the slag becomes liquid and, as the term suggests, flow freely. In this study, the free running temperatures were defined according to the viscosity-temperature curves (Figures 5 and 6) as shown in Figure 7a,b: our method for doing so involved drawing a tangent line at 45 degree angle to the X axis, where the temperature of the tangent point on the curve is the free running temperature of the slag [21]. Figure 7 shows where increasing quaternary basicity by addition of CaO decreases the free running temperature until reaching a threshold at about 0.95 basicity, at which free running temperature increases. Free running temperature increases as quaternary basicity increases by addition of MgO.

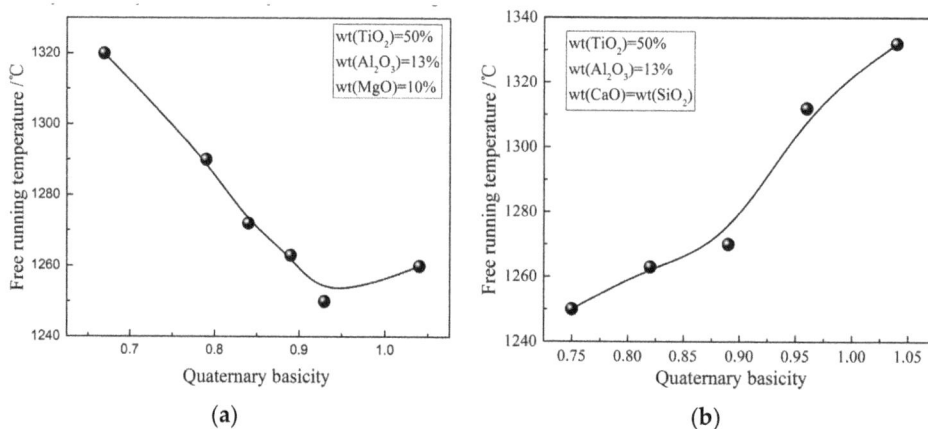

(a)   (b)

Figure 7. (a) Effects of quaternary basicity by addition of CaO content on free-running slag temperature; and (b) effects of quaternary basicity by addition of MgO content on free-running slag temperature.

Briefly, our thermodynamic analysis of the titanium slag indicated that the suitable quaternary basicity range is 0.9–1.2 and that MgO addition decreases titanium slag viscosity but increases its melting temperature and free-running temperature. Accordingly, the effects of quaternary basicity (0.9–1.2) and MgO content (8%–14%) were the focus of our subsequent investigation.

4. Results and Discussion

In this study, effects of quaternary basicity and MgO content in slag were investigated through adding different contents of CaO and MgO, and contents of Al_2O_3 and SiO_2 were not changed. Table 2 shows the design of experiments and the smelting parameters.

Table 2. Titanium slag composition and smelting parameters.

No.	Binary Basicity	Quaternary Basicity	MgO/%	Smelting Temperature/°C	Smelting Time/min	Reducing Agent/%
1	0.8	0.98	11.38	1550	20	2
2	0.9	0.99	11.38	1550	20	2
3	1.0	1.04	11.38	1550	20	2
4	1.1	1.10	11.38	1550	20	2
5	1.2	1.16	11.38	1550	20	2
6	1.1	0.88	7.97	1550	20	2
7	1.1	0.96	9.79	1550	20	2
8	1.1	1.13	12.75	1550	20	2
9	1.1	1.10	11.38	1550	20	5

4.1. Effects of Basicity

The effects of quaternary basicity on the smelting of vanadium titanomagnetite ore were investigated by varying the quaternary basicity level by addition of CaO under the following conditions (No. 1–No. 5): constant MgO content (11.38%), smelting temperature (1550 °C) and smelting time (20 min) with 2% reducing agent. Figure 8 shows that lower basicity levels with constant MgO content had significant effect on the recovery rates of Fe and V, as well as the Ti grade in the iron. As showed in

Figure 8a, the recovery rates of Fe and V increased as basicity increased, indicating that higher level of basicity benefits the reduction of iron oxides and vanadium oxides. Figure 8b indicates that titanium content dropped as iron decreased with the increasing quaternary basicity. In short, the results show that the basicity must be kept above 1.0 to ensure favorable results.

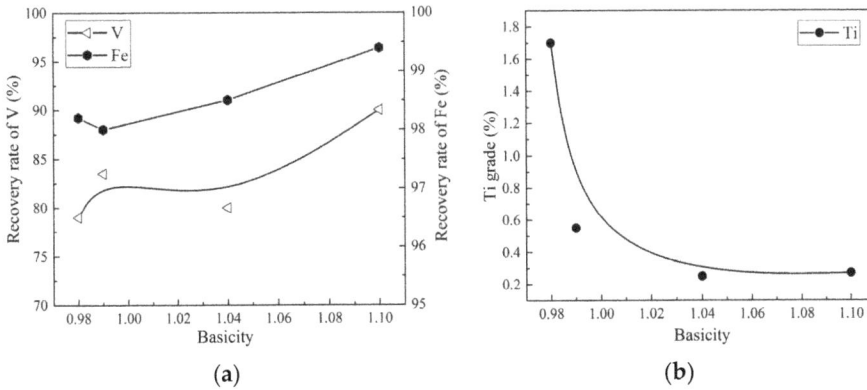

Figure 8. (a) Effects of quaternary basicity by addition of CaO on Fe and V recovery rates; and (b) effects of quaternary basicity by addition of CaO on Ti grade in iron.

The XRD pattern shown in Figure 9 demonstrates that the main phase compositions in the samples were perovskite ($CaTiO_3$), anosovite ($MgTi_2O_5$), and magnesium-aluminum silicate ($Mg_3Al_2CaSi_2O_6$). As basicity increased, the content of $MgTi_2O_5$ decreased while the contents of $CaTiO_3$ and $Mg_3Al_2CaSi_2O_6$ increased. Further, the results of thermodynamic analysis (Equations (4) and (6)) suggested that, increases in $CaTiO_3$ and decrease in $MgTi_2O_5$ inhibit the thermodynamic trend of titanium reducing to iron, and cause the Ti grade in the iron to decrease (Figure 8b). Moreover, the appearance of TiC particles in the titanium slag is known to raise its viscosity [22], hence the increasing $CaTiO_3$ having decreased the reduction trend to TiC in the titanium slag. Excessive basicity, however, increased the melting temperature because $CaTiO_3$ is a high-melting-point material.

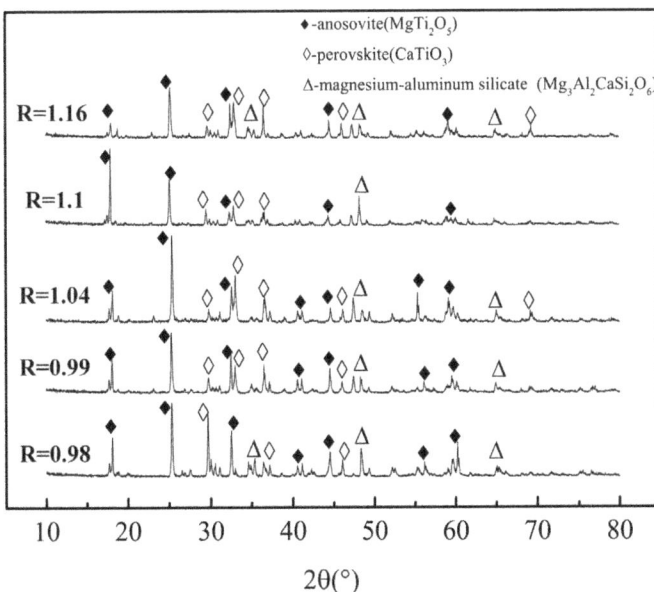

Figure 9. XRD patterns of titanium slag with different quaternary basicity levels by addition of CaO.

Figure 10 shows macroscopical pictures of the separation of iron and titanium slag with different basicity conditions; the iron is marked by the red line and the titanium slag is marked by the green line. Clearly, the separation was incomplete at a basicity of 0.98 as evidenced by the numerous iron shots (red circles) in the titanium slag. There were also several iron shots in the slag at a basicity of 1.04. Separation was optimal at a basicity of 1.10, where there were the fewest iron shots. Figure 11 shows the yield of residual iron in magnetic separation decreases firstly as basicity increases, then increases when basicity above 1.10. In short, our analysis of the effects of basicity on Fe, V, and Ti separation behaviors in iron and titanium slag indicated that the optimum quaternary basicity is about 1.10.

Figure 10. Pictures of iron and titanium slag separation with varying quaternary basicity by addition of CaO.

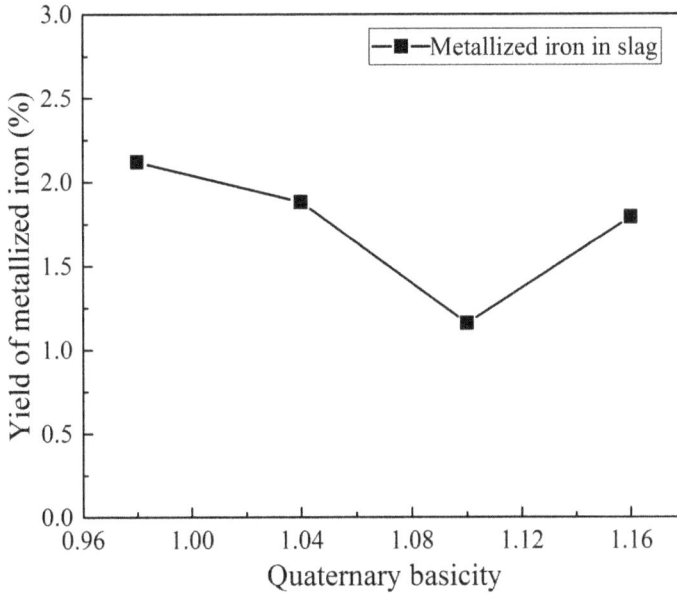

Figure 11. Effects of quaternary basicity by addition of CaO on yield of residual metallized iron in slag.

4.2. Effects of MgO

The effects of MgO on the smelting process were studied in the mass range of 7.97% to 12.75% with constant CaO. Figure 12 shows that as the content of MgO increased, the recovery rate of Fe increased slightly and the recovery rate of V increased considerably. It can be inferred that the grades of Ti in molten iron were at a low level and did not change significantly at MgO content below 11%, but increased as MgO content increased to 12.75%. Basically, increasing MgO content in the slag is beneficial, but only within a suitable range; otherwise, the reduction of titanium oxide will increase.

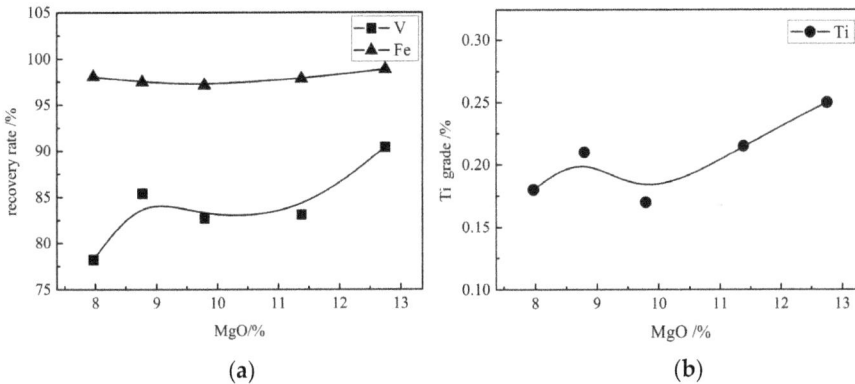

Figure 12. (a) Effects of MgO on Fe and V recovery rates; and (b) effects of MgO on the Ti grade in iron.

The XRD pattern shown in Figure 13 demonstrates that main phase compositions were $CaTiO_3$, $MgTi_2O_5$, and $Mg_3Al_2CaSi_2O_6$. As MgO content increased, the content of $MgTi_2O_5$ likewise increased, while the contents of $CaTiO_3$ and $Mg_3Al_2CaSi_2O_6$ decreased. Again, based on thermodynamic analysis (Equations (4) and (6)) of titanium oxide reduction, the thermodynamic tendency of the reduction from $MgTi_2O_5$ to TiC occurs more readily than that of $CaTiO_3$, so as MgO content increased beyond 12.75%, the Ti grade in the iron increased. Previous research has also indicated that Ti-bearing blast furnace slag viscosity increases as the perovskite phase increases [23], so reduction in titanium slag viscosity facilitates the reduced diffusion of Fe and V. The same thermodynamic analysis mentioned above also suggests that decrease in $CaTiO_3$ and increase in $MgTi_2O_5$ increase the thermodynamic trend of titanium reducing to iron and increase the Ti grade in iron (Figure 12b).

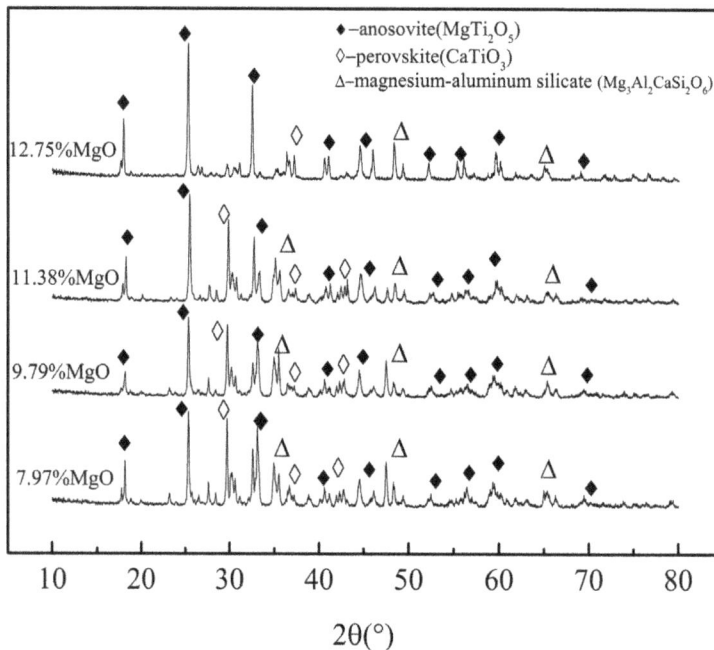

Figure 13. XRD patterns of titanium slag with varying MgO content.

The separation pictures of iron and titanium slag with varying MgO content are shown in Figure 14. Clearly, different MgO contents had different effects on separation. When MgO content was 7.97%, the separation was not optimal (*i.e.*, there were iron shots in the titanium slag), and when the MgO content increased to 12.75%, iron and titanium slag mixed together and the separation of iron and slag was again incomplete. Separation was better as MgO content increased from 9.79% to 11.38%, as the degree of polymerization among Si–O and Al–O decreased as MgO content increased; as the O^{2-} ions were impacted by the addition of MgO, the network structures of their formation were destroyed and transformed into simple single- and double-tetrahedron structures [21]. Thus, MgO reduced slag viscosity and promoted aggregation and settlement of molten iron, ultimately favoring the separation of iron and titanium slag. Figure 15 shows the yield of residual metallized iron in magnetic separation decreased gradually as MgO content increased. In short, it shows where about 11.38% MgO was optimal for the effective separation of iron and titanium slag.

In summary, the optimum quaternary basicity and MgO content were 1.10 and 11.38%, the vanadium-bearing iron and titanium slag were produced under the following conditions (No. 8): smelting at 1550 °C and 20 min, the reducing agent is 5%. Tables 3 and 4 show the main elements in iron and the main compositions of titanium slag. The recovery rates of Fe and V were 99.59% and 95.52%, respectively, and the titanium slag contained 50.34% TiO_2. The vanadium-bearing iron can be applied to recover iron and vanadium by the converter steelmaking process, and the titanium slag can be used to recover titanium by the acid leaching method [1,24].

Figure 14. Separation photos of iron and slag with varying MgO content.

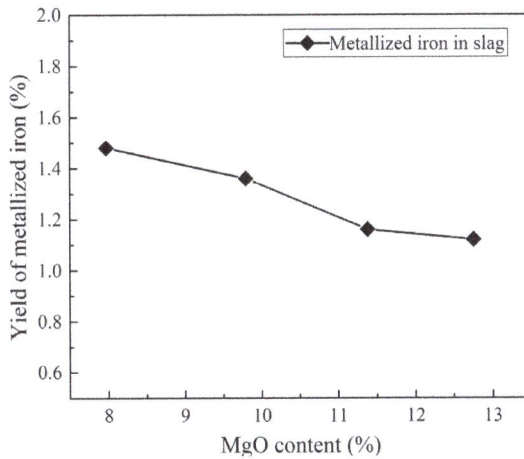

Figure 15. Effect of MgO content on yield of residual metallized iron in slag.

Table 3. Production index of vanadium-bearing iron.

Element Grade in Iron (wt. %)				Recovery Rate/%	
TFe	V	Si	Ti	V	Fe
98.09	0.683	0.199	0.296	95.52	99.59

Table 4. Main composition of titanium slag (wt. %).

TiO_2	V_2O_5	TFe	Al_2O_3	MgO	CaO	SiO_2
50.34	0.162	1.051	13.05	11.38	11.79	11.68

5. Conclusions

Effects of basicity and MgO content on reduction behaviors and iron-slag separation during smelting vanadium titanomagnetite ore by the direct reduction-electric furnace process were investigated in this paper.

(1).   The related thermodynamic analysis indicated that the suitable basicity range was 0.9–1.2. The addition of MgO decreased titanium slag viscosity but increased its melting temperature and free-running temperature.

(2).   A higher level of basicity benefits the reduction of iron oxides and vanadium oxides, and titanium content dropped in molten iron with the increased basicity. The optimum basicity was about 1.1. As the content of MgO increased, the recovery rate of Fe increased slightly, but the recovery rate of V considerably increased. The grades of Ti in molten iron were at a low level without significant change when MgO content was below 11%, but increased as MgO content increased to 12.75%. MgO content of 11.38% and basicity of 1.1 were the optimum conditions in this study.

(3).   The vanadium-bearing iron and titanium slag were produced at 1550 °C and 20 min, with 5% reducing agent. The recovery rates of Fe and V were 99.59% and 95.52%, respectively. The titanium slag contained 50.34% TiO_2. The vanadium-bearing iron can be applied in the converter steelmaking process and titanium slag can be utilized by the acid leaching method.

Author Contributions: Tao Jiang, Yufeng Guo and Shuai Wang conceived of and designed the experiments. Shuai Wang and Fuqiang Zheng performed the experiments. Shuai Wang and Yufeng Guo analyzed the data. Yufeng Guo contributed materials. Shuai Wang wrote the paper. Yufeng Guo and Feng Chen reviewed and contributed to the final manuscript.

Conflicts of Interest: The authors declare no conflict of interest.

References

1.   Zheng, F.Q.; Chen, F.; Guo, Y.F.; Jiang, T.; Travyanov, A.Y.; Qiu, G.Z. Kinetics of Hydrochloric Acid Leaching of Titanium from Titanium-Bearing Electric Furnace Slag. *JOM* **2016**, *68*, 1476–1484. [CrossRef]

2.   Taylor, P.R.; Shuey, S.A.; Vidal, E.E.; Gomez, J.C. Extractive metallurgy of vanadium-containing titaniferous magnetite ores: A review. *Miner. Metall. Proc.* **2006**, *23*, 80–86.

3.   Du, H.G. *Principle of Blast Furnaces Melting Vanadium-Titanium Magnetite*; Science Press: Beijing, China, 1996; pp. 6–8.

4.   Tan, Q.Y.; Chen, B.; Zhang, Y.S.; Long, Y.B.; Yang, Y.H. Characteristics and Current Situation of Comprehensive Utilization of Vanadium Titanomagnetite Resources in Panxi Region (In Chinese). *Multipurp. Util. Miner. Resour.* **2011**, *6*, 6–10.

5.   Chen, S.Y.; Chu, M.S. A new process for the recovery of iron, vanadium, and titanium from vanadium titanomagnetite. *J. South. Afr. Inst. Min. Metall.* **2014**, *114*, 481–488.

6.   Samanta, S.; Mukherjee, S.; Dey, R. Upgrading Metals via Direct Reduction from Poly-metallic Titaniferous Magnetite Ore. *JOM* **2015**, *67*, 467–476. [CrossRef]

7.   Fu, W.G.; Wen, Y.C.; Xie, H.E. Development of intensified technologies of vanadium-bearing titanomagnetite smelting. *J. Iron Steel Res. Int.* **2011**, *18*, 7–18. [CrossRef]

8.   Liu, X.J.; Chen, D.S.; Chu, J.L.; Wang, W.J.; Li, Y.L.; Qi, T. Recovery of titanium and vanadium from titanium–vanadium slag obtained by direct reduction of titanomagnetite concentrates. *Rare Met.* **2015**. [CrossRef]

9. Guo, Y.F. Study on Strenthening of Solid-State Reduction and Comprehensive Utilization of Vanadiferous Titanomagnetite (In Chinese). Ph.D. Thesis, Central South University, Changsha, China, July, 2007.

10. Jones, J.A.; Bowman, B.; Lefrank, P.A. Electric Furnace Steelmaking. Available online: http://jpkc.gsut.edu.cn/upload/20120523/20120523181249992.pdf (accessed on 6 May 2016).

11. Tang, J.; Chu, M.; Xue, X. Optimized use of MgO flux in the agglomeration of high-chromium vanadium-titanium magnetite. *Int. J. Miner. Metall. Mater.* **2015**, *22*, 371–380. [CrossRef]

12. Liu, Z.; Chu, M.; Wang, H.; Zhao, W.; Xue, X. Effect of MgO content in sinter on the softening-melting behavior of mixed burden made from chromium-bearing vanadium-titanium magnetite. *Int. J. Miner. Metall. Mater.* **2016**, *23*, 25–32. [CrossRef]

13. Chen, M.; Raghunath, S.; Zhao, B. Viscosity Measurements of SiO_2-"FeO"-MgO System in Equilibrium with Metallic Fe. *Metall. Mater. Trans. B* **2014**, *45*, 58–65. [CrossRef]

14. Gao, Y.; Wang, S.; Hong, C.; Ma, X.; Yang, F. Effects of basicity and MgO content on the viscosity of the SiO_2-CaO-MgO-9wt%Al_2O_3 slag system. *Int. J. Miner. Metall. Mater.* **2014**, *21*, 353–362. [CrossRef]

15. Li, P.; Ning, X. Effects of MgO/Al_2O_3 Ratio and Basicity on the Viscosities of CaO-MgO-SiO_2-Al_2O_3 Slags: Experiments and Modeling. *Metall. Mater. Trans. B* **2016**, *47*, 446–457.

16. Fan, J.J.; Cai, M.X.; Zhang, H. Study on effects of Al_2O_3 content and MgO content on melting temperature at BF slag (In Chinese). *Shanxi Metall.* **2007**, *30*, 22–23.

17. Shankar, A.; Rnerup, M.R.G.; Lahiri, A.K.; Seetharaman, S. Experimental Investigation of the Viscosities in CaO-SiO_2-MgO-Al_2O_3 and CaO-SiO_2-MgO-Al_2O_3-TiO_2 Slags. *Metall. Mater. Trans. B* **2007**, *38*, 911–915. [CrossRef]

18. He, H.Y.; Wang, Q.X.; Zeng, X.N. Effect of MgO content on BF slag viscosity (In Chinese). *J. Iron Steel Res.* **2006**, *18*, 11–13.

19. Li, J.; Zhang, Z.; Liu, L.; Wang, W.; Wang, X. Influence of Basicity and TiO_2 Content on the Precipitation Behavior of the Ti-bearing Blast Furnace Slags. *ISIJ Int.* **2013**, *53*, 1696–1703. [CrossRef]

20. Bale, C.W.; Bélisle, E.; Chartrand, P.; Decterov, S.A.; Eriksson, G.; Hack, K.; Jung, I.H.; Kang, Y.B.; Melançon, J.; Pelton, A.D.; *et al.* FactSage thermochemical software and databases—Recent developments. *Calphad* **2009**, *33*, 295–311. [CrossRef]

21. Huang, X.H. *Principles of Iron and Steel Metallurgy*; Metallurgy Industry Press: Beijing, China, 2013; pp. 60–440.

22. Zhen, Y.; Zhang, G.; Chou, K. Viscosity of CaO-MgO-Al_2O_3-SiO_2-TiO_2 Melts Containing TiC Particles. *Metall. Mater. Trans. B* **2015**, *46*, 155–161. [CrossRef]

23. Zhang, L.; Zhang, L.N.; Wang, M.Y.; Lou, T.P.; Sui, Z.T.; Jang, J.S. Effect of perovskite phase precipitation on viscosity of Ti-bearing blast furnace slag under the dynamic oxidation condition. *J. Non-Cryst. Solids* **2006**, *352*, 123–129. [CrossRef]

24. Yang, S.L. *Non-Blast Furnace Smelting Technology for Vanadium Titanomagnetite Ore*; Metallurgy Industry Press: Beijing, China, 2012.

18

Effect of Sb Addition on the Solidification of Deeply Undercooled Ag-28.1 wt. % Cu Eutectic Alloy

Su Zhao *, Yunxia Chen and Donglai Wei

School of Mechanical Engineering, Shanghai Dianji University, Shanghai 200240, China; chenyx@sdju.edu.cn (Y.C.); wdonglai@163.com (D.W.)
* Correspondence: wellzs@163.com

Academic Editor: Hugo F. Lopez

Abstract: Ag-28.1 wt. % Cu eutectic alloy solidifies in the form of eutectic dendrite at undercooling above 76 K. The remelting and ripening of the original lamellar eutectics result in the formation of the anomalous eutectics in the final microstructure. The addition of the third element Sb (0.5 and 1 wt. %) does not change the growth mode, but enlarges the volume fraction of anomalous eutectics because of the increasing recalescence rate. The additional constitutional supercooling owing to the Sb enrichment ahead of the eutectic interface promotes the branching of the interface and as a result fine lamellar eutectic arms form around the anomalous eutectics in the Sb-added Ag-28.1 wt. % Cu eutectic alloy.

Keywords: eutectic alloys; undercooling; growth modes; microstructure

1. Introduction

Binary eutectic alloys usually solidify into regular lamellar (rod) eutectics under equilibrium solidification conditions [1–3]. With the increase of undercooling, the solidification rate increases and the solidification structures can be composed of a mixture of lamellar eutectics and anomalous eutectics or full anomalous eutectics [4–8]. Furthermore, the rapidly solidified alloys can have a series of good properties, such as high strength and toughness, high temperature creep resistance, electromagnetism and corrosion resistance, *etc.*, because the properties of the as-solidified materials are determined by the conditions under which the liquid solidifies. Therefore, much attention has been paid to the solidification at high undercooling [9–12]. As a small amount of third element is added to a binary eutectic alloy, both a transverse and longitudinal diffusions of the added element in the liquid ahead of the solid-liquid interface is established, due to which the growth behavior changes. However, there have been few experimental investigations on the solidification of eutectic alloys containing a small amount of third element at high undercooling. In this paper, the effect of trace Sb addition on the crystal growth mode and the microstructure of Ag-28.1 wt. % Cu eutectic alloy at high undercoolings will be systematically investigated.

2. Materials and Methods

Three alloys, Ag-28.1Cu, (Ag-28.1Cu)-0.5Sb and (Ag-28.1Cu)-1Sb (weight percent), were investigated. The mass of each alloy sample is 6 g. The high undercooling of the experimental alloys was achieved by glass fluxing method in a high-frequency induction facility [13]. In each experiment, the raw materials (99.999% purity Ag, 99.999% purity Cu and 99.999% purity Sb) were put in a quartz crucible and covered with 0.6 g glass flux. The glass flux consists of 12.3 wt. % B_2O_3, 17.7 wt. % $Na_2B_4O_7$ and 70 wt. % Na_2SiO_3, and was dehydrated at 1273 K for 6 h in advance.

Then, the crucible was placed in the vacuum chamber of a high-frequency induction facility. The vacuum chamber was evacuated and backfilled with ultrapure argon. The raw materials were melted, superheated and cooled through adjusting the input power of the induction coil. To acquire a desired undercooling, it was necessary to repeat the heating-cooling cycle many times with an overheating degree of about 400 K and a holding time of 2–5 min. The power supply was turned off, when the desired undercooling was achieved. Then the sample solidified and cooled to room temperature. All the experimental samples were nucleated spontaneously without any manual triggering. The thermal history of the alloy was monitored by a two-color infrared pyrometer with an accuracy of 1 K and a response time of 1 ms. The sample solidified is a short column with a diameter of about 10 mm.

The phase transformation with temperature was analyzed by differential scanning calorimetry (DSC, Netzsch, Bavaria, Germany). The calibration of temperature and energy scale was performed with pure In and Zn. An argon purge of 80 mL/min was employed for all measurements. Both the heating and cooling rates were 10 K/min.

The phase constitution of the alloy was examined by X-ray diffraction (XRD, Thermo, Waltham, MA, USA). The surface structure of the sample was directly examined without any etching by a scanning electron microscope (SEM, Hitachi, Tokyo, Japan). Then the sample was cut, mounted in epoxy, polished, and etched with a mixture of 4% alcoholic picric acid. The internal structure of the sample was observed by an optical microscope (OM, Olympus, Tokyo, Japan).

3. Results

3.1. Thermal History Analysis

A temperature-time profile recorded in the experiment is shown schematically in Figure 1. When melt is cooled, a temperature decreases to T_N (nucleation temperature) (Stage 1), and at this time, nuclei form, and $\Delta T = T_E - T_N$ (ΔT, undercooling; T_E, eutectic temperature). The latent heat of crystallization released during rapid solidification is absorbed by undercooled melt, which leads to an increase of the temperature, *i.e.*, the so-called recalescence (Stages 2 and 3). During rapid recalescence (Stage 2), the temperature rising rate is faster, crystals grow rapidly and dendrite skeleton forms. Then the coarsening of fine dendritic structures makes the slow recalescence appear (Stage 3). After recalescence, the remaining melt solidifies under near-equilibrium conditions. The solidification during Stage 4 is controlled by the heat transfer from the sample to the environment. Finally, the sample cools to room temperature (Stage 5). Recalescence time, degree and rate refer to the time interval, temperature rise and the rising rate from the start to end of the rapid recalescence (Stage 2 in Figure 1), respectively [14]. Moreover, with the increase of undercooling, recalescence degree decreases and the recalescence time becomes shorter and shorter during rapid recalescence.

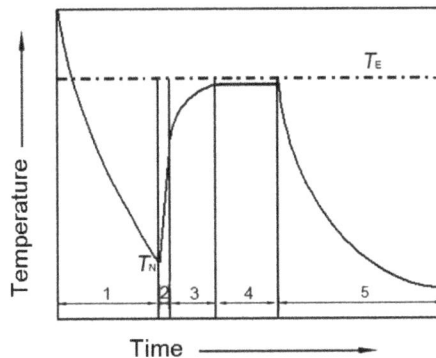

Figure 1. Schematic illustration of a cooling curve during the solidification of undercooled eutectic melt.

3.2. DSC Analysis

At 1054 K the eutectic phase transformation, L → α-Ag + β-Cu, occurs in Ag-28.1Cu alloy [15], indicated by only one endothermic peak during heating and only one exothermic peak during cooling (Figure 2a). With 0.5 and 1 wt. % Sb added to the base alloy, the melting point temperature gradually decreases to 1045 K and 1034 K, respectively, while the heating and cooling curves of the (Ag-28.1Cu)-0.5Sb and (Ag-28.1Cu)-1Sb alloys still exhibit the similar feature with those of the Ag-28.1Cu eutectic alloy (Figure 2b,c).

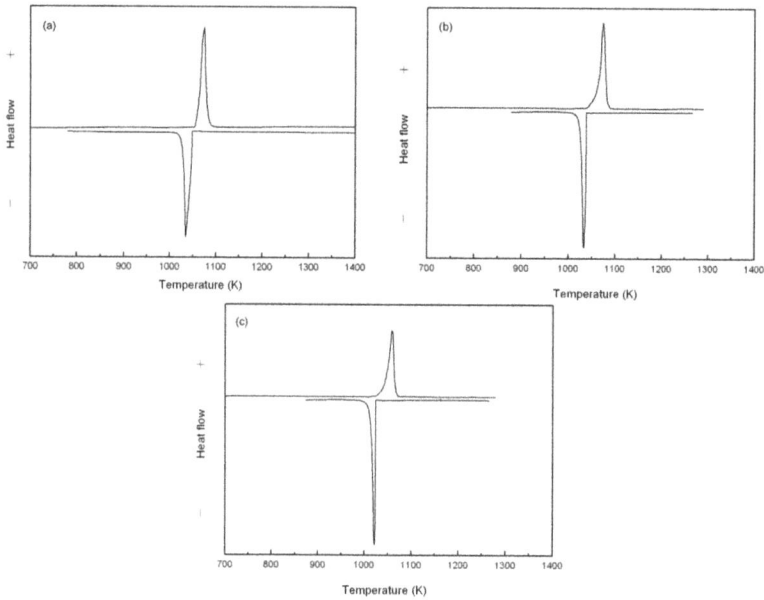

Figure 2. DSC curves of different alloys, **(a)** Ag-28.1Cu, **(b)** (Ag-28.1Cu)-0.5Sb, and **(c)** (Ag-28.1Cu)-1Sb.

3.3. XRD Analysis

Most of Sb can be dissolved in the Ag-rich (α) and Cu-rich (β) phases during solidification to form substitutional solid solutions [16]. Figure 3a,b show that only α-Ag and β-Cu face-centered cubic phases exist in the solidification structures of (Ag-28.1Cu)-0.5Sb and (Ag-28.1Cu)-1Sb alloys, no matter how large the undercooling is. This means that the addition of Sb does not influence the phase constitution of Ag-28.1Cu eutectic alloy.

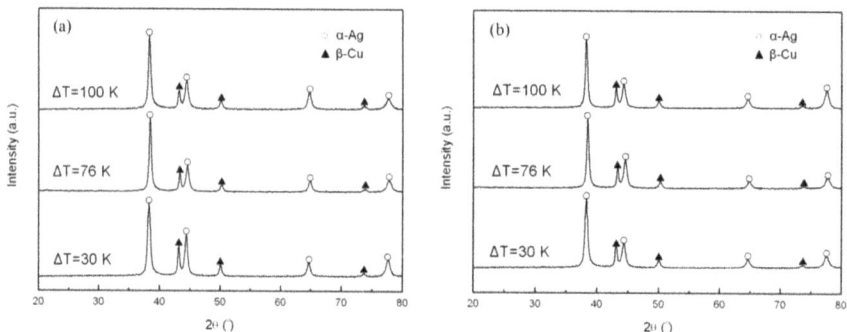

Figure 3. XRD patterns of the alloys Sb-containing solidified at different undercoolings, **(a)** (Ag-28.1Cu)-0.5Sb, and **(b)** (Ag-28.1Cu)-1Sb.

3.4. Microstructures

A large undercooling level up to 100 K can be achieved experimentally in the (Ag-28.1Cu)-xSb (x = 0, 0.5, 1.0) eutectic alloys. Taking 76 K as a critical undercooling, the alloys solidify in a form of cellular eutectics, cellular dendritic eutectics and the undeveloped dendritic eutectics at lower undercooling but dendritic eutectics at larger undercooling. The solidification of the alloys undercooled no more than 76 K has been reported previously [16]. Here, we pay attention to the solidification of the alloys undercooled above 76 K. For Ag-28.1Cu eutectic alloy, developed dendrites on the sample surface are observed under the SEM (Figure 4a). At high magnification, lamellar eutectics are found in the dendrite arms (Figure 4b). Such solidification structures are referred to as eutectic dendrites so as to distinguish them from the dendrites consisting of single phase. For the internal structure, a dendritic morphology is revealed, and the dendrite arms observed are roughly parallel to each other (Figure 4c) and the bright phase has been determined to be rich in Ag, and the dark phase rich in Cu under the OM. But now the dendrite arms are composed of anomalous eutectics and lamellar eutectics around the former (Figure 4d). And the anomalous eutectics refer to the granular particles and fine lamellar fragments [17]. The microstructures of (Ag-28.1Cu)-0.5Sb and (Ag-28.1Cu)-1Sb eutectic alloys are similar to those of Ag-28.1Cu eutectic alloy, as shown in Figures 5 and 6. Their arms are composed of lamellar eutectics on the sample surface but a mixture of anomalous eutectics and lamellar eutectics inside the sample, as occurred in the base alloy, except that the dendrite arms become thinner (Figures 5a and 6a).

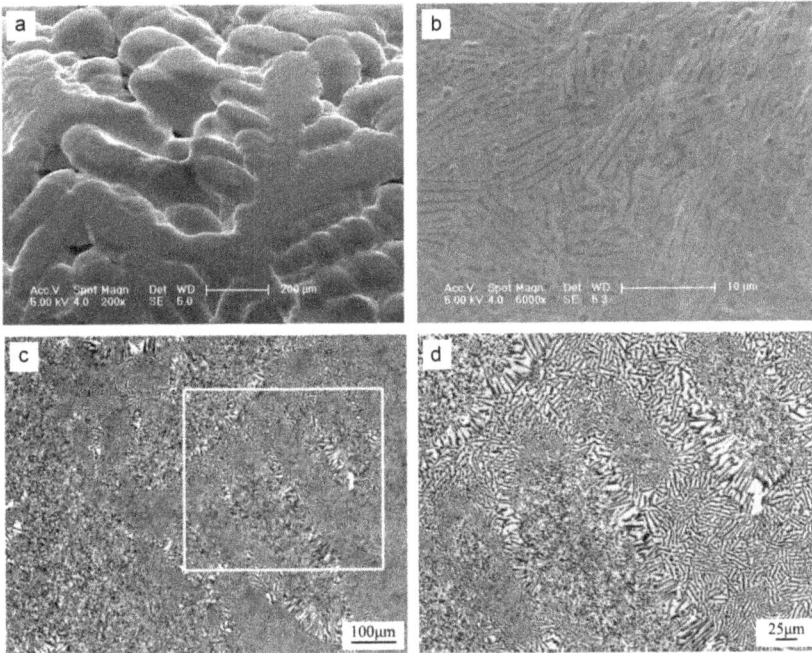

Figure 4. Microstructures of Ag-28.1Cu eutectic alloy undercooled by 85 K, (**a**) the surface structure with dendritic morphology at low magnification, (**b**) lamellar eutectics in the surface dendrite arms at high magnification, (**c**) the internal structure with dendritic morphology at low magnification, and (**d**) anomalous eutectics and the lamellar eutectics around them in the quadrilateral region at high magnification.

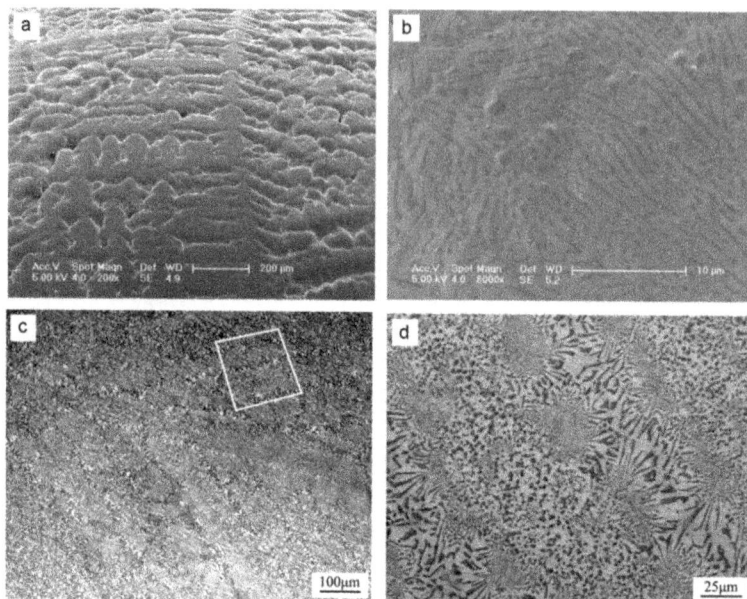

Figure 5. Microstructures of (Ag-28.1Cu)-0.5Sb alloy undercooled by 85 K, (**a**) the surface structure with dendritic morphology at low magnification, (**b**) lamellar eutectics in the surface dendrite arms at high magnification, (**c**) the internal structure with dendritic morphology at low magnification, and (**d**) anomalous eutectics and the lamellar eutectics around them in the quadrilateral region at high magnification.

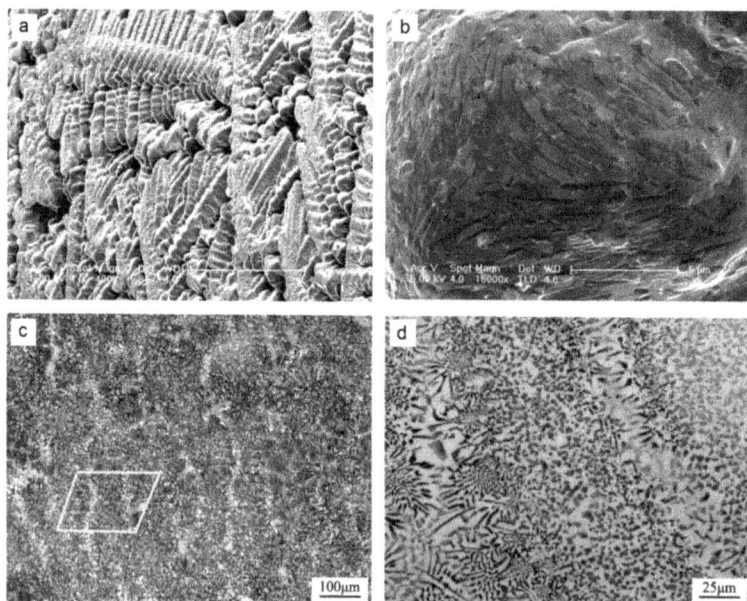

Figure 6. Microstructures of (Ag-28.1Cu)-1Sb alloy undercooled by 85 K, (**a**) the surface structure with dendritic morphology at low magnification, (**b**) lamellar eutectics in the surface dendrite arms at high magnification, (**c**) the internal structure with dendritic morphology at low magnification, and (**d**) anomalous eutectics and the lamellar eutectics around them in the quadrilateral region at high magnification.

The volume fraction of anomalous eutectics at each undercooling was evaluated by measuring the area of the anomalous eutectic zone on the sample cross-section, and the results are shown in Figure 7. It can be seen that the volume fraction increases with the increasing undercooling and the increasing content of Sb.

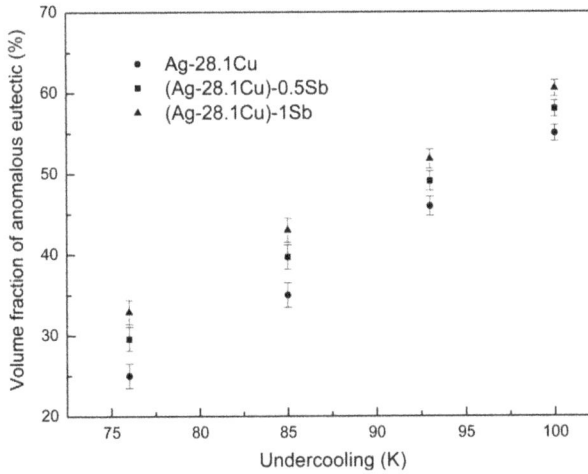

Figure 7. Volume fraction of anomalous eutectic *versus* undercooling for different Sb contents.

3.5. Recalescence Rate

Recalescence time, degree and rate can be used to describe the rapid solidification behavior of undercooled alloy melts. Among them the recalescence rate is directly proportional to the crystal growth rate. The recalescence rates as a function of undercooling at different contents of Sb are shown in Figure 8. It can be seen that the recalescence rate increases with increasing undercooling for the three alloys and at the critical undercooling of 76 K, there is a jump. Furthermore, the higher the Sb content, the larger the recalescence rate.

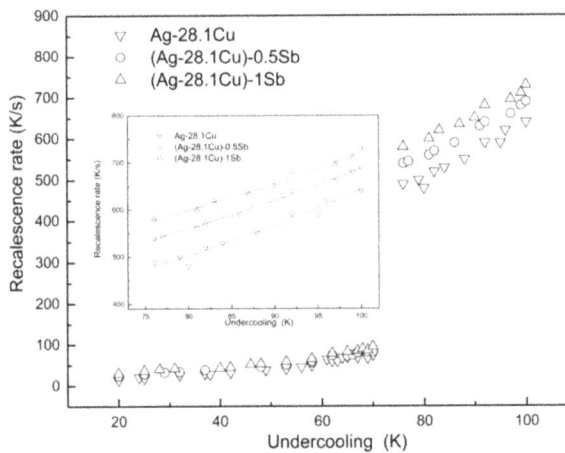

Figure 8. Recalescence rate *versus* undercooling for different Sb contents. The inset is a local amplification.

4. Discussion

4.1. Effect of Sb Addition on the Crystal Growth and the Formation of Anomalous Eutectic

The single-phase dendrite growth in undercooled melts has been well described in the BCT model [18]. According to the BCT model, the physical data of the Ag-Cu eutectic alloy (Table 1) were used to calculate the dendrite growth velocities of α-Ag and β-Cu *versus* undercooling and the results are shown in Figure 9. It can be seen that β-Cu should grow as the primary phase once decoupled growth occurs. For the eutectic dendrite growth in the undercooled Ag-Cu eutectic melt, the growth velocity during rapid solidification can be calculated by the eutectic growth model [19] that has taken the solute trapping into account and been established on the basis of the LZ model [20].

Table 1. Physical parameters of Ag-Cu eutectic alloy [21,22].

Parameter	Value
Latent heat of fusion for α-Ag/β-Cu (J/mol)	11,326/13,027
Specific heat of the liquid (J/mol· K)	32.21
Interdiffusion coefficient of solutes (m²/s)	$2.42 \times 10^{-7} \exp(-48{,}886/8.314T)$
Volume fraction of α-Ag in the eutectic	0.74
Thermal diffusivity in the liquid (m²/s)	5.44×10^{-5}
Gibbs-Thomson parameter for α-Ag/β-Cu (m· K)	$1.5 \times 10^{-7}/1.3 \times 10^{-7}$
Eutectic temperature (K)	1053
Liquidus slope for α-Ag/β-Cu (K/at. %)	−4.56/5.07
Kinetic parameter for α-Ag/β-Cu (m/s· K)	0.61/0.71
Equilibrium solute partition coefficient	0.353

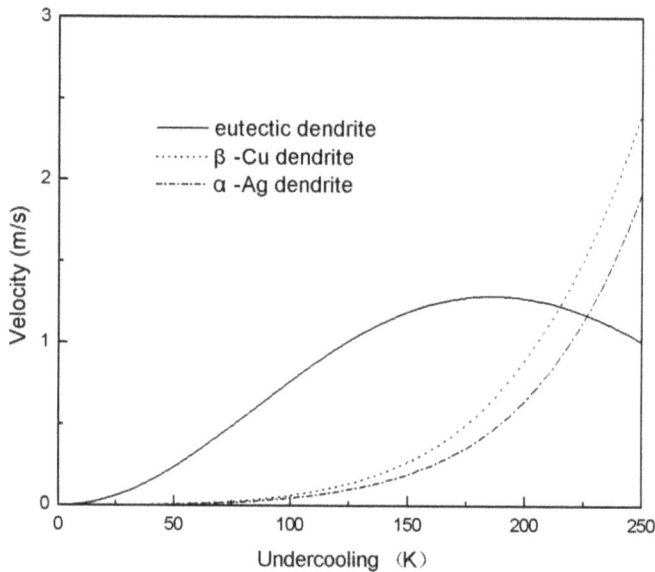

Figure 9. Calculated growth velocities in Ag-28.1 wt. % Cu eutectic melt when eutectic dendrite, β-Cu dendrite or α-Ag dendrite is assumed to form.

The bulk undercooling ΔT at the dendrite tip can be expressed as four terms:

$$\Delta T = m_v a_v^L \frac{1}{\lambda} + m_v \left(V\lambda/D\right) Q_v P \left(f, P_e, k_v\right) + m_v \frac{V}{\mu} + \left(\frac{\Delta H_f}{C_p}\right) Iv \left(P_t\right) \tag{1}$$

Where the parameters are as follows:

$$\frac{1}{m_v} = \frac{1}{m_\alpha^v} + \frac{1}{m_\beta^v} \tag{1a}$$

$$Q_v = \frac{1 - k_v}{f(1 - f)} \tag{1b}$$

$$a_v^L = 2\left[\frac{\Gamma_\alpha \sin\theta_\alpha}{fm_\alpha^v} + \frac{\Gamma_\beta \sin\theta_\beta}{(1 - f)m_\beta^v}\right] \tag{1c}$$

$$P(f, P_e, k_v) = \sum_1^\infty \frac{1}{(n\pi)^3}[\sin(n\pi f)]^2 \frac{p_n}{\sqrt{1 + p_n^2} - 1 + 2k_v} \tag{1d}$$

$$\frac{1}{\mu} = \frac{1}{m_\alpha^v \mu_\alpha} + \frac{1}{m_\beta^v \mu_\beta} \tag{1e}$$

where m_i^v is the liquidus slope of the α or β phase under non-equilibrium conditions; Γ_i, the Gibbs-Thomson coefficient; λ, the lamellar spacing; f, the volume fraction of the α phase; θ_i, the contact angle; D, the solute diffusion coefficient in the liquid; k_v, the solute distribution coefficient under non-equilibrium conditions; $P_n = 2n\pi/P_e$, where solute Peclet number $P_e = V\lambda/2D$; μ, the kinetic parameter; $\Delta H_f = f\Delta H_f^\alpha + (1 - f)\Delta H_f^\beta$, where ΔH_f^i ($i = \alpha$, β) is the heat of fusion; C_p, the specific heat of liquid phase; $I_v(P_t) = P_t\exp(P_t)E_1(P_t)$, the Ivantosv function of thermal Peclet number $P_t = VR/2\alpha$, where α is the thermal diffusion coefficient in the liquid, and R is dendrite tip radius.

Then we calculate the eutectic growth velocities using the physical parameters of Ag-Cu eutectic alloy and the results are also shown in Figure 9. The velocities for β-Cu dendrite and eutectic growths are equal at about 230 K, below which the eutectic dendrite grows faster than β-Cu dendrite. That means the coupled growth of lamellar α-Ag and β-Cu rather than single-phase dendrite growth should take place at the present experimental undercoolings. So the dendritic lamellar eutectics are observed on the sample surface, as shown in Figure 4a,b. Inside the sample, the severe superheating in the primary eutectic dendrites results in the appearance of anomalous eutectics (Figure 4c,d) [23]. When the melt contacts with the crucible wall or glass purifier, the recalescence is depressed greatly. So the eutectic dendrites on the sample surface are retained. On the contrary, inside the sample the temperature can be recalesced close to the eutectic temperature during rapid solidification. Excess solute is trapped in the original lamellar eutectics during rapid solidification because of the large growth velocity and the deviation of the solidification temperature at the solid-liquid interface from the equilibrium eutectic temperature [17,23]. As a result, parts of the original lamellar eutectics are remelted under the action of severe superheating during recalescence. Furthermore, after the recalescence, the broken lamellae decay into anomalous eutectics with granular morphology under the driving force of a reduction of the interfacial energy stored in the fine lamellar eutectics [17].

It is also shown in Figure 9 that when the single-phase dendrite growth occurs, the growth velocity of single-phase dendrite is much greater than that of eutectic dendrite, and for the same growth modes, their growth velocities should rise in the same law with the increase of undercoolings [24,25]. In the present experiment, it is difficult to determine the crystal growth path, so the recalescence rate rather than the growth velocity was measured. The recalescence rate is directly proportional to the growth velocity, so from the results in Figure 8 it is known that the growth velocity rises gradually with the increase of Sb content for a fixed undercooling. Furthermore, the addition of Sb does not change the variation trend of the growth velocity along with undercooling (Figure 8). In addition, perfect eutectic dendrites are formed on the sample surface, but the internal structure are anomalous eutectics and the lamellar eutectics which compose the dendrite arms (Figures 5 and 6), just like the structures of Ag-28.1Cu eutectic alloy (Figure 4). Therefore, it is believed that eutectic dendrite growth also takes place for the eutectic alloys containing Sb. And it is shown in Figure 8 that the growth velocity rises abruptly at the critical undercooling of 76 K. This is in good agreement with the microstructural evolutions: the cellular eutectics (Ag-28.1Cu), cellular dendritic eutectics (Ag-28.1Cu-0.5Sb) and the undeveloped dendritic eutectics (Ag-28.1Cu-1Sb) [16] change into the developed dendritic eutectics at 76 K. The developed eutectic dendrites grow faster, because they have smaller tip radius, which favors the dissipation of the latent heat.

The reduction of tip radius can lead to the increase of recalescence rate [16]. Thus, the composition of the primary eutectics deviates more severely from the equilibrium value, and the more solute is trapped in the primary lamellar eutectics with the rise of recalescence rate [26]. As a result, the volume fraction of anomalous eutectics increases with the increase of both Sb content and undercooling (Figure 7).

4.2. Effect of Sb Addition on the Formation of Lamellar Eutectics

Normally the negative temperature gradient ahead of the solid-liquid interface during solidification in undercooled melt leads to the formation of dendritic morphology [20,27], while planar interfaces are observed in the directional solidification for a pure eutectic alloy [28]. With the proceeding of solidification, the temperature of the undercooled melt increases gradually until it is close to the equilibrium eutectic temperature. The eutectics formed at the stage of slow solidification cannot be remelted and keeps the morphology of regular lamellae because of the fall or disappearance in the supersaturation degree of solute and superheating. So the lamellar eutectics grow outward around the anomalous eutectics for the Ag-28.1Cu eutectic alloy (Figure 4d). For the eutectic alloys containing Sb, however, because the equilibrium partition coefficients of Sb with respect to the both α-Ag and β-Cu phases phases are about 0.46 and 0.31, respectively [15], part of Sb atoms are rejected into the liquid during solidification. As a result, the new solute boundary layer with the enrichment of Sb atoms exists in front of the solid-liquid interface, and a zone of constitutional supercooling is formed. The enrichment of Sb atoms ahead of the interface must be diffused along the direction perpendicular to the interface during solidification, thus destabilizing the planar interface. Therefore, lamellar eutectics with dendritic morphology growing outward around the anomalous eutectics are observed (Figures 5d and 6d).

5. Conclusions

(1)   The eutectic dendritic growth takes place in the eutectic alloys containing Sb at the undercoolings ranging from 76 to 100 K. The remelting and ripening of the original lamellar eutectics result in the formation of anomalous eutectics.

(2)   For a fixed undercooling, the recalescence rate and the volume fraction of anomalous eutectics increase with the increase of Sb content. For a fixed Sb content, the recalescence rate increases with the increase of undercooling, and there is a jump at 76 K, indicating that the developed eutectic dendrites grow much faster.

(3)   The additional constitutional supercooling in the liquid ahead of the eutectic interface caused by Sb addition results in the dendritic growth of lamellar eutectics outward from the anomalous eutectics at the stage of slow solidification of the eutectic alloys containing Sb.

Acknowledgments: This work was supported by National Natural Science Foundation of China, No. 50571068, Research and Innovation Project of Shanghai Municipal Education Commission, No. 14YZ159, Climbing Peak Discipline Project of Shanghai Dianji University, No. 15DFXK02, Shanghai Young College Teacher Training and Financial Assistance Scheme, No. ZZSDJ12006.

Author Contributions: Su Zhao conceived and designed the experiments, analyzed the data and wrote the paper. Yunxia Chen and Donglai Wei performed the experiments.

Conflicts of Interest: The authors declare no conflict of interest.

References

1. Zhu, M.F.; Hong, C.P. Modeling of microstructure evolution in regular eutectic growth. *Phys. Rev. B* **2002**, *66*, 1554281–1554288. [CrossRef]
2. Folch, R.; Plapp, M. Towards a quantitative phase-field model of two-phase solidification. *Phys. Rev. E* **2003**, *68*, 106021–106024. [CrossRef] [PubMed]
3. Xu, J.F.; Liu, F.; Xu, X.L.; Dang, B. Undercooled solidification of Ni-3.3 wt. % B alloy and cooling curve description. *Mater. Sci. Technol.* **2013**, *29*, 36–42. [CrossRef]
4. Han, X.J.; Wei, B. Microstructural characteristics of Ni-Sb eutectic alloys under substantial undercooling and containerless solidification conditions. *Metall. Mater. Trans. A* **2002**, *33*, 1221–1228. [CrossRef]
5. Yang, C.; Gao, J.; Zhang, Y.K.; Kolbe, M.; Herlach, D.M. New evidence for the dual origin of anomalous eutectic structures in undercooled Ni-Sn alloys: *In situ* observations and EBSD characterization. *Acta Mater.* **2011**, *59*, 3915–3926. [CrossRef]
6. Liu, L.; Wei, X.X.; Huang, Q.S.; Li, J.F.; Cheng, X.H.; Zhou, Y.H. Anomalous eutectic formation in the solidification of undercooled Co-Sn alloys. *J. Cryst. Growth* **2012**, *358*, 20–28. [CrossRef]
7. Clopet, C.R.; Cochrane, R.F.; Mullis, A.M. The origin of anomalous eutectic structures in undercooled Ag-Cu alloy. *Acta Mater.* **2013**, *61*, 6894–6902. [CrossRef]
8. Wei, X.X.; Lin, X.; Xu, W.; Huang, Q.S.; Ferry, M.; Li, J.F.; Zhou, Y.H. Remelting-induced anomalous eutectic formation during solidification of deeply undercooled eutectic alloy melts. *Acta Mater.* **2015**, *95*, 44–56. [CrossRef]
9. Li, M.; Nagashio, K.; Ishikawa, T.; Yoda, S.; Kuribayashi, K. Microtexture and macrotexture formation in the containerless solidification of undercooled Ni-18.7 at. % Sn eutectic melts. *Acta Mater.* **2005**, *53*, 731–741. [CrossRef]
10. Xu, J.; Liu, F.; Zhang, D. *In situ* observation of solidification of undercooled hypoeutectic Ni-Ni$_3$B alloy melt. *J. Mater. Res.* **2013**, *28*, 1891–1902. [CrossRef]
11. Mullis, A.M. The origins of spontaneous grain refinement in deeply undercooled metallic melts. *Metals* **2014**, *4*, 155–167. [CrossRef]
12. Herlach, D.M.; Feuerbacher, B. Non-equilibrium solidification of undercooled metallic melts. *Metals* **2014**, *4*, 196–234. [CrossRef]
13. Liu, N.; Yang, G.C.; Liu, F.; Chen, Y.Z.; Yang, C.L.; Lu, Y.P. Grain refinement and grain coarsening of undercooled Fe-Co alloy. *Mater. Charact.* **2006**, *57*, 115–120. [CrossRef]
14. Wei, B.; Yang, G.C.; Zhou, Y.H. High undercooling and rapid solidification of Ni-32.5% Sn eutectic alloy. *Acta Metall. Mater.* **1991**, *39*, 1249–1258.
15. Subramanian, P.R.; Perepezko, J.H. The Ag-Cu (Silver-Copper) system. *J. Phase Equilibria* **1993**, *14*, 62–75. [CrossRef]
16. Zhao, S.; Li, J.F.; Liu, L.; Zhou, Y.H. Solidification of undercooled Ag-Cu eutectic alloy with the Sb addition. *J. Alloy. Compd.* **2009**, *478*, 252–256. [CrossRef]
17. Goetzinger, R.; Barth, M.; Herlach, D.M. Mechanism of formation of the anomalous eutectic structure in rapidly solidified Ni-Si, Co-Sb and Ni-Al-Ti alloys. *Acta Mater.* **1998**, *46*, 1647–1655. [CrossRef]
18. Boettinger, W.J.; Coriell, S.R.; Trivedi, R. Principles and technologies. In *Rapid Solidification Processing*; Mehrabian, R., Parrish, P.A., Eds.; Claitors Publications: Baton Rouge, LA, USA, 1988; p. 13.
19. Zhao, S.; Li, J.F.; Liu, L.; Zhou, Y.H. Effect of solute trapping on the growth process in undercooled eutectic melts. *Acta Metall. Sin.* **2008**, *44*, 1335–1339.
20. Li, J.F.; Zhou, Y.H. Eutectic growth in bulk undercooled melts. *Acta Mater.* **2005**, *53*, 2351–2359. [CrossRef]
21. Walder, S.; Ryder, P.L. Critical solidification behavior of undercooled Ag-Cu alloys. *J. Appl. Phys.* **1993**, *74*, 6100–6106. [CrossRef]
22. Wang, N.; Cao, C.D.; Wei, B.B. Rapid solidification of Ag-Cu eutectic alloy by drop tube processing. *Acta Metall. Sin.* **1998**, *34*, 824–830.
23. Li, J.F.; Li, X.L.; Liu, L.; Lu, S.Y. Mechanism of anomalous eutectic formation in the solidification of undercooled Ni-Sn eutectic alloy. *J. Mater. Res.* **2008**, *23*, 2139–2148. [CrossRef]
24. Liu, X.R.; Cao, C.D.; Wei, B. Microstructure evolution and solidification kinetics of undercooled Co-Ge eutectic alloys. *Scr. Mater.* **2002**, *46*, 13–18. [CrossRef]

25. Yao, W.J.; Han, X.J.; Wei, B. Microstructural evolution during containerless rapid solidification of Ni-Mo eutectic alloys. *J. Alloy. Compd.* **2003**, *348*, 88–99. [CrossRef]

26. Aziz, M.J. Model for solute redistribution during rapid solidification. *J. Appl. Phys.* **1982**, *53*, 1158–1168. [CrossRef]

27. Goetzinger, R.; Barth, M.; Herlach, D.M. Growth of lamellar eutectic dendrites in undercooled melts. *J. Appl. Phys.* **1998**, *84*, 1643–1649. [CrossRef]

28. Kurz, W.; Fisher, D.J. *Fundamentals of Solidification*, 4th ed.; Trans Tech Publications Ltd.: Dürnten, Switzerland, 1998; p. 108.

Permissions

The contributors of this book come from diverse backgrounds, making this book a truly international effort. This book will bring forth new frontiers with its revolutionizing research information and detailed analysis of the nascent developments around the world.

We would like to thank all the contributing authors for lending their expertise to make the book truly unique. They have played a crucial role in the development of this book. Without their invaluable contributions this book wouldn't have been possible. They have made vital efforts to compile up to date information on the varied aspects of this subject to make this book a valuable addition to the collection of many professionals and students.

This book was conceptualized with the vision of imparting up-to-date information and advanced data in this field. To ensure the same, a matchless editorial board was set up. Every individual on the board went through rigorous rounds of assessment to prove their worth. After which they invested a large part of their time researching and compiling the most relevant data for our readers.

The editorial board has been involved in producing this book since its inception. They have spent rigorous hours researching and exploring the diverse topics which have resulted in the successful publishing of this book. They have passed on their knowledge of decades through this book. To expedite this challenging task, the publisher supported the team at every step. A small team of assistant editors was also appointed to further simplify the editing procedure and attain best results for the readers.

Apart from the editorial board, the designing team has also invested a significant amount of their time in understanding the subject and creating the most relevant covers. They scrutinized every image to scout for the most suitable representation of the subject and create an appropriate cover for the book.

The publishing team has been an ardent support to the editorial, designing and production team. Their endless efforts to recruit the best for this project, has resulted in the accomplishment of this book. They are a veteran in the field of academics and their pool of knowledge is as vast as their experience in printing. Their expertise and guidance has proved useful at every step. Their uncompromising quality standards have made this book an exceptional effort. Their encouragement from time to time has been an inspiration for everyone.

The publisher and the editorial board hope that this book will prove to be a valuable piece of knowledge for researchers, students, practitioners and scholars across the globe.

List of Contributors

Shi-Hong Zhang and Hong-Wu Song
Institute of Metal Research, Chinese Academy of Sciences, Shenyang 110016, China

Neng-Yong Ye and Ming Cheng
Institute of Metal Research, Chinese Academy of Sciences, Shenyang 110016, China
State Key Laboratory of Advanced Processing and Recycling of Non-ferrous Metals, Lanzhou University of Technology, Lanzhou 730050, China

Hong-Wei Zhou and Ping-Bo Wang
State Key Laboratory of Advanced Processing and Recycling of Non-ferrous Metals, Lanzhou University of Technology, Lanzhou 730050, China

Huei-Sen Wang and Pei-Ju Hsieh
Deptartment of Materials Science and Engineering, I-Shou University, Kaohsiung 84001, Taiwan

Hiroshi Suzuki
Quantum Beam Science Center, Japan Atomic Energy Agency, Tokai, Naka, Ibaraki 319-1195, Japan

Rui Yamada and Junji Saida
Frontier Research Institute for Interdisciplinary Sciences, Tohoku University, Sendai, Miyagi 980-8578, Japan

Shinki Tsubaki and Muneyuki Imafuku
Faculty of Engineering, Tokyo City University, Setagaya, Tokyo 158-8857, Japan

Shigeo Sato
Graduate School of Science and Engineering, Ibaraki University, Hitachi, Ibaraki 316-8511, Japan

TetsuWatanuki and Akihiko Machida
Quantum Beam Science Center, Japan Atomic Energy Agency, Sayo, Hyogo 679-5148, Japan

Jinghua Jiang
College of Mechanics and Materials, Hohai University, Nanjing 210098, China
Jiangsu Key Laboratory of Advanced Structural Materials and Application Technology, Nanjing 226000, China

Fan Zhang, Aibin Ma, Dan Song, Jianqing Chen, Huan Liu and Mingshan Qiang
College of Mechanics and Materials, Hohai University, Nanjing 210098, China

Nakarin Srisuwan and Nantawat Kreatsereekul
Thai-French Innovation Institute, King Mongkut1s University of Technology North Bangkok, Bangsue, Bangkok 10800, Thailand;

Krittee Eidhed
Faculty of Engineering, King Mongkut1s University of Technology North Bangkok, Bangsue, Bangkok 10800, Thailand

Trinet Yingsamphanchareon and Attaphon Kaewvilai
Department of Welding Engineering Technology, College of Industrial Technology, King Mongkut1s University and Technology North Bangkok, Bangsue, Bangkok 10800, Thailand
Welding Engineering and Metallurgical Inspection, Science and Technology Research Institute, King Mongkut1s University and Technology North Bangkok, Bangsue, Bangkok 10800, Thailand

Takeshi Egami
Department of Materials Science and Engineering, Joint-Institute for Neutron Sciences, University of Tennessee, Knoxville, TN 37996, USA
Department of Physics and Astronomy, University of Tennessee, Knoxville, TN 37996, USA
Oak Ridge National Laboratory, Materials Science and Technology Division, Oak Ridge, TN 37831, USA

Yang Tong
Department of Materials Science and Engineering, Joint-Institute for Neutron Sciences, University of Tennessee, Knoxville, TN 37996, USA
Department of Mechanical and Biomedical Engineering, City University of Hong Kong, Hong Kong, China

Wojciech Dmowski
Department of Materials Science and Engineering, Joint-Institute for Neutron Sciences, University of Tennessee, Knoxville, TN 37996, USA

Mohd Halim Irwan Ibrahim, Azriszul Mohd Amin, Rosli Asmawi and Najwa Mustafa
Advanced Manufacturing and Materials Center, University Tun Hussein Onn Malaysia (UTHM), Parit Raja, Batu Pahat, Johor 86400, Malaysia

Hikari Nishijima
Department of Applied Beam Science, Graduate school of Science and Enginieering, Ibaraki University, 4-12-1 Naka-narusawa, Hitachi 316-8511, Japan

Yo Tomota
Graduate school of Science and Engineering, Ibaraki University, 4-12-1 Naka-narusawa, Hitachi 316-8511, Japan

Yuhua Su and Wu Gong
Japan Atomic Energy Agency, 2-4 Shirane Shirakata Tokai, Ibaraki 319-1195, Japan

Jun-ichi Suzuki
Comprehensive Research Organization for Science and Society, 162-1 Shirakata, Tokai, Ibaraki 319-1106, Japan

Seung-Baek Yu and Mok-Soon Kim
Division of Materials Science and Engineering, Inha University, Incheon 22207, Korea

Hsin-Hsiung Huang
Metallurgical and Materials Engineering, Montana Tech, Butte, MT 59701, USA

Xiaomeng Luo, Lizhan Han and Jianfeng Gu
Institute of Materials Modification and Modeling, School of Material Science and Engineering, Shanghai Jiaotong University, 800 Dongchuan Road, Shanghai 200240, China

Cheng-Yi Chen, Fei-Yi Hung, Truan-Sheng Lui and Li-Hui Chen
Department of Materials Science and Engineering, National Cheng Kung University, Tainan 701, Taiwan

Seda Çam, Vedat Demir and Dursun Özyürek
Manufacturing Engineering of Technology Faculty, Karabuk University, 78100 Karabuk, Turkey

Darren J. Goossens
School of Physical, Environmental and Mathematical Sciences, University of New South Wales, Canberra ACT 2600, Australia

Pavel Strunz
Nuclear Physics Institute ASCR, CZ-25068 Řež near Prague, Czech Republic

Martin Petrenec
Institute of Physics of Materials of the ASCR, CZ-61662 Brno, Czech Republic

Jaroslav Polák
Institute of Physics of Materials of the ASCR, CZ-61662 Brno, Czech Republic
CEITEC Institute of Physics of Materials of the ASCR, CZ-61662 Brno, Czech Republic

Urs Gasser
Laboratory for Neutron Scattering, PSI, CH-5232 Villigen, Switzerland

Gergely Farkas
Department of Physics of Materials, Faculty of Mathematics and Physics, Charles University, Ke Karlovu 5, 121 16, Prague 2, Czech Republic

Fei Wang, Lintao Zhang, Anyuan Deng, Xiujie Xu and Engang Wang
Key Laboratory of Electromagnetic Processing of Materials (Ministry of Education), Northeastern University, No. 3-11,Wenhua Road, Shenyang 110004, China

Tao Jiang, Shuai Wang, Yufeng Guo, Feng Chen and Fuqiang Zheng
School of Minerals Processing and Bioengineering, Central South University, Changsha 410083, China

Su Zhao , Yunxia Chen and Donglai Wei
School of Mechanical Engineering, Shanghai Dianji University, Shanghai 200240, China

Index

www.ingramcontent.com/pod-product-compliance
Lightning Source LLC
Chambersburg PA
CBHW061948190326

41458CB00009B/2818